Zumwalt

ZUMWALT

The Life and Times of
Admiral Elmo Russell "Bud" Zumwalt, Jr.

LARRY BERMAN

HARPER

An Imprint of HarperCollins*Publishers*
www.harpercollins.com

HarperCollins books may be purchased for educational, business, or sales promotional use. For information, please write: Special Markets Department, HarperCollins Publishers, 10 East 53rd Street, New York, NY 10022.

FIRST EDITION

Designed by Fritz Metsch

Frontispiece photograph courtesy of the Naval History and Heritage Command, Washington, D.C.

Library of Congress Cataloging-in-Publication Data has been applied for.

ISBN: 978-0-06-169130-0

12 13 14 15 16   OV/RRD   10 9 8 7 6 5 4 3 2 1

For my grandchildren, Isabel and Ian

"Father of the Modern Navy" will remain
a unique epitaph to this man.

—VICE ADMIRAL EARL FRANK "REX" RECTANUS
IN A LETTER TO MOUZA ZUMWALT, JANUARY 10, 2000

CONTENTS

ZUMWALT

CONSCIENCE OF THE NAVY

*I have been called controversial. I am glad that this is so
because the requirement was to be as Robert Frost phrased it,
"And I have miles to go before I sleep."*[1]

With its copper-clad dome soaring two hundred feet above the ground,
the historic chapel at the center of the yard can be seen from every approach to the United States Naval Academy in Annapolis. Built upon
a cornerstone laid in June 1904 by the legendary Admiral of the Navy
George Dewey, the chapel serves as the navy's cathedral. A below-
ground crypt holds the remains of John Paul Jones, the great naval
leader of the American Revolution who gave the navy its earliest traditions of heroism and victory.[2]

On the cold and overcast morning of January 10, 2000, over two
thousand people filled the wooden pews of the chapel to bid farewell to
Admiral Elmo Russell Zumwalt, Jr., whose life exemplified the words
etched above the massive bronze entrance doors: NON SIBI SED PATRIAE,
meaning "Not for self but for country."

Bud Zumwalt, which is what just about everyone called him throughout his life, is remembered as a trailblazer who reformed the navy and
as a champion of the men and women who served in it. He was the icon-
oclastic admiral who brought a navy drifting toward the shoals back
into the channel of the twentieth century and prepared it for the new
millennium. Bud Zumwalt forced the navy to think more deeply and
objectively about things that should have been self-evident. By doing
so, he became a sailor's admiral, often referred to as Zorro, fighting for
the rights of oppressed navy men and women. In 1970 he landed on the

cover of *Time* magazine, which called the charismatic chief of naval operations (CNO) "the Navy's most popular leader since World War II."[3]

The press quickly dubbed it the Mod-Squad Navy led by its "psychedelic admiral." Sailors began sporting longer hair, beards, and sideburns. The infamous "Z-grams" attempted to meld the traditions of navy service with the needs of a nation in turmoil and a culture in transition. The navy was never the same. Beer dispensers were allowed in enlisted men's barracks, acid rock blared from service clubs, and women were going to sea. Traditionalists, mostly white, retired admirals, ridiculed the reforms as the Three Bs—beer, beards, and broads, deriding "Zumwaltism" for undoing navy discipline and leading to mutinies at sea.

Bud Zumwalt requested that his funeral be held at the Naval Academy, the place of many memories and milestones in his distinguished career. At Bancroft Hall, his plebe class had been sworn in as midshipmen, taking their oath under Commodore Perry's flag and Captain Lawrence's words "Don't Give Up the Ship." Graduating cum laude with the wartime class of 1943, the young ensign made his way to the Pacific, soon reporting to the destroyer USS *Robinson*, where his performance as head of the ship's Combat Information Center during night torpedo attacks in the historic Battle of Leyte Gulf earned him the Bronze Star for valor in combat. It was the start of a career characterized by service to the nation and commitment to the welfare of those under his command.

Inside the chapel were the country's highest dignitaries, the most prominent being the nation's commander in chief President William Jefferson Clinton and First Lady Hillary Rodham Clinton. Two years earlier, in the East Room of the White House, the president had presented his dear friend with the nation's highest civilian honor, the Presidential Medal of Freedom.[4] At the ceremony, Clinton joked that "these days, Elmo 'Bud' Zumwalt introduces himself as 'a former sailor.' That's sort of like calling Henry Ford a former car salesman." When the laughter subsided, the president described Zumwalt as a "great patriot" and "one of the greatest models of integrity and leadership in genuine humanity our nation has ever produced." The Medal of Freedom citation provided the most appropriate accolade: "In both wartime and peacetime, Elmo Zumwalt has exemplified the ideal of service to our nation. A distinguished veteran of World War II and Korea, he served as Commander of U.S. Naval Forces in Vietnam and rose to become the Navy's youngest chief of Naval Operations in 1970. As CNO, he worked vigorously to improve our sailors' quality of life and devoted himself to eliminating discrimination in the Navy. In a life touched by tragedy, he became a great champion of veterans afflicted by ailments related to service in Vietnam. For his dedication, valor, and compassion, we salute Bud Zumwalt."

Bud Zumwalt and Bill Clinton first met when Governor Clinton consulted Zumwalt on the disposal of Agent Orange barrels in his home state of Arkansas. After Clinton's election, Bud sought the president's support for Vietnam-era veterans who had been exposed to Agent Orange, a topic whose importance Clinton grasped immediately. The president designated Robert G. Bell, a member of the National Security Council staff, to liaise directly with Zumwalt on all Agent Orange issues.

The president sought Bud's advice on normalizing relations with Vietnam, strategic-arms agreements, and cabinet-level appointments. In 1995 Clinton appointed Bud to the president's Foreign Intelligence Advisory Board, which focused on threats to our nation's national security involving weapons of mass destruction, biological, chemical, and nuclear. Bud and the president worked closely together to pass the Chemical Weapons Convention. Zumwalt then served as an honorary member of the Veterans for Clinton/Gore National Steering Committee in 1996.

The "unique chemistry" between Bud Zumwalt and Bill Clinton took root at the annual Renaissance Weekends in Hilton Head, South Carolina, and one Renaissance gathering forever solidified their friendship.[5] Speaking on the topic "If These Were My Last Words," Bud directed his comments to the nation's first daughter, Chelsea Clinton. He spoke about the reasons Chelsea could be proud of her parents' accomplishments with respect to the country's national security and domestic strength. Bud received a resounding standing ovation from the audience. It was a moving tribute at a time of great personal family crisis for the Clintons. Bill and Hillary never forgot how Bud Zumwalt stood up for them in their time of need.[6]

An early arrival at the chapel was Bud's younger brother Jim. Neither man could have asked for a more caring or loving sibling, although Jim's strong antiwar views and liberal leanings often tested the patience of his more conservatively inclined brother. A July 1968 personal letter of congratulations on the occasion of Bud's promotion to the rank of vice admiral and new assignment as head of the Brown Water Navy in Vietnam expressed Jim's heartfelt sentiments: "In spite of all the political differences, I have never stopped loving you as a brother, nor have I questioned your integrity or sincerity. If ever a military figure had compassion, you have it."[7] Two years later, on the occasion of Bud's nomination as the country's youngest CNO, Jim wrote that "in spite of some political and philosophical differences with you, I have nothing less than the most profound respect for the genius and hard work which has brought you [to] the pinnacle of your career."[8] Bud once asked his brother to stand in for him at a reunion of his World War II shipmates from the *Robinson*. As the evening drew to a close, one sailor said, "My greatest hero in life is Admiral Zumwalt because he was an enlisted man's sailor. He always treated us with respect."[9] Jim was especially proud that Bud was beloved by enlisted men, women, and minorities.

After he passed through a metal detector, secret service agents ushered Jim into a private vestibule for family members. With Bud gone, Jim and Saralee Crowe were the surviving children raised by two country doctors on the "avenue of the sycamores" in Tulare, California. Saralee was the eldest, followed by Elmo, Jr.; Bruce Craig; and James Gregory. As Saralee embraced Jim, her mind raced through a lifetime of memories that included Crowe family visits to Mare Island, where after dinner at the officers' club, they listened to Bud make Tarzan calls from a tree house and went to sleep in a sun-heated Quonset hut. Her favorite memory was the time Bud drove across the country to visit the Crowes at Wawona Lodge in Yosemite. Bud arrived at the gate exhausted and out of money, but the Ranger would not allow any servicemen to enter the park free who were not in uniform. Bud got his suitcase from the car and changed right there "in front of God and everybody."[10]

First Lady Hillary Clinton soon entered the room, embracing Bud's wife, Mouza. Looking at the family, Mrs. Clinton said, "Bud and Mouza have been friends of ours for several years. We felt compelled to come down here today to offer our condolences and sympathy to all of you." Before she could say anything else, the president arrived. When embracing Mouza, family members saw tears in the president's eyes. "Bud was a good friend. I want to tell you that Hillary and I loved Bud so much. It was a great honor for us," said Clinton.[11]

Fifty-four years earlier, twenty-four-year-old Bud Zumwalt, two years removed from the Naval Academy and a prize crew captain, charted and sailed up the Yangtze River to Shanghai, clearing Japanese mines and joining with iconic Admiral Milton E. "Mary" Miles in disarming the surrendering Japanese forces. In Shanghai he met Mouza Coutelais-du-Roché, the daughter of White Russians from Harbin, Manchuria. It was love at first sight. Bud asked Mouza to teach him Russian and in return promised to help her with English. A week later he proposed; two weeks later they were married, with *Robinson* crew serving as honor guard.

Bud ignored the uniform phalanx of people offering advice that marriage to a foreign-born spouse was a career killer for advancing to higher ranks in the navy, and Mouza ended up making immense contributions to Bud's career. Bud was at sea for fourteen years of their fifty-four-year marriage. "I can't imagine what life would have been without her," wrote Bud.[12] Mouza was always HIS STRENGTH.[13] Within a short time, alone with a new baby, she adapted to a new culture. During her husband's three commands, she served as an indispensable "house mother" to waves of young officers and their wives and the wives of enlisted men. Filling the role of both parents during Bud's long deployments, Mouza moved their four children more than forty times. During the Vietnam War, with two sons and her husband in harm's way, she was a regular at Clark Air Force Base hospital, counseling and comforting those wounded in combat. "She so loved our Navy, and you knew she loved our Navy. And we so loved her," said Admiral Mike Mullen. "She was extraordinary in so many ways; extraordinary in how she was to others, and quick to notice those in need. She personified empathy and compassion."[14] When Bud became CNO, Mouza played an important contributing role in retention study groups and programs aimed at improving the quality of life for navy people and their families. "Mouza possessed that rare quality that allowed her to be in the public shadow of a larger-than-life figure," recalled Ambassador Philip Lader.[15]

Bud understood that a crow or a star on a man's sleeve or collar did not make him a leader. He possessed a deep, abiding passion for those entrusted to his care, adhering to the belief that a person who has command of forces in war has a lifetime responsibility for their well-being. Sitting in the chapel was Joe Muharsky, who had just made the seven-hour drive from Cleveland. Joe was one of Big Z's Swift Boat sailors in Coastal Squadron One, who had fought the Vietcong along tributaries of the Cua Lon River during the time Bud headed the Brown Water Navy that patrolled the rivers and coast of Vietnam.[16] Dressed in a black suit pinned with medals awarded by Zumwalt, Joe had come to honor the man who "fought for us when the battle was over and others chose to forget."[17] A few days earlier, Joe had written a heartfelt letter to Mouza. "I never met one sailor who was not proud that he had served under your husband's command in Vietnam. He gave us something we never had in the Navy until then. It may not have meant much to others but to us it meant the world. There is a simple word for it, it's called respect."

The distinguished elder statesman Paul Nitze was sitting in the first row of the chapel. Decades earlier, as assistant secretary of defense for international security affairs, Nitze spotted Zumwalt at the National War College and lost no time offering him a position. Zumwalt moved up to director of arms control and contingency planning for Cuba, and the two men worked side by side during the tense days of the Cuban Missile Crisis and later in negotiations leading to the Nuclear Test Ban Treaty. When Nitze was appointed secretary of the navy in 1963, he made Zumwalt his executive assistant and senior aide.

Normally, before a captain is promoted to flag rank, he needs a major sea command, which for a surface officer meant a destroyer squadron or a cruiser. Nitze's fitness report recommended that the requirement be waived, and Bud received the second star of rear admiral two years ahead of his Naval Academy class. The promotion sent Bud to San

Diego as commander of a cruiser/destroyer flotilla in the First Fleet, but the tour was cut short when Nitze and CNO Admiral David McDonald created the Division of Systems Analysis for the navy and appointed Zumwalt as chief, a position he held for two years preceding his tour in Vietnam.

Under Nitze's tutelage, Bud Zumwalt expanded his intellectual scope and sharpened his administrative skills. "I earned what I think of as a Ph.D. in political-military affairs," recalled Zumwalt, who often described his relationship with Nitze as "Plato and Socrates."[18] Nitze owned a 2,200-acre farm on the Potomac near Port Tobacco, Maryland. The Zumwalt family spent weekends at the farm, where they had their own guest house. When Nitze became deputy secretary of defense during the Johnson presidency, Zumwalt became privy to the Vietnam policy thinking of Secretary of Defense Robert McNamara and the president. Whatever Nitze knew, he shared with Zumwalt. Together their doubts grew concerning the viability of military victory in Vietnam. It was not long before Zumwalt's boss, Chief of Naval Operations Admiral Thomas Moorer, came to resent his subordinate's access to this information. Moorer decided to ship Zumwalt out of town; the farthest place he could find was Vietnam. Moorer saw it as win-win: he would break the Nitze-Zumwalt axis, and no one would ever hear of Bud Zumwalt again.

Bud was expected to be in Vietnam by September 1968, two months before the American presidential election. Believing that the war had already been lost politically, therefore making it impossible to prevail militarily, Bud got cracking on shifting responsibility to the South Vietnamese navy and devising innovative strategies for the Brown Water Navy. If Bud Zumwalt had never become CNO, he would be remembered for his planning and conduct in the management of naval forces in an insurgency environment.

Also in the chapel was Qui Nguyen, whose uncle, Vice Admiral Chung Tan Cang, had been the South Vietnamese navy's final CNO. When South Vietnam fell, in April 1975, Qui's family was one of two sponsored by Bud Zumwalt, who pledged to provide "for their needs until they are ready to take their place in society."[19] In July 1975 the eleven-member family took a bus from Indiantown Gap, Pennsylvania, to Washington, D.C., where they were met by Bud and his son James. "We had our first dinner together that evening with his wife and children," recalled Qui. The family lived in the Zumwalts' basement for the next six months. Mouza drove them around for job interviews and introduced them to the rudiments of shopping, keeping a checkbook, enrolling the children in school, and making the transition to a new home. "They have basically always looked out and kept in touch with us for twenty-six years watching us grow as Americans. The Admiral inspired us with his inner strength," said Qui.[20] Qui's older brother Phu never forgot what the entire Zumwalt family did for them. "The Admiral and his wife helped us to see life in the United States and helped transform our life into good American citizens. Our family wanted to make Admiral Zumwalt proud of our accomplishments, which would not have been possible without their support."[21]

Sitting near Qui in the chapel was James Reckner, professor of history and director of the Vietnam Center at Texas Tech University. Jim's navy career spanned twenty years; he started as an enlisted man and later became a commissioned officer. Reckner served two tours on some of the most hazardous rivers in Vietnam. Two decades later, the professor came to see Bud Zumwalt with a request that the retired admiral serve as chairman of the board to preside over the creation of the Vietnam Center and Archive at Texas Tech. The two men had never met. "I went to his office in Arlington, Virginia, knees knocking somewhat, as I've never recruited a four-star admiral before," Reckner said.

Bud Zumwalt embraced the assignment as a labor of love. "With Jim Reckner's wartime knowledge and his experience with historical research, he was keenly aware of the need to start early the retrieval and assembly of all documents from all sources and all sides if later historians were to be able to write definitive and accurate histories," said Zumwalt.[22] Between 1989 and today, the Vietnam Archive has emerged as a world-class repository of memorabilia and documents relating to the Vietnam era. Admiral Zumwalt chose the archive as the custodian for his personal papers and encouraged all Vietnam veterans to follow his lead.[23]

———

Ties to Vietnam defined so many aspects of Bud Zumwalt's life, most especially with his firstborn son and namesake. Lieutenant Junior Grade Elmo Russell Zumwalt III served proudly and bravely in Vietnam, commanding Swift Boat PCF-35—maneuvering his lightly armed vessel along jungle-choked streams in the Mekong Delta from June 1969 to July 1970. Elmo survived Vietnam, returning home in 1970, only to learn in 1983 that he had developed herbicide poisoning.

The navy had been taking heavy casualties in Vietnam, which meant that the average young naval person had a high probability of being killed or wounded in a year's tour. Vietcong snipers preyed on sailors from their hiding spots along the riverbank, perhaps ten to fifteen feet from their targets. When Admiral Zumwalt asked what could be done, experts in the Pentagon advised that Agent Orange defoliation offered the promise of moving those snipers back a thousand yards. He was assured that the herbicide was nontoxic and not dangerous to human or animal life. The jungle terrain was stripped bare. Zumwalt was unaware that the chemical companies producing these herbicides had evidence that the dioxin used in the manufacturing process was carcinogenic to humans. There was the cruel irony that Bud Zumwalt was responsible for ordering the spraying along the rivers and canals that his son and crew patrolled.

Elmo fought his illness for five years, succumbing in 1988 at age forty-two.[24] Elmo's son Russell, Bud's grandson, was born with severe sensory integration dysfunction attributed to his father's exposure. Bud and Elmo coauthored a bestselling book, *My Father, My Son,* followed with a made-for-TV movie starring Karl Malden as Admiral Zumwalt and Keith Carradine as young Elmo.

Bud Zumwalt paid a deep personal price for his decision to use Agent Orange. In a pledge to his dying son, Bud promised accountability from the government and private industry as to who knew what about the poison that had killed his son and so many of his sailors. Bearing responsibility for the loss of a son, Bud dedicated the rest of his life to studying the medical linkages between exposure and illnesses. No service chief ever demonstrated such a continuing and selfless commitment and loyalty to the men and women who served under his command. His son was dead, but by spearheading a citizen education and mobilization effort, he could help others in securing reparations for thousands of Vietnam veterans and their children whose lives had been permanently damaged.

For several years, Bud could not visit the Vietnam Memorial without feeling the pain of losing so many under his command. He found peace only when the government accepted responsibility for taking care of his valiant warriors and their families. Visiting the wall for the first time and pressing his hand firmly against the black granite, Bud could finally "then envision Elmo's hand reaching out to touch mine."[25]

Bud Zumwalt believed that government had a legal and moral responsibility to those who fought for their country. He used his position and reputation to fight for those who lacked the power and resources to do so. Bud founded the Marrow Foundation and served as chairman in order to assist in registering donors as well as raising the funds to help those needing transplants. "Without him, the National Marrow Donor Program could not have survived," wrote Bob Hartzman. "His determination carried the program through its birth and its most contentious and difficult years. Without him, support for those injured by Agent Orange would not have happened; without him, the studies of medical effects following Chernobyl would have been impossible; and, without him, we could not now be developing ways to protect our people from potential exposure to biological, chemical and radiologic weapons."[26]

Bud's membership on the Special Oversight Board for Investigations of Gulf War Chemical and Biological Incidents enabled a new generation of servicemen and -women to benefit from his leadership, his expertise, and his compassion.[27] In the words of President Clinton, "He never stopped fighting for the interests, the rights and the dignity of those soldiers and sailors and airmen and marines and their families."

Nothing illustrated Bud's commitment and compassion more than his friendship with Zoltan Merszei, who was also in the chapel. Bud had asked the retired president, CEO, and chairman of the Dow Chemical Company, which produced Agent Orange, to serve as a founding trustee on the board of the Marrow Foundation, a nonprofit organization formed in 1986 in support of a national registry of bone marrow donors. Zoltan and Bud became close friends. "I can attest to never having met another individual with the admiral's leadership skills, charisma, and dedication," said Merszei. "If I could turn back the clock and rewrite my own history, I would make sure that it included military service to this extraordinary country of ours under the exemplary direction of Admiral Elmo R. Zumwalt, Jr."[28]

Jerry Wages was one of several honorary pallbearers. In March 1969, Captain Clarence J. Wages assumed duties as senior advisor and commander of the Rung Sat Special Zone in Vietnam. The majority of his time in-country was spent commanding a river task group. His job was to protect ships coming up the Saigon River to discharge their cargo. The river patrol group commanded by Wages was later awarded the Presidential Unit Citation by President Nixon for extraordinary heroism. Zumwalt considered Wages "one of the finest wartime unit commanders in Vietnam."[29] Tragically, like many of his fellow sailors exposed to Agent Orange, Wages developed leukemia and prostate cancer, earning a 100 percent disability status from the Veterans Administration.[30] When Bud told Jerry that President Clinton "had done much more for the Vietnam veterans than either of the past two presidents," Jerry offered to head a "Vietnam Veterans of Agent Orange Casualty for Clinton" group.[31]

In 1970, at the age of forty-nine, Bud Zumwalt was appointed chief of naval operations. The choice of Defense Secretary Melvin Laird and Navy Secretary John Chafee, Zumwalt was deep-selected over the heads of thirty-three more senior admirals. The youngest CNO in history was expected to create a more egalitarian navy as well as transition it into a smaller and modernized fleet that would be better prepared to cope with the burgeoning Soviet naval threat. Bud Zumwalt took over a navy whose capabilities were deteriorating at precisely the time in history when the world's rival navy was growing in quantity and quality. "It amuses me a little that I am known mostly as the CNO who allowed sailors to grow beards, wear mod clothes, and drive motorcycles," wrote Zumwalt. "In truth, I spent almost all of my time pondering upon and seeking to make a contribution to American society with respect to the U.S.-Soviet maritime balance, strategic arms limitation, naval modernization, and a number of other matters that most people would agree have more bearing on the fate of the nation than what a sailor wears to supper."[32]

Bud understood that the navy had reached a point in its history where it could no longer drift with the tides and winds of change, totally oblivious to the needs of civilian society and the dignity of its personnel. In a March 1970 letter to the editor of *Time* titled "Camelot in Blue & Gold," Dick Rose compared Zumwalt to King Arthur in recognizing that enlisted men were human, too. Rose warned that like Arthur, the navy's barons—Mustangs and senior officers—would sabotage him.[33] Traditionalists charged that good order and discipline were no longer valued in a fighting force. Bud often quoted Winston Churchill's alleged reply when told that an initiative flouted the traditions of the Royal Navy. "I'll tell you what the traditions of the Royal Navy are . . . rum, sodomy, and the lash."[34] Bud was not going to saddle the majority of responsible, conscientious, and mature navy people with restrictive regulations designed to restrain the few who would abuse privileges. His goal as a leader was to reform, not destroy the institution. Traditions that aided in the achievement of the navy's mission and added to its pride, esprit de corps, and morale would be retained. Arbitrary, unduly or unnecessarily irritating regulations and those demeaning to dignity or counterproductive of mission would be eliminated or modified. "I'm sure that when flogging was abolished in the Navy there were those (in uniform and out) who regarded that as a fatally 'permissive' move," Bud wrote to one critic.[35]

What set Bud Zumwalt apart from senior officers in the navy as well as from most military leaders was that he was a maverick with the courage of a lion. Bud Zumwalt dedicated his life to the qualities of leadership espoused in Admiral Ernest J. King's definition of military discipline: "The intelligent obedience of each for the effectiveness of all."[36] To achieve this meant having the most dedicated, perceptive, and imaginative leaders possible, leaders who possessed and utilized the virtues of firmness, fairness, compassion, and a sense of judgment.

In order to save the navy, Bud Zumwalt embraced the redrafting and reshaping of its social contract. His aim was to cast the navy in a more humane and more just light. In doing so, he was essentially telling the navy it was grossly behind the country, and certainly the other services, in regard to equality of opportunity. He brought his enormous intellectual powers to this new social contract, and also the courage and the strength to take all the heat from those who didn't want to see that contract rewritten.[37]

One of Bud's closest associates, Admiral Worth Bagley, thought that "he showed—probably more than anybody else—that the navy's not as great as it thinks it is."[38] Bud's greatest challenge as the navy's leader was in combating institutional racism. To no one's surprise, the ancient sons of Neptune did everything possible to construe *integration* as a synonym for *permissiveness.*

Sitting in one of the pews was another honorary pallbearer, Captain William Norman, a black naval officer who taught Bud Zumwalt what it was like to be black in the white navy. In 1956 Norman thought of applying to the Naval Academy, but the recruiter dissuaded him because "blacks were not ready."[39] In 1967 Norman had been teaching political science at Annapolis, where no one would rent him an apartment. Seared in Norman's memory was the experience of going to an Officers' Club in Meridian, Mississippi, and being told by the executive officer that he should not come to dinner because other officers and guests would be embarrassed by his presence. Norman was an officer, but whites of inferior rank did not salute him.

Norman detested the treatment so much that he decided to leave the navy. "I knew I could not continue in the navy to serve with the kind of pride and dignity that I was a part of because I still was comparatively ashamed of being a naval officer. I was ashamed every time I walked into a wardroom and looked around me and saw myself as just about the only black face in there except those that were serving food or the Filipinos."[40] When Bud learned that one of the navy's most promising black officers was resigning his commission, the new CNO requested a meeting. Norman arrived with a one-page list of proposals and ideas for dealing with the navy's deeply rooted institutional racism, including tackling the racially segregated rating system that kept Filipinos as stewards and manservants for the white admirals, addressed as "Cook," "Stew," or "TN" (table navigator). It was a cozy system wherein navy recruiters steered Filipinos to steward and mess jobs despite the fact that twenty-four other positions were available to them. After listening to Norman, Bud pledged that as CNO he would be the agent of social change, but said he needed Norman at his side.

All of the ratings were soon opened to minority enlisted personnel, an action that was met with extraordinary resistance and fierce opposition, even from close friends. Bud refused to waver, because in his heart he knew it was the right thing to do for the future of all career service members and their families. Z-gram 66, one of his famous memos, represented this commitment to eradicate demeaning areas of discrimination. "Ours must be a navy family that recognizes no artificial barriers of race, color or religion. There is no black navy, no white navy—just one navy—the United States Navy."

Former deputy assistant secretary of defense Dov Zakheim was upset that he was not able to be in the chapel. He was at the JFK School of Government, about to deliver a major address to senior Russian and American officers on major global military trends. Setting aside his lecture notes, Dov first spoke about his friend Bud Zumwalt: "He recognized that no military is stronger than its personnel. And that a military that discriminated against its own was a military doomed to defeat. So he cleaned up the Navy. He eliminated institutionalized racism. He brought women in to serve. He offered promotions to blacks and Jews and other minorities where none had previously been available. And the Navy grew stronger, not weaker. . . . He fashioned the Navy's relevance in today's world, a quarter-century after he left service. . . . There was nothing in the Navy of 1970, when Bud became its highest-ranking officer, that pointed to the Navy of today. Nothing. It was a troubled force consisting of demoralized men and aging ships, plagued by bitterness, discrimination and hatred. Zumwalt turned the Navy around. He reversed the trends. All great leaders do."[41]

———

Dave Woodbury first met Bud Zumwalt in 1972 when Jerry Wages brought him into the CNO's office to introduce him as Jerry's relief as naval aide to the CNO. Sitting in the chapel pew, Dave remembered that he had last been to the Naval Academy for Bud's controversial change-of-command ceremony some twenty-six years earlier. An honorary pall-bearer, Dave had seen his boss for the final time a few months earlier at a meeting to discuss medical issues related to veterans' exposure to Agent Orange. At the time, he thought Bud looked unusually tired, but Bud tried putting him at ease by saying that he and Mouza were driving the next morning from their home in Arlington, Virginia, to Cary, North Carolina, in order to spend time with their younger daughter, Mouzetta, and her husband, Ron. He would have plenty of time to relax during the family visit.[42]

When they arrived in North Carolina on September 22, Bud felt ill and was taken to a local hospital, where the initial diagnosis was a collapsed lung. Three days later, Bud was transferred to Duke University Hospital, where tests revealed mesothelioma—an incurable cancer of the lung lining related to asbestos exposure. Bud returned to Duke on October 13 for a radical procedure known as extrapleural pneumonectomy—removal of the lung, the lining around the lung sac, and the upper half of the diaphragm. It was a "Hail Mary," especially because such surgery is not normally performed on patients over fifty-five years of age. Bud's physical condition was so good that the attending physicians felt he could undergo the very dangerous operation.[43]

In the days prior to the surgery, hundreds of letters and cards arrived. Scott Davis of the Fred Hutchinson Cancer Research Center opened a meeting of the International Consortium for Research on the Health Effects of Radiation in Russia with the announcement that Zumwalt was ill. Every member of the Russian delegation signed a message wishing "our warm feelings which are coming from our souls and hearts . . . we know you as a person who can foresee the dangers facing humanity and find ways to overcome it."[44]

Rosemary Bryant Mariner, one of the first eight women selected for naval flight training and a retired naval aviator, wrote, "Your leadership as CNO, not just in opening pilot training for women but in forcing the Navy to deal with the reality of an all-volunteer force, made a profound difference in my life and for the national security of our great republic. I am proud to have served under you in the U.S. Navy and as an American citizen."[45]

Joe Ponder sent a gift of "dancing flowers" to the hospital along with a note that "you've brightened my day many times by caring, giving and simple listening."[46] Over thirty years earlier, on November 27, 1968, Zumwalt had been at Joe's bedside in the U.S. Army's Twenty-ninth Medical Evacuation Hospital in Binh Thuy, in the heart of the Mekong. A gunner's mate serving aboard Patrol Craft Fast 31, Joe had been on special missions from Ha Tien near the Cambodian border all the way down to the southernmost tip of the Ca Mau peninsula. On November 24, 1968, his Swift Boat was on a search-and-destroy mission that was part of Zumwalt's Operation SEALORDS, an acronym for Southeast Asia Lake Ocean River Delta Strategy. Joe's boat was soon engaged in armed conflict with enemy forces along the Bo De River. In a fierce and blistering gun battle with VC forces, Joe was seriously wounded in the right leg. As he regularly did, Bud Zumwalt showed up to personally present the Purple Heart, but what Joe remembered most is that the top dog pulled up a chair to speak for thirty minutes. When it was time to leave, Zumwalt maneuvered his way around Joe's tubes to give his sailor a hug, whispering "God Bless you, son. I wish you the very best." Joe thought it was all a dream and almost pinched himself to prove the admiral's visit wasn't. "I will *NEVER* get over the fact that a Vice Admiral, the Commander of all U.S. Naval Forces in Vietnam, Elmo R. Zumwalt took the time out of his very busy and important schedule to visit such a low ranking person!"[47]

Returning to headquarters, Bud penned a personal letter to Rebecca Ponder: "I visited your husband yesterday at the hospital in Binh Thuy. The opportunity for me to talk with such a fine, brave young American was one I will long remember." Bud reassured her that Joe was recovering and looking forward to getting back to his unit. "Your husband told me you were expecting a child. You and your baby should certainly be proud of your husband."[48]

Retired admiral Harry Train wanted his friend to know "you are my all-time hero . . . what you have done for this nation, what you have done for our navy and what you have done for its people is beyond description."[49] Retired admiral Arthur Price, one of Bud's component wartime commanders in Vietnam, wrote, "Remember it was 'you' who did so much to improve conditions for the little guy and his dependents in the Navy."[50]

John Ryland wanted "Big Z" to know that "you will always be remembered as 'The Sailor's Admiral.'" Les Garrett, a Swift Boat sailor on PCF-73, recalled that "at Christmas time in 1969, you opened your Villa to us sailors who were in Saigon at the time. I have never been able to tell you, until now, how *special* that gesture was to us enlisted sailors, who were far from home. You gave us cheer & joy and a little touch of home for a wonderful moment that day. You took the time to give personal encouragement to each and every one of us, something we all remember and pass on to others, just as you did that day. Thank YOU for that special day set against turbulent times. Thank you for being a Commander who never forgot his sailors."[51]

Before becoming ill, Bud had been scheduled to speak at the thirtieth reunion of the Black Ponies, the navy's only permanently based fixed-wing attack squadron in Vietnam and its last combat unit there. Dozens of get-well thoughts arrived, all echoing the words of retired captain Larry Hone, "Admiral Z, thank you for all you've done for the Navy and this nation. Your leadership was an inspiration for us all. Fight hard, get well and make the next VAL-4 reunion."

Registered at Duke under the name of his deceased brother Bruce Craig, Bud was ready for the battle ahead. Minutes before being taken into surgery, S. Scott Balderson, the physician's assistant who developed a profound respect for his patient, came into the room: "Admiral, it's time to say goodbye." Mouza was in her wheelchair, the result of excruciating back and neck pain. Scott thought that Bud "looked like he could still run laps around the nursing station." As Bud bent over, Mouza grabbed her husband's face between her hands to kiss him good-bye. "I love you," she said. Bud took charge as he always did, replying "I love you," and then looked at Scott. "OK, let's go!" After Bud had been wheeled into surgery, Scott noticed that the attendant who had spent the morning preparing Bud for surgery was standing at near attention outside the OR. Scott asked Al, who was older than most of the attendants, why he had not left. "That's my admiral in there," replied Al, who had served under Zumwalt years earlier.[52]

Zumwalt developed complications during surgery.[53] The final eighty-two days of Bud Zumwalt's life were spent in the intensive care unit. His condition remained critical but relatively stable; Bud was usually alert and attentive enough to read or have letters read to him. He was unable to speak, but communicated through the use of a tablet and marker. As he was celebrating his seventy-ninth birthday in the ICU, special notes arrived. Paul Nitze encouraged his friend to be strong and stay the course so that they could resume their lunches at the Metropolitan Club. "We can reminisce about all the good and exciting things we were able to get done together."[54] Another note came from President Clinton, who had called a few days earlier. "I know things have been difficult for you lately—but I want you to know I'm still thinking about you and pulling for you. Hang in there! Hillary and I are holding you in our prayers."[55]

Bud soon took a turn for the worse. He had brain-stem damage, and major organs were shutting down. He had prepared for this moment by executing a health-care proxy and designating son-in-law Dr. Michael Coppola to carry out his final wishes. Years earlier, Michael had served as Elmo's proxy. Bud's desire was to follow procedures similar to those Elmo had established prior to his own death. Bud left clear instructions with Michael that his life should not be prolonged by life support. He trusted Michael to advise the family when that time arrived.

Sitting at her husband's bedside and realizing his end was imminent, Mouza asked Michael and Ann to bring a letter to the annual New Year's Renaissance gathering. The host of Renaissance, Ambassador Philip Lader, read Mouza's letter aloud. "As I sit here at Bud's bedside in the intensive care room, the walls covered with pictures of our life together, my emotions are mixed. There is sadness over the thought of losing my partner of 54 years but also joy over the terrific life we enjoyed. As I stroke his hair, it is not the gray that has set in over the past many years that I see—it is the ever-young, dashing Navy lieutenant who swept me off my feet when I first met him in Shanghai, China, in October of 1945. Despite the fact I then spoke very little English and he very little Russian, we did speak the international language of love which resulted, as many of you are aware, in our getting married after only knowing each other three weeks. As I look at Bud, I do not see a bed-ridden man now dependent upon others during his last days of life—I see a man who championed the causes of so many throughout that life. I see a white knight who knew he had so much to accomplish yet so little time in which to do it. I see a man who in nearly eight decades of life was no stranger to tragedy, suffering the loss of a young mother and brother in his early years and a son in his later ones, yet using these personal tragedies as a foundation for helping others."

After a moment of silence, 1,600 attendees spontaneously stood to salute their longtime friend.[56] "It was not the Admiral that we saluted," recalled Lader. "It was the man."[57]

It was at Hilton Head that Michael made the decision to take Bud off the ventilator. He knew Bud's wishes, and as proxy his job was to determine when Bud's specifications and criteria had been met. After sitting down with the immediate family, the decision was made to set sunrise January 2, 2000, as the time for bidding farewell. In the words of their daughter Mouzetta, "sunrise seems to represent the beginning of a new day synonymous with hope and promise—which was everything Dad represented."

That morning, with the men dressed in jacket and tie and the ladies in their best finery, family members assembled in the hospital room for their personal good-byes. The shade of the hospital window was removed so that sunrise could be observed. At 6:30 a.m., a 36 × 24 color photo of young Elmo and his father walking down a wooded path was brought into the room. Bud had previously confided and written that he hoped to soon be reunited with Elmo and his parents. Mouzetta felt that "one had the sensation that perhaps Elmo was greeting him."

Grandson Elmo Russell Zumwalt IV, born with learning disabilities that his father and grandfather attributed to Elmo's exposure to Agent

Orange, came earlier to say good-bye. Weeping into his grandfather's ear, Russell could be heard saying, "You are a good sailor, Granddad."

Each family member then took a few minutes in private, although it was all too much for James Zumwalt, the surviving son of Bud and Mouza. A combat-hardened marine who had served tours in Vietnam and Iraq, years earlier he had held his brother Elmo in his arms as the valiant warrior took his last breath. Jim chose not to bear witness to his father's passing.

The attending physicians and family looked to Michael for the nod that life support was to be turned off. As the nurse began to administer slow-drip morphine, life-support lines were removed. Family members saw Bud's eyes open to look at the picture of Elmo. Granddaughter Lauren held his hand, whispering words of love into his ears. "I will never forget laying my cheek on yours and stroking your soft hair. . . . I couldn't let go of your hand those last minutes of your remarkable life. . . . As I watched and held your hand in a tight clutch I saw you take your last deep breath for air in life. I looked up at the machine above your bed, your heart rate went to zero. Laying there you looked so peaceful and I knew you were in a better place."[58]

The nurses who had grown especially fond of their patient had brought a small wooden cross that they placed in Bud's hands. The cross, as well as notes written by the family, was later sealed in the coffin. Colonel Mike Spiro, who served as the CNO's marine aide, came to the hospital just to be close to "the greatest person I have ever known in my life." Mike made certain that Bud's remains were handled respectfully and escorted his boss's body from North Carolina to the funeral home in Washington, D.C., then from the funeral home to the Navy Chapel, and then on to his final resting place at Annapolis. He returned weekly to make certain that the grave site was maintained to the standards set by the U.S. Marines. "I honor him with no limitations and no bounds" is what Mike wanted Bud to know.[59]

Bud Zumwalt lived to see only two days of the new millennium, but he had played a major role in shaping the future. As news of his death spread, the impact of this remarkable man manifested itself in several ways. William Franke, who served with Elmo on the Swift Boats, wrote Mouza that he "was an inspirational man . . . history will treat him with honor." In a mass mailing, Rick Hind of Greenpeace informed his fellow activists that their great ally in the fight against dioxin was gone. "He came to that fight after seeing his own son die from exposure to Agent Orange used in Vietnam. This tragedy was compounded by the fact that as commander of naval forces in Vietnam he ordered the use of Agent Orange to 'save lives.'" It was Zumwalt who stood up to the "company docs" as he called them, who denied Agent Orange's dioxin toxicity. "He was one of a kind and will be sorely missed but always remembered."[60]

In a personal letter to Mouza, President Bill Clinton described Bud as "a great patriot, leader, husband, father and friend. Hillary and I will always cherish the times we shared, the work we did on war-related health problems and the opportunity I had to award him the Medal of Freedom."[61]

Most revealing of Bud Zumwalt's character was a letter written to Mouza by a complete stranger. Cindy Wofford spoke of her eight-year-old daughter, who had relapsed with leukemia. Someone had given her Bud Zumwalt's name as a person who had information on bone marrow treatment. Wofford was desperate when she called Zumwalt, who spent time speaking with Cindy, demonstrating caring and compassion. "Even more amazing was the way he included your son in our discussion. He was so proud of your son and so thankful of the time they shared. I was humbled by the thankfulness in his voice. It was evident that your husband was so honored by his role as a father. Your son was such an important part of his life. Very openly, your husband shared the human side of the transplant procedure and stressed the fact that although your son died post-transplant that he had always viewed the transplant as successful. The procedure had given them more time to be together. He shared the closeness they shared towards the end and the bond that would never be broken. . . . Your husband amazed me with his empathy, his ability to share details of his life which are far too often guarded from others—the ramblings of the heart, the love of a child, and the grace in knowing how to let go when it is time."

Cindy ended the letter by telling Mouza that Bud's words provided the strength needed as she held her daughter until she took her last breath.

Bud requested to be buried in his blue uniform with service ribbons, hat by his side, leaving to Mouza the decision of an open coffin. Reflecting his high esteem for enlisted sailors, the current master chief petty officer of the navy was to present the flag to the next of kin. Two songs were to be played, "The Impossible Dream" and "Lara's Theme," the same music that had been played four years earlier at Bud and Mouza's fiftieth wedding anniversary.

At ten minutes after two o'clock, the Zumwalt family entered the chapel in the company of the president of the United States, who slowly pushed Mouza's wheelchair. Four midshipmen, four chief petty officers, and four enlisted sailors sat nearby, representing "the heart, soul and future of the greatest Navy in the world." In accordance with the admiral's final instructions, Master Chief Petty Officer of the Navy James L. Herdt presented the flag to Mouza.

The flag-draped coffin was set in front of the altar, adjacent to a stained-glass window depicting David Farragut lashed to the rigging of his flagship, the USS *Hartford*, while navigating the minefields of Mobile Bay, with the Archangel Michael showing the way. The Reverend Burton Shepherd read sentences from Scripture. Rear Admiral Shepherd's naval career spanned over thirty years; he was one of the most highly decorated aviators of the Vietnam War and executive assistant to Bud as CNO. "He called me from North Carolina just before the surgery to ask if I would do the funeral service," recalled Shepherd. "These were his final instructions to me."[62] Shepherd especially admired what Bud had done for others throughout his life. He "was always trying to do something for the underdog; he touched the lives of so many."

The first speaker was Chief of Naval Operations Admiral Jay Johnson. Focusing on Zumwalt's "visionary leadership and unswerving commitment to improving the lives of our sailors," Johnson noted that his friend had "profoundly changed and enhanced the character and culture of our Navy." In Johnson's words, Bud Zumwalt was "the epitome of humility, dignity and grace, a gentleman and reformer." These words did not sit well with Admiral Thomas Moorer, sitting in one of the pews alongside another former chairman of the Joint Chiefs of Staff, Admiral William Crowe. "I sat beside Admiral Moorer at the Zumwalt funeral, and all these eulogies started talking about how Zumwalt saved the Navy. And I thought, 'God, I can't look at Moorer. He's madder and madder and madder.'"[63]

The CNO read a passage from the New Testament, which captured the essence of how Bud Zumwalt lived his life and what he imbued in his sailors: "Whatever is truth, whatever is noble, whatever is right . . . whatever you have learned . . . from me . . . put it into practice. And the God of peace will be with you."[64] Johnson closed with the words, "The God of Peace is with Admiral Bud Zumwalt—and his son Elmo. Father and son—both men of the sea—both surface warriors—reunited again after eleven years."

Richard Schifter, former assistant secretary of state for human rights and former deputy ambassador of the United States to the United Nations, spoke of his friend's engagement as a citizen in matters of public concern at the individual level. "He demonstrated his ability as a leader and trendsetter in public life, both in the Navy and as a civilian. But all of his work played out against the backdrop of his humanity, his concerns for people and for our country, not just in the aggregate, but as they related to individuals whose lives he would touch, whether it was a sailor who did not have to change his clothes before he could have a meal, or a veteran who had contracted a war-related disease."

Over the years, Schifter and Bud worked tirelessly in getting clearances and sponsorship for dozens of Vietnamese to enter the United States, culminating with the arrival on December 9, 1992, of Admiral Tran Van Chon, who spent fifteen years imprisoned by the Vietnamese government. Bud called in every favor and used his extensive government contacts in trying to get Chon released so that he could be reunited with a man he considered a brother.

Schifter and Bud spent numerous weekends together at Renaissance, becoming the closest of confidants. Schifter admired Bud Zumwalt for being a great humanitarian, speaking up about the dangers posed by a Soviet arms buildup when the president of the United States wanted to look the other way. He felt that Zumwalt played a key role in launching an awareness effort that led to the adoption of policies that brought the Cold War to a successful conclusion for the cause of freedom. Schifter also knew the critically important role Zumwalt played during the October 1973 invasion of Israel by its Arab neighbors. The Israelis had not counted on the Soviet supply of weapons to the Arab countries and quickly ran out of supplies during the first days of the war. There was hesitation in the United States government as to whether to move quickly to resupply Israel. It was the chief of naval operations who recognized the danger to Israel's survival and the U.S. national interest in preventing its collapse. Bud informed Senator Henry "Scoop" Jackson that Israel was in dire straits and in desperate need of a resupply of arms. The problem was that Henry Kissinger was opposed to an immediate resupply, saying "let them bleed a little." Senator Jackson immediately got in touch with Secretary of Defense James Schlesinger, warning him that he would go public in pointing out the serious problem if Schlesinger did not act without delay. Schlesinger, who had been uncertain as to what to do, got President Nixon to authorize the resupply.[65]

The next speaker was the U.S. ambassador to the Court of St. James's, Philip Lader, who, with his wife, Linda, was cofounder of the annual Renaissance Weekends. Lader spoke eloquently of a "model of the life well lived." It was "the man, not the Admiral; the story, not his achievements, that most touched us." By this Lader meant "not simply of sailors or Americans, but of handicapped children in Vietnam, victims of radiation in Russia, marrow transplant recipients." These commitments and caring for an "extended family" characterized the dominant trait of Bud's entire life.

Bill Clinton had last been in the chapel for the funeral of Admiral Arleigh Burke. If Burke had been the spirit of the navy, Clinton said, "Bud Zumwalt was its conscience." The president told the audience that as he was getting dressed that morning, his navy steward said of Zumwalt,

"He's the best we ever had. He was for us." The president noted that "Americans could always count on Bud Zumwalt to do the right thing. . . . He sailed through rough waters more than once. . . . When we struggled through the racial tensions of the sixties and seventies, he worked in the face of wilting criticism and a highly resistant institutional culture to make the Navy do the right thing and make the Navy one of the most color blind institutions in our entire Nation. I know it was a special point of pride for him that the very first African-American admiral earned his star on Bud Zumwalt's watch. At a time when morale and enthusiasm were at an all-time low, he had the vision to see a great future for the Navy." The president noted that "thousands of naval leaders like former Secretary of the Navy John Lehman have said that they actually made the decision to stay in the Navy because Bud Zumwalt made the Navy exciting again."

President Clinton recalled the time that Vietnamese refugees had been placed in temporary housing in his home state of Arkansas. To stay, they needed a sponsor. The only name they knew was Admiral Zumwalt. Bud made arrangements for them to fly to North Carolina, where they stayed in his son Elmo's home because the admiral already had families living with him. "When Bud Zumwalt made a commitment, he stuck with it," said the president.

The president spoke about how Bud had given "honest, caring, steadfast friendship. His letter to our daughter about what her parents tried to do for America is one of our family's most cherished possessions. It is the symbol of everything he was as a man, a leader, and a friend. And so today we say goodbye to the sailor who never stopped serving his country, never stopped fighting for the men and women in uniform, never stopped being *the conscience of the Navy.*"[66]

After the service, with the U.S. Navy Band leading the way, the procession walked from the chapel across the Naval Academy grounds to the cemetery. The sun broke through the chapel windows just before they departed, and Jim Zumwalt told those close by "it was if God had showed my dad where the climate control switch was and Dad pulled it." As sailors fired nineteen guns at five-second intervals in salute, Joe Muharsky thought that Elmo had asked God to turn off the rain in order to say, "Welcome home, Dad."

Then, near the crest of a hill overlooking the blue waters of the Severn River, where he had first learned seamanship in the class of '43, Bud Zumwalt was laid to rest. The single-word inscription on his tombstone encapsulated his life, REFORMER.

The next morning, far from the pomp of the previous day's remembrance, Bud's brother Jim returned to the cemetery in order to deliver a personal eulogy. Turning to the grave site and surrounded by his close-knit family, Jim saluted his brother one last time with the words "Bravo Zulu, Admiral Zumwalt. Well done!"[67]

CHAPTER 2

THE ROAD FROM TULARE TO ANNAPOLIS

*Few people in my home town would have predicted in 1938
that the Class Valedictorian of Tulare Union High School,
locally famous for his humorous pranks and hell-bent ways,
would someday achieve unparalleled acclaim as the Navy's
top officer.*

—JAMES G. ZUMWALT, BUD'S BROTHER[1]

The attending physician at Stanford University Hospital joked about the nine-pound, thirteen-ounce newborn having the largest set of eyebrows he had ever seen.[2] Elmo Russell Zumwalt, Jr., the son of two physicians, Dr. Elmo Russell Zumwalt and Dr. Frances Frank Zumwalt, entered the world on November 29, 1920. "I well remember the day in 1920," recalled his father, "following a long drive to San Francisco, and viewing a very ruddy faced, compact, chesty individual with a dark head of hair soon to be named by your sister, who was then twenty-months old, who being told 'this is your brother,' enunciated 'Bud-dy.' "[3]

Saralee's nickname stuck. While in primary school, her brother went by the moniker Buddy, but by the time he reached junior high school, Elmo Russell Zumwalt, Jr., introduced himself as Bud and asked others to "call me Bud." This usually posed little problem, although it generated criticism from old-guard retired admirals who objected to such informality. Representative of the traditionalists' viewpoint was retired rear admiral Colby Guequierre Rucker, who criticized Admiral Zumwalt for using "Bud" when signing communications to the naval community, especially those junior in rank to him. "I can assure it does not arise from a misplaced desire on my part to be overly 'buddie' with anyone," replied Bud. "With a name like Elmo, as you may have found in *your own experience*, one must look for alternatives where he can."[4]

Coincidentally, St. Elmo is the patron saint of sailors, but in this case the name Elmo was selected by Bud's grandmother for his father from one of the most popular books of the nineteenth century, written by Augusta Jane Evans. The protagonist in the novel *St. Elmo* transformed himself from a rather contemptible individual into a dashing hero. The book was so popular that dozens of southern plantations, local schools, cocktail drinks, and newborn children were given the name Elmo.[5]

Raised in the city of Tulare in the fertile San Joaquin Valley, Bud and his siblings enjoyed an almost idyllic pastoral life. Founded in 1872 by the Southern Pacific Railroad as headquarters for the valley, Tulare was a small town of approximately seven thousand proud and close-knit people, who despite differences in background were vitally interested in and supportive of one another.[6] In a letter to a former teacher, Bud recalled how growing up in a small town provided him "the basic serenity with which to survive."[7]

Bud was a fourth-generation Californian. His great-grandfather had been one of the area's first settlers, and the Zumwalt name was a familiar one in Tulare County. Genealogists trace all Zumwalts in America as descendants of Andreas zum Wald, who boarded the SS *Virtuous Grace* in the port of Rotterdam, disembarking in the port of Baltimore in 1737.[8] After marrying a Swiss-born wife, Mary, Andreas farmed a few acres near York, Pennsylvania. Five children were born of that marriage, but after Mary's death, Andreas sent back to Switzerland for another wife. Eventually settling in Virginia, Andreas and Ann Regina had six children of their own, one daughter and five sons. One of those sons was named Jacob, from whom Bud Zumwalt is descended.

Following the war for independence, Jacob moved to Missouri in 1798, building a fort northwest of the Missouri River that provided refuge for many families during the War of 1812. The ruins of Fort Zumwalt, near O'Fallon, Missouri, are today preserved as a National Monument. In 1849 one of Jacob's great-grandchildren, James Brown Zumwalt, orphaned at the age of three and a blacksmith by trade, joined a wagon train to California, eventually settling in the Sacramento Valley. After working in the mines at Murder's Bar on the Middle Fork of the American River, James settled in Red Bluff, where he opened a blacksmith shop. In 1860, James married Lydia DeWitt, who had also crossed the continent by wagon, from Kentucky in 1852.[9]

In 1864 the family moved to Grand Island, Colusa County, purchasing 160 acres of land for farming. Fourteen years later, in search of flood-free fertile farmland, the family packed their belongings and traveled by wagon to Tulare, where James purchased 900 acres of land a few miles northwest of the city line for five dollars an acre. Within a few years, he was farming alfalfa and fruit trees, primarily raisin grapes. The land was also home to 350 head of horses, cattle, and hogs and a dairy of 80 cows. During the summer, they grazed the cattle in the Sierras in an area identified today on topographical maps as Zumwalt Meadows, along the banks of the Kings River in Kings Canyon National Park.[10] His plantation-style home in Tulare became a showplace, known as the Palace Ranch of the San Joaquin Valley.[11]

James and Lydia needed a large home for their eleven children, one of whom was James Eleazar Zumwalt, who later married Mabel Ford. Their only son, Elmo Russell Zumwalt, Bud's father, was born at the Palace Ranch in 1892.[12] James Zumwalt worked as a schoolteacher, earning a salary of $75 a week, and young Elmo began school in the small town of Dixon, near Davis and Sacramento. "Dad had a school out in the country about five miles and used a bike to make the trip," recalled Elmo.[13] The family lived without electricity or an indoor bath. "I remember the first stationary bathtub brought into the house and the thrill on the day the house was wired for electricity that had been brought to town and we mounted an old Edison carbon light bulb that gave about a twenty watt light, a tremendous advance over the coal oil lamps."

When James was offered a better teaching job, the family relocated to Richmond, just across the bay from San Francisco. James later became principal of Lincoln High School, the first of its type in Richmond. Elmo was one of seven students in the third graduating class from Richmond High. He intended to study dentistry, which required going to San Francisco for a three-year course of study, but his father's counsel was that his son would be much better off at UC Berkeley. Elmo accepted the advice and enrolled as a freshman in 1911. Required by the registrar to put down a field of study, Elmo wrote *medicine*, because it was the field closest to his previous career interest. After two years of premed classes, Elmo needed a break and dropped out of college, taking a job working on the Santa Fe Railroad in Richmond. Fourteen months later, he returned to Berkeley as a premed student, buttressed by a savings account of $500 from a monthly wage of $76 working for the railroad.

While on the Berkeley campus, Elmo met Frances Frank, another premed student, whose family resided in Los Angeles. By all accounts, they took an immediate liking to each other and quickly fell in love. Intent on finishing their respective medical degrees, Elmo attended the University of California San Francisco Medical School while Frances returned to Los Angeles to attend the University of Southern California Keck Medical School. Frances and Elmo wrote each other regularly but did not see each other until Frances started her residency at Children's Hospital in San Francisco in 1917. In that same year, Elmo was commissioned second lieutenant in the Army Medical Corps and sent to Fort McDowell on Angel Island.

Lieutenant Elmo Russell Zumwalt and Frances Frank were married in Richmond's First Christian Church on Elmo's birthday, February 7, 1918. It was the city's first military wedding, attended by over 150 guests, although noticeably absent were the bride's parents. With a large American flag spread above the pulpit, the couple approached the altar to the music of Mendelssohn's "Wedding March," exchanged vows to the soft strains of "Sunshine of Your Smile," and departed to the "Wedding March" from *Lohengrin*.[14]

Following a brief honeymoon, the newlyweds moved into a cottage on Angel Island, where Elmo was stationed. The two doctors started their medical careers in San Francisco, but Frances soon became pregnant. The tranquility of Tulare appealed to them, as did the idea of returning to the place of Elmo's birth to practice medicine. "Neither your mother or I enjoyed SF and we cast our lot in Tulare for a few years," wrote Elmo to his children. "I remember driving there with a new Buick, no money, and two weeks later Saralee arrived."[15]

Dr. Elmo Zumwalt's arrival in Tulare was big enough news to make the front page of the *Daily Tulare Register*—NEW DOCTOR TO LOCATE HERE: DR. ZUMWALT.[16] Setting up practice in the old Ryan Building, "Doc Zumwalt" was a general practitioner who became one of the most respected and beloved men in Tulare. Frances Zumwalt did not practice in a formal sense, but mothers started bringing their children to see "Doc Frances" as a pediatric doctor at home. Before long, mothers and babies seemed to always be in their home.

Devotion to public service characterized Doc Zumwalt's life. "I well recall that more evenings than not he returned home from a difficult day to have a quick meal with us before departing to meet with one of two school boards, to meet with the scouting council, or to attend to miscellaneous community chores," recalled Bud.[17] Elmo served as city health officer, county health officer, school board member, Rotarian, and Boy Scout leader. He was also elected mayor of Tulare. He served his country in two wars, volunteering to return to army service in World War II, reentering as a captain and returning to Tulare in 1919 as a full colonel. Tularians recognized him as the complete citizen. "Tulare can offer everlasting thanks to its lucky star that Elmo R. Zumwalt, Sr., saw fit to return in 1919 to the place of his birth—and then spent much of the rest of his life helping make it the fine community it is today."[18]

One of the cherished moments in Bud's life came when he and his seventy-nine-year-old father returned to Tulare for "Zumwalt Day." The city renamed a park in honor of both of them, even unveiling a special Z-gram for the occasion. Four thousand citizens turned out to acknowledge the family's lifelong service to the community and nation. "I have searched the world over and have never found another more worthy, in my judgment, of being called The Good Samaritan," said Bud of his father.[19] In Zumwalt Park are two plaques, side by side, dedicating the park to Dr. Elmo R. Zumwalt, Sr., and Admiral Elmo Russell Zumwalt, Jr. Following Elmo's death, the inscription on his plaque read: "He dedicated his life to his fellow man, to his country and to his profession. With his ashes scattered over this land he loved so well."[20]

Little was known about Frances Frank except for what she told people. That story began with the birth of Frances Pearl in Burlington,

Vermont, on December 12, 1892, to French-Canadian parents. When she was three months old, her entire family died from smallpox. Frances was taken in by the neighbors and raised under their surname, Frank. By the early 1900s, she had moved west with the Franks to Los Angeles.[21] The Zumwalt children recall their mother being very guarded about her childhood, always managing to divert conversation. "She seldom spoke of her early childhood," recalled Saralee, "but I did learn that when she was a toddler, both her parents died in some kind of epidemic. She was taken in by a family who had befriended her folks at the time they came to Vermont."[22]

The story that Frances was born to French-Canadian parents in Vermont who died in a smallpox epidemic and was then adopted by the Frank family in Burlington is false. Extensive research conducted by Saralee's son, Richard Crowe, combined with DNA testing, has solved that mystery, but spawned others. The Frank family was Jewish and immigrated to America in 1890 from a part of Russia that is now Lithuania. Frances was one of six children born to Julius and Sarah Frank.

For the first ten years of her life, Frances lived in Burlington, Vermont. The Franks then moved for a short time to Canada before settling in a newly established Jewish community in Los Angeles known as Boyle Heights. The Franks were deeply religious orthodox Jews, and Yiddish and English were spoken in their home. When Frances and Elmo married in 1918, the Frank family was not present. The marriage began a lifelong estrangement in which Frances shut the Frank family out of her Zumwalt life. The estrangement appears to center on the Franks being tradition-bound religious conservatives who told their daughter that if she married outside the faith, she would be dead to them. Sarah Frank's 1933 will, updated in 1940, probated upon Sarah's death, sent a clear message: "To my daughter, Frances Zumwalt, of Tulare, Calif., the sum of Ten Dollars ($10.00). I am purposely leaving no more to her for personal and sufficient reasons, and may God forgive her."[23]

Away from home for the first time at Berkeley, Frances found freedom from the strict religious upbringing that she found so stifling. By marrying Elmo, Frances divorced herself from her parents and chose to hide her Jewish roots forever. It seems that Frances made up the entire adoption story and never discussed her Frank family—nor introduced them—to her Zumwalt family, even though they lived just hours apart in the same state. Saralee occasionally asked her mother about the Frank family and her life as a child. "Frances curtly replied that she was disowned when she married Elmo and there was nothing to discuss."[24] Both Frances and Elmo went to great lengths to hide the truth from their children. Even after Frances's death and his own remarriage, Elmo never told his children the truth. There is ample evidence that Frances's siblings were aware of their Zumwalt family. On January 10, 1971, Bud received a letter from Anna Rich, Frances's younger sister, who wrote that "as your mother's younger sister, I feel that I stood in for her in sharing the rewards of your success."[25]

The DNA results raise an interesting issue with respect to Bud's navy career. He remains the nation's youngest CNO, but because his mother was Jewish, he is a Jew by birth and can be considered the navy's first Jewish CNO, replacing his dear personal friend Admiral Jeremy Michael Boorda, who also married out of his faith. It was Bud's call to Boorda at the behest of Dov Zakheim that began the process of creating a privately funded Jewish Center and chapel at the United States Naval Academy.[26] "Helping to create a Jewish chapel for the Naval Academy was typical of Bud. It was another case of reversing trends and creating new ones," observed Zakheim.

Until Bud was about eight years old, the family resided in a small two-bedroom, single-bath bungalow at 136 North H Street, across from the Congregational Church in Tulare. Jim recalled the house being surrounded by huge trees and shrubbery, with a single large brown turkey fig in the yard, along with a grove of bamboo.[27] The family of six quickly outgrew the H Street home and at the beginning of the Depression in 1929 moved to a much larger home on the other side of town.[28]

Sycamore Avenue was the nicest avenue in Tulare, and 854 Sycamore was a stately residence with a trimmed Bermuda lawn and colorful flower beds with zinnias, roses, azaleas, and irises. Built in 1906 at a time when workers took personal pride in craftsmanship, the home reflected the architectural style of the era—a large screened front porch, separate garage, hardwood floors of oak, redwood, and cedar, a large living room and formal dining room, separate reading room, large kitchen and eating nook, three bedrooms, a huge fireplace, a cellar, and an attic.

The children loved their new home, especially in spring and early summer as the sycamores' green leaves cast a wide swath of shade over the street. From their bedroom, they could smell the distinctive aroma from Giannini Winery two miles north. Sundays were reserved for formal family dinners. With guests sitting around the huge mahogany dining table on mahogany chairs stuffed with horsehair and covered with gold upholstery, dinner was overseen by Frances. Out came the fine china and silverware, otherwise stored in a matching armoire and chest with intricate soft-green inlaid figures representing classical Chinese scenes. A bell was placed alongside Frances's setting, "which she rang at the end of each course," recalled Jim. The family's longtime housekeeper and maid, Sebelia Hamilton, then cleared the table for the next offering.

Frances had more than food on her agenda for these family gatherings. She was determined that her children learn proper dinner manners and social etiquette, and they were schooled in the art of carrying on dinner conversation. Years later, after attending a special dinner at the home of the Naval Academy's chaplain, Bud proudly reported that all the hours practicing manners and social graces at Sunday dinner had paid off. He impressed everyone, most especially the young lady he was courting.

Elmo was a staunch Republican and one of President Herbert Hoover's strongest supporters in the elections of 1928 and 1932. In November 1935, local Republican leaders brought Hoover, then an ex-president, through Tulare and arranged for a dinner at the Zumwalt home.[29] In the days preceding the retinue's arrival, the home was full of people cleaning rugs, walls, dinnerware, waxing the hardwood floors and polishing the silver. Gardeners manicured the lawn, eradicating all weeds and crabgrass. The ex-president's group included Mrs. Hoover, their son, and a number of aides and spouses. Bud and Saralee were invited to meet the ex-president. Tulare was a staunchly Democratic community, and Hoover was extremely unpopular among locals. Following the visit, Bud and Jim were ridiculed at school, suffering the taunts of children that they were rich capitalists with little empathy for the common people.[30]

The first of several tragedies that would shape Bud's approach to life occurred when he was ten years old. As a youngster, Bruce Craig was the most daring and independent of the four siblings. The most frequent word used on report cards to describe the boy was *rebellious*. His mother, Frances, thought that "this particular boy represented angel and devil—the thunder and lightning and sunshine—the flowers and trees and the mountains—the noise and yells and scuffling, the dear little sly ways sensitive boys have of showing affection—the airplanes, worms, bugs, butterflies-birds, etc., etc., that he loved. I see him swishing through every puddle, his face turned to the skies and hear him say, 'Let it pour.'"[31]

The family loved spending summers at the Sea Garden apartments in Pismo Beach, where they dug for clams, swam in the surf, or drove to the sand dunes a few miles away at Oceana. Doc Elmo would remain at his medical practice in Tulare during the week and make the five-hour drive on Friday in his 1925 Packard. One day after returning from the beach, Bruce Craig became ill and was taken by emergency transport to a hospital in San Francisco. The original diagnosis was infantile paralysis with a chance of tubercular meningitis. The symptoms closely resembled each other and were atypical for either. Elmo and Frances flew in specialists whom they trusted and respected. The consensus was that the serum had to be given for infantile paralysis because if it were tubercular meningitis there would be no harm done. There was a glimmer of hope that infantile paralysis could be avoided if the serum was given early enough. If it proved to be the other, there was none.

Bruce Craig's illness proved to be tubercular meningitis. "Everything was done that could be done," explained Frances. "To say we are reconciled is to be hypocritical. I know we are going to have to bear it but both of us will always rebel that it had to be."[32] With their parents in Los Angeles consulting with physicians, Bud and Jim were staying with the Hopkins family in Tulare while Saralee was with Frances's close friends, Ida and Otto Parlier. One day Fred Hopkins took Bud aside to say that his brother had passed away. With tears in his eyes, Bud approached his younger brother. "Jim, Bruce Craig is dead. Don't you understand? Bruce is dead; he's dead!"

A pall of gloom descended over the household. Bruce Craig's funeral was held at the family's Sycamore home and his body cremated. His ashes were later scattered over Tulare with those of his mother. Frances wanted no grave or tombstone. "I don't think my mother ever stopped grieving for him," said Jim. Doc Elmo was adrift as well. For years he kept on his desk a crude lantern made a few days before Bruce Craig became ill. It was a tin can with ten ragged holes punched out with a blunt tool. A candle had been fused with melted wax to the bottom. The family never returned to Pismo again. "My mother always associated Pismo to that tragedy. The incubation period of his fatal disease meant that he had contracted it during the time we had been vacationing there."[33]

By his late teens, Bud had discovered girls, leading to major conflict with his father and creating much tension within the household, which was compounded by his mother's diagnosis of cancer. "The overriding trauma of my youth was the slow demise of my mother," recalled Bud. "She had had a very minor lump in a breast, which was to have been biopsied at the time of the death of my brother. But the tragedy of that loss led these two doctor parents to defer the removal of the lump for about a year. By the time it was removed, it was malignant. The breast was removed, and then she nearly made it. Five years later, almost to the day, the metastasis had apparently taken place, and there was a recurrence so that in my senior year in high school she began the terminal period."[34]

Most of Bud's attention was directed at classmate Anita Whistler, whom he almost married during his senior year. On one occasion, the couple left Tulare for Nevada to get married. "We got as far as Fresno—40 miles away—and I persuaded her that we should turn around and go home. I just couldn't face what it would do to my parents."[35] Elmo threatened to throw Bud out of their home if he continued seeing Anita.

Bud's socializing convinced his parents that West Point could provide the discipline needed for assuming greater individual responsibility. "The lack of strict discipline and the temptation of youth led me to flirt very seriously with a less responsible life," recalled Bud.[36] Elmo devised a "high-pressured" effort for doing everything possible to secure a spot in one of the military service academies.[37] "I guess there was always a dichotomy in my own mind, with my father and, I think, my mother also, really clearly preferring that I go to West Point. And I was kind of fighting it in the early years," recalled Bud. "I think, in my own mind, I really visualized the optimal life being the achievement of a medical degree and going into practice with my parents, because I was very devoted to them both."[38]

Bud had little say in the matter, especially since his mother felt that her son needed more discipline to avoid the temptations of youth, especially exposure to women. Elmo and Bud hit the road, trying to meet as many influential people as possible, particularly in political circles. Elmo recalled "the campaign to 'reach' the Congressman as we toured his six county districts and the interesting people we contacted."[39] The plan took a detour when a family friend and advisor Bart Longan arranged for Bud to meet his brother, P. M. Longan, who maintained that the navy offered more exciting opportunities than the army. "He went back and described the adventures of going to sea—going all the way back to the days of his whaling captain forebears. He described the difference between the sea campaigns and the land campaigns, and the ability of an individual to be an individual in command of ships and in command of fleets. I found it all very exciting and persuasive. With just that much thought, I shifted from competing for West Point to the Naval Academy examination. I never asked my father whether he agreed with that, but I think he did. Although he himself was Army, I think he was also mesmerized by the description."[40]

Bud was selected as the valedictorian speaker for his June 7, 1938, Tulare High School graduation. In spite of her pain, Frances managed to attend her son's graduation, providing the family with their final memorable time together. Within a month, Frances would be bed-bound. In Bud's graduation speech we see the roots of themes and commitments that would define his later career—compassion for the less fortunate and a strong line against dictatorships. Addressing the audience, Bud reflected on the factors that made for a satisfactory life. Canvassing the problems facing mankind—dictatorship, poverty, crime, and class struggle, Bud offered solutions for neutralizing them. "Surely by nourishing all that is ideal and beautiful in life, we can hope to attain the quintessence of progressive civilization." Lamenting the millions of Americans living in poverty, suffering from scarcity of food and clothing, Bud urged fellow graduates to consider "humanitarian opportunities" as "the prime requisite considered in choosing an occupation. Too many have lost their youth in a struggle for material wealth, only to discover in old age that they had misunderstood the true values of life."

The valedictorian also lashed out at the world's dictatorships, seeing that they not only threaten democracy, but that the "imposition of dictatorial aims on humanity produces the complete degradation of hope" and "disintegration of human spirit." Invoking the pioneer spirit that allowed previous generations to survive, the seventeen-year-old said, "We face the wilderness of the future with the strength of a pioneer birthright. We stand at the threshold of a strange, new world. With the light hearts of youth, with the joy of righteous struggle, we shall plunge into the intangible wilds, resolving that courage, eagerness and intelligence—the heritage from a pioneer past—shall continue the progressive civilization of our America."

Over three decades later, in a 1974 *Playboy* interview, Bud proffered, "I think every young generation's approach to the world is to generalize idealistically—dissatisfied with what they see—hoping for a better world." The essence of growth is learning how to make decisions and compromises "without giving up one's fundamental beliefs and aspirations. . . . When people achieve positions of importance, the real test, for naval officers or anybody is whether they recall those youthful aspirations and measure themselves against those early ideals, modified by maturity, but hopefully not too much."[41]

Following his high school graduation in June 1938, Bud spent a year at Rutherford Preparatory School in Long Beach. His congressman did not have an appointment available, so in order to better prepare for the

Naval Academy, Bud enrolled at Rutherford where he was mentored by the school's founder. "For some reason or other, I was designated to sit on the right hand of Mr. Rutherford throughout that year and had many long talks with him at breakfast and dinner. Mr. Rutherford became a rather impressive influence in my life. He talked to me at great lengths about the Navy and the Navy system, about his adventures in the naval civil service, and about the problems that one has with bureaucracy and about how hard it is to move things in a bureaucratic system."[42]

Bud hedged his bet for a political appointment by joining the enlisted Naval Reserve in Long Beach in order to compete for a presidential appointment. He took the competitive exam for Senator Hiram Johnson's appointment, which was awarded on merit, not political connections. On the day of the exam, Bud was ill with the flu and running a high fever. He was disconsolate in a phone call to his dad, "I know I flunked it."[43] The rest is history. Bud's appointment to Annapolis was announced in the Tulare newspaper on Friday, September 23, 1938.

A few weeks prior to his departure for the Naval Academy, Bud found himself in deep trouble. Along with his friend Gordon White, Bud had perfected the technique of dive-bombing by driving a car at full speed toward hitchhikers and then letting go with a barrage of eggs at the petrified pedestrians. Unfortunately for Bud and Gordon, their license plate was written down and shortly therafter the police showed up with a warrant. They were taken to the police station and charged with malicious misdemeanor. "My father reacted more strongly than I have ever seen him react about that one," recalled Bud. Elmo did not want Bud's mother, who by this time was quite ill, to hear about it. Elmo made Bud go down to the local paper and plead with the local editor to keep it out of the papers, unaware that Elmo had already arranged this outcome. Next, Elmo persuaded the local judge to haul both boys into court to give them the scare of their lives. Judge Smith raked Bud over the coals, asking how being sent to jail would look on the record of someone admitted to the Naval Academy. Bud feared that he might not be able to attend the academy. With Elmo's encouragement, the judge sentenced Bud to several hours of work detail seven days a week until he went to the academy.[44]

On the day he was scheduled to depart for the Naval Academy, Bud went upstairs to say good-bye to his mother. Frances was in bed, under the care of a nurse. They both tried keeping everything casual, putting on a brave front for each other, but by the time he had reached the front door, Bud was overcome by emotion. He could not leave like that. Returning to the bedroom, he found Frances crying. They embraced and shared tears. "We sat there in each other's arms, motionless, for what seemed like hours." The silence was interrupted when Frances made her son promise that he would not return for her funeral. She emphasized how rugged the academic discipline of plebe year was going to be, and she wanted Bud to remember her as she was then. "I promised to abide by her wishes, kissed her and departed," knowing this was the last time he would see his mother.

Frances was dead ten weeks later.

CHAPTER 3

EDUCATION OF A NAVAL OFFICER

To Bud, the only two things of any importance in this life are
women and women. But when he did take time off from his
amorous pursuits, he could do amazing things in other fields
as well.

—NAVAL ACADEMY GRADUATING CLASS YEARBOOK,
LUCKY BAG.[1]

In June 1939, three months before the German invasion of Poland, eighteen-year-old Bud Zumwalt walked toward the main gate of the Naval Academy. In a letter home, he described the academy as "a beautiful sight."[2] It had been an exhausting train ride across the country from Corcoran, California, to Annapolis, Maryland. His mother's mounting medical expenses ruled out buying an airline ticket or first-class train accommodations. Unable to secure a sleeping berth, he sat the entire four days. Arriving at the Annapolis train station, Bud was physically tired and emotionally apprehensive about what awaited him.

The Gate, as midshipmen called it, was the dividing point between the world of youth and the path of a career naval officer. Entering the Yard for the first time, Bud traversed Stribling Walk, named for a former superintendent of the Naval Academy, Rear Admiral Cornelius Kinchiloe Stribling, who served during the War of 1812, the Second Barbary War, the Mexican-American War, and the Civil War. At the head of Stribling Walk, he saw the bronze replica of Tecumseh, the Indian chief of the Battle of Tippecanoe, whose likeness once served as the figurehead of the USS *Delaware*. Tecumseh was revered by midshipmen as the "God of 2.5" or "God of C"—the academy's passing grade. Tecumseh awaited incoming midshipmen, who cast pennies before him and paid tribute with a left-handed salute, so they might have enough good luck in examinations to receive a 2.5. Midshipmen decorated him with war paint three times a year, before athletic contests against Army.

Bud soon reached Tecumseh Court, the scene of daily brigade formations for meals, rallies, and other events. In the center of the court, a plaque marked the spot where Commander Franklin "Buck" Buchanan had assembled the instructors and midshipmen at eleven o'clock on the morning of October 10, 1845, to read a letter of instruction from Navy Secretary George Bancroft authorizing the establishment of a naval school, which ultimately became the United States Naval Academy.

He next arrived at Bancroft Hall, home to all midshipmen, including over eight hundred first-year plebes, who were scheduled to be sworn in the next day, taking their oath at the center of Bancroft, under Commodore Oliver Hazard Perry's flag embroidered with Captain James Lawrence's immortal words, DON'T GIVE UP THE SHIP. Meandering through Bancroft's seemingly endless marble corridors and following a path up its rotunda to Memorial Hall, Bud stopped to gaze at the trophies, rolls of honor, and portrayals of great American naval victories. Reaching the plebe dormitory, Bud found his plainly furnished room, where each piece of furniture had its correct place. His roommate Skipper Dean was already there.

Plebes had been instructed to deposit their luggage and head for physicals and issuance of their white working uniforms, shoes, books, and other prescribed items. Bud and Skip went together for their regulation haircut and to the tailor to be measured for blue serge uniforms and multibuttoned, high-jacketed full-dress attire. Returning to their room, the roommates carefully reviewed their plebe instruction sheet for stowage. Each item from handkerchief to clothes had a designated shelf.[3]

Each of the 813 plebes had been told that if he had any reservations whatsoever, he should not take the oath at the next day's induction ceremony. "When I arrived at Annapolis as a young and somewhat frightened California boy too many years ago," Bud later mused, "I concluded that the setup was not as I had expected and, on the day I was about to become a Midshipman, I called my father to tell him I was coming home, instead."[4] When his father heard the words "Dad, I can't take it,"

Elmo could only say "Okay, then come home."[5] Whether it was a case of not liking the vibes or agonizing over being so far from home during his mother's final weeks, Bud wanted out. Yet when he heard the disappointment in his father's voice, he knew there could be no backtracking. Whatever trepidation Bud may have felt took second place to feelings toward a father he could not disappoint. "I hung up the phone and without looking back took my place in the Class of 1943."

Joe Warren Stryker, a 1925 graduate of the U.S. Naval Academy, was one of those in charge of swearing in the plebe class that summer. In a 1970 congratulatory letter to Bud on the occasion of his appointment as chief of naval operations, Rear Admiral Stryker reminded Bud that he was "pointing at random at the lot of you saying that you, and you, and you would probably be admirals some day and not to forget it."[6] Bud would never have imagined that Stryker was speaking about his future, because Annapolis was the means to pursue a medical career. The academy offered a free education, which was important, given the large stack of medical bills associated with his mother's illness. "In the final analysis, in my own mind, in making the decision to go to the Naval Academy, initially West Point and then the Naval Academy, I was merely postponing until after the war was over going to medical school."[7]

Plebe summer is designed to lay the foundation of the life of a naval officer. Classwork did not begin until the full brigade of midshipmen returned from summer activities. For the new midshipmen, this began a period of intense training, marching, shooting, sailing, rowing, and learning knots. Parents received a letter from the superintendent recommending that they refrain from sending money during plebe summer in order to avoid providing outside temptation. Instead, plebes spent their time learning how to keep a room neat and memorizing myriad academy and navy regulations. Within days, Bud knew how to wear his uniform, salute, and carry out basic plebe tasks. In the nearby waters of the Severn, he learned to judge wind and currents, as well as the rudiments of navigation and boat handling. He also started to develop his own ideas about leadership for when and if his own time came to command.

When the brigade returned at the end of the summer, plebes were quickly put to the test by upperclassmen, especially by the second-year youngster class, who took full advantage of no longer being at the bottom rung of the academy's ladder. Youngsters took special delight in hazing rituals that included ordering plebes to walk down the center of corridors with eyes in the boat (looking straight ahead), memorizing and repeating meal menus, squaring all corners at right angles, addressing upperclassmen as Mister, performing pushups for minor infractions of regulations, or "shoving out"—sitting without a chair at meals. *Reef Points* was the plebe bible, containing all the information needed to survive; every word of it had to be memorized. At any time of day or night, a plebe could be asked to recite a poem, sing a naval song, or provide details on a particular battleship history. All of this was geared to instill discipline and learn how to accept authority and punishment with a stiff upper lip.

The glossary for the 1939 *Reef Points* provided a list of terms to be used by all midshipmen: *Bull* was the nickname for subjects like English and history. *Chico* referred to Filipino mess boys. *Dago* was a reference to any foreign language. To *French out* was to take unauthorized leave or go *over the wall*. *Steam* referred to engineering; *skinny*, to physics; *math*, to analytic and applied mechanics, later calculus. *Spanish athlete* was a nonathlete. *Moke* referred to a "colored corridor boy or mess attendant." To *drag* was to have a date, usually for the *hop*.[8]

Bud survived plebe summer, enjoying the physical challenges as well as the new structure. "I think I always enjoyed the discipline. In part there is a certain challenge in measuring up, and there's a certain challenge in figuring out how to beat the system too."[9] Classes had just started when the inevitable news arrived on September 13, 1939, that Bud's mother was dead. With Bud at the academy and Saralee at Berkeley, fourteen-year-old Jim bore the brunt of Frances's final weeks of morphine-induced nightmares. "By the end of August, she had been crying out for death while I held her hand. Mercifully, she slipped into a coma a few days later," recalled Jim. "Now her suffering was over."

Elmo had anticipated the moment since Bud had departed for the academy. In a letter to the superintendent, Elmo explained that "for two years his [Bud's] mother has fought a lingering illness buoyed up by the hope that she could know he had attained this goal. Now that this wish is consummated, like a ship that has slipt [*sic*] its anchor, we are seeing her drift rapidly downstream. Your plebe knows this situation and I have tried to prepare him for the zero hour."[10] Elmo requested only that when the time arrived, he could send a telegram to the superintendent, and "perhaps a few words of advice may help him to carry on in the true tradition of the navy."

The call came just two weeks into the start of the academic semester. Bud had braced himself for the moment but found the news difficult to accept. "For the first time I could remember, my life seemed to be without meaning or purpose. It would take the better part of a world war for me to regain that sense of purpose."[11] By Bud's own admission, he underwent "a reexamination of my direction and goals in life." Three thousand miles from home and without any family support, Bud questioned the role of God and religion in life. His brother and mother had died so young. How was it possible to believe in a just God?[12] It was truly the most traumatic experience of his life at that point. He questioned why such a warm, compassionate, and understanding woman "who had given so much to others was taken during the prime of her life, after a slow and painful demise." His mother's death "left me bitter. The goals I had set to serve my fellow man in the field of medicine no longer seemed so important to me. That long term goal was replaced by a short term one that required I only do now that which I needed to get me through until tomorrow."[13]

Bud honored his mother's wishes by not returning home for the funeral service at the family's Sycamore home. With over two hundred Tularians present, Tulare High School music teacher Cyril White played "Auld Lang Syne" on the violin, the same melody Frances had requested for the funeral of Bruce Craig.[14]

Bud's remaining anchor was now his father. "I worshipped the man. I consulted with him at every step of my career," Bud said of the man he considered "his hero, ideal, and father."[15] In a 1972 letter to his father written en route to his hospital bed, Bud encapsulated his father's influence upon him: "They say that fathers should not try unduly to

influence their sons in a career. Yet it was clearly your influence that carried me into the Naval Academy and it was, in the early post-war years,
my knowledge of how much it meant to you that kept me in the naval
service. The fact that I have had the opportunity to serve my country in
a top post is, then, directly attributable to the love and respect I have for
you and for all to which that has led, I am grateful to you. But beyond
that, the way you passed on as a faith, in my youth, of public service,
devotion to country, love of family, courage to face the issues squarely
and to dare to deal with them forthrightly, disinterest in wealth, dedication to the pursuit of excellence in leadership and in one's profession,
provided a guidepost which I have tried to use in my life."[16]

Nowhere was this bond between father and son more evident than
in the dozens of letters exchanged between 1939 and 1943, which reveal
the joys and trepidations, the ups and downs, and the shaping of a
worldview that would define Bud's later life.[17] Early letters document
the travails of academic struggle, the pressures to succeed, the tension
between yearning to live a life on the edge by testing academy rules
and the demands of military conformity. There was the incessant fear
of losing his place in class standing, of not being able to cut it, of letting
his father down. These anxieties were later replaced by a growing self-
awareness that he possessed the right stuff to succeed. The letters also
reflect a young naval officer's evolving views on the war in Europe, the
rise of fascism, and the prospect of dying in battle. At bottom, the letters reveal a process whereby a young man discovers his commitment
to duty, honor, and service.

Bud found it difficult getting back to the rigorous schedule of the
academy. Reeling from his mother's death, he sought emotional support
from a former steady girlfriend from Long Beach, Geraldine Chapman,
attending school in the Washington, D.C., area. She had known about
Frances, and it was understandable that Bud would turn to her. The one
problem he faced was that plebes were not permitted off the Yard for
dating. Plebes could, however, sign out for cross-country hiking. Bud
came up with a plan whereby he would sign out for the hike but, once
across the railroad trestle bridge, he'd meet his girlfriend. In a letter to
his father, Bud bragged about beating the system, under which plebes

were not allowed to drag. He began by packing two picnic lunches and walking with a carefully screened upperclassman until he saw Geraldine, at which time she would join them. "Anyone seeing them would think that two others were dragging," wrote Bud. The lowly plebe was just tagging along. Once out of sight of the academy, the upperclassman would go his own way so that Bud and Geraldine could have lunch along the Severn. They would then walk back until they saw Naval Academy personnel and go their separate ways, "as tho [sic] we were total strangers. It was a good strategy," bragged Midshipman Zumwalt.[18]

There were times when Bud partied too long with his girlfriend. "We had a wonderful time but when Sunday nite came around I was so tired and worn out that I couldn't see straight," Bud wrote to his dad.[19] He reported having no energy and a low-grade fever, but feared academy medics would kill him. Bud preferred a diagnosis from "an old country doctor from a cow town" rather than from those "who would likely remove his spine clear up to medulla oblongata" when all he had was a cyst.

This was all rather disheartening for Elmo to hear. Memories flooded back of Bud's senior year in high school, when they argued about Bud's amorous pursuits. In no uncertain terms, Elmo advised Bud that he was "painting the social trail too vigorously." Elmo challenged Bud to get into shape by exercising.[20] The Naval Academy became, in retrospect, boot camp for a life of multitasking and burning the candle at both ends. Bud insisted that he could wake up forty-five minutes before reveille to catch up on classwork and also study Saturday afternoons, if necessary. He preferred socializing and sleeping to studying and intended to keep at it. "I console myself with this logic—Plebe year counts only a minute fraction toward our final standing. The next three years increase in importance. In Sept we can discuss the whole situation to decide if I should turn on more steam. Until then, I shall give em both barrels at my present rate."[21]

With this kind of attitude, it is not surprising that the plebe-year letters reveal Bud's apprehension about staying afloat academically. "I have just completed the hardest academic week I have had," Bud informed his father.[22] After taking his math exam, Bud reported that he "went into the exam room . . . and blew up." He feared losing 200 places in the class standing. "I feel like a man who has just been knocked out. Sorry I don't have the capacity both you and I thot [sic] I had."[23] When first half-year grades were posted, Bud "hit a new low" in math, in which he plummeted from 49th to 111th. "All because I received a 2.61 in my final. In short, two hours of 'blow up' cost me 62 numbers."[24] He struggled in other subjects as well. "In Skinny I dopt [sic] from 9 to 17 which makes it look as though I am on the skids pretty bad. All I can say is that I know I can do better—that I can exert more push and goddamit—I am going to show them a thing or two."[25] The one consolation was "125 plebes are trying to bilge out, almost 1/6th of class . . . this is certainly 'survival of fittest' contest."[26]

By mid-February of the new year, Bud was finding his ground, exercising, and focusing on how to relax under pressure. "I think sometimes that I am a temperamental nitwit," he wrote in a candid self-appraisal. "As you know (better than anyone else) I am a mixture of high strung foolishness at times."[27] For the first time since his mother's illness, Bud forgot his father's birthday. "No gift that I could have sent you, *on time*, could express the thought, no yearly eulogy could convey the continued year round day by day love, that your son bears for you," wrote Bud in apology. "Rarely a night passes by, but that my pre-slumber thots [sic] turn to you and the family. Always in these moments, there is thankfulness for the Providence that gave me my Dad. . . . And tho I confess with shame to have forgotten your birthday, I point with pride to the countless hours of every day that are devoted to thinking of you and your strength."[28]

The drums of war occupy a large part of the correspondence, offering a window into world events as they unfolded and the shaping of a worldview. By March 1940, Finland had been at war with Russia for four months, beseeching the Allies for assistance amid the bitter fighting.[29] On March 10, 1940, Bud wrote, "I hope like hell that England and France send troops to Finland, but I doubt that they will. That British caution is certainly a trying thing." A day after he mailed the letter, Britain and France announced "all available resources" to help the Finns,[30] but the pledge arrived too late, and Finland soon signed a pact ceding substantial territories, as well as economic and military privileges, to the Soviet Union.[31]

"The European situation has got me down," Bud wrote two weeks later. "England's blockade is a failure. Russia and Germany are working like demons to exchange their agriculture and industry respectively. They have licked the blockade, thus making France and Eng take the offensive." Bud was also concerned with war looming in the Near East. The *New York Times* reported on March 24, 1940, that the region was "bristling with troops and armament" and filled with the growing belief that Europe's war heading their way was "almost inevitable and perhaps not far off."[32] Maxime Weygand, commander in chief of Allied Near Eastern Forces, had assembled a large force of at least two hundred thousand troops. Bud envisioned geographic difficulties that would likely complicate if not crush chances for Allied victory. "If war starts in the Balkans, Weygand in spite of his forces will be operating far from a home base and supplies," wrote Bud. "He will lose to a German army operating close to home. Neither side has a chance in the west—so there you are. At best a stalemate, which is tantamount to an Allied defeat."[33]

Bud held hopes for Maurice Gustave Gamelin, commander in chief of the allied troops in France.[34] "My only hope lies in Gamelin. His is a genius. I hope to see some great strategy from him. Do you know it would be a break for the Allies to have Italy declare war, either FOR or AGAINST. The FOR is obvious. AGAINST, Gamelin would smash the Italians in nothing flat. The mountain passes from France to Italy run from a French focal point to spread out in Italy (like your fingers spread from your hand). With Italy as a base, a new front against Germany thru Brenner pass can be opened up."

Concluding with a personal note, Bud admitted, "This week for some reason, I have been homesick and dissatisfied. I can't get over the feeling that I have let people down with my record."[35] A week later he reported, "We finished our exams today. I am dissatisfied with the results. . . . There are only 68 days more before we finish this horrible year." His grade in Steam placed him 611, leading Bud to think that his professor "is a Jew (I am almost certain) and hates my German name. . . . Now this MAY BE imagination, but I feel sure my work is better than the grade."[36]

Meanwhile in Europe, Gamelin led a combined British, French, Belgian, and Dutch force as German troops entered the Low Countries, but when German forces broke through French lines, he was replaced by General Maxime Weygand.[37] "War has begun like a hell at last," wrote Bud. "All these things have broken into the routine of our life here. I am so keyed up over this war that life here seems almost inconsequential."[38] "The worst hell of history has begun, I think. I pray to God that the Tommies and Frenchies can hold them."[39]

After listening to an international news hookup depicting the raging battle from Arctic Norway to the Swiss border, he wrote, "I can't leave the radio, I am so entranced." By now Bud favored the Allies sending their Asia Minor army of 750,000 up through Greece and into the Balkans. "If they disrupt the southeastern industry, they have Germany licked. Russia is not going to molest them. She is going to 'sit it out.'" He also favored a U.S. declaration of war: "At last this country has gotten into a friendly state of cooperation with the Allies. It's about time. I think that in spite of the losses we would incur, that it would be wise to declare war, protect the Dutch Indies, join the British blockade and gear our industry to produce enough planes to blast the Germans from the face of the earth. Our planes could throw air power to the allies in a big hurry."[40]

Bud's plebe year took a dramatic turn when he attended a meeting of the Quarterdeck Society, the academy's oratorical and debating club. He was not intending to speak, but after sizing up the three midshipmen participating in the preliminary round, Bud signed up for the main event, having little doubt he could beat the other speakers. Bud had supreme confidence in his communication skills, having honed them in high school and with his father. One of Bud's second-year roommates, fellow Californian Ray Angelo, liked to recall the quiet Sunday afternoon in Bancroft when Bud first demonstrated his oratorical talents. They had received a crate of old mushy grapes that Ray's dad had shipped from California. Bud decided to give their next-door dorm mates a wine-making lecture. "The silver tongued Zumwalt" strode into the room barefooted and "immediately launched into an impromptu ten minute discourse on the history and art of wine making in old Italy. His eloquence held the amazed audience spellbound and he then announced that they would be treated to an actual demonstration. The grapes were dumped on the polished floor, stomped upon." Bud and his roommates then fled. "Thus was born a great naval orator."[41]

Bud's plan for participating in the academy's debate challenge seemed well thought out until seven new participants showed up, including the academy's most experienced and skilled orators. Looking at the competition, "My heart sank," he later wrote.[42] Being designated to speak sixth provided Bud with a chance to estimate adversaries and compose himself. "I was in a perfect emotional pane to deliver my efforts," Bud reported to his father. Addressing the coming war, Bud chose to attack two myths that had become accepted as truths: that strict neutrality could safeguard national existence and that aloofness from foreign alliances offered a similar protection from aggressors. "Until six weeks ago, a large part of the world existed in a realm of imagination—a land of make believe" because they had accepted these myths. "In past weeks, the world had been forced to awaken from this fantasy and face a stark reality that a neutral Norway had been swallowed up. . . . Today that government has ceased to exist. That brave contented people has become mute with horror and dumb at the sudden terrible catastrophe to their nation. Almost like a giant octopus, the monogarchy [sic] of Central Europe has stretched its tentacles into a death grip on the land of the Norsemen."

By Bud's line of reasoning, these events had transpired "because Norway was a myth. Because Norway's very existence and her attitude toward the outer world was sublimely ridiculous, because Norway believed in THE GOLDEN RULE. Hitler built a Trojan Horse within the land of the Viking . . . they stood idly by while the Germans marched in—and with Norway, perished the first of the two great myths—the fantasy that blind, idealistic neutrality can secure a national safety."

Farther south, Belgium and Holland had not claimed strict neutrality, so they had eliminated Norway's Achilles' heel, that is, they had armed themselves, had fortified themselves, "but it was as two Boy Scouts 'Being Prepared' against some monstrous Frankenstein."

Norway was overrun and so many killed because "these countries cast aside that time worn adage, 'United We Stand.' They chose to stand alone . . . with them fell the Second Great Myth." Rattling off a list of the fallen countries, Austria, Czechoslovakia, Finland, Norway, Luxembourg, Holland, and Belgium, Bud posited that the lesson for democracies was that there can be no neutrality, no escape from alliances—"The only defense against totalitarian Blitzkrieg is the defense of collective security and 100% preparedness."

Turning to the homeland, Bud asked, "Are we in America safe from these tragedies?" The answer was no. "We must become fully alive to the genuine threat of subversive elements within our own borders." Closing his speech with a call for preparedness and mobilization, Bud urged the audience and judges never to think it can't happen in the homeland. "When we have awakened from the stupor engendered of these myths—when we have signified by complete preparedness and by our alliances, that we have thrown off their throttling grasp—when we can say, not with blind Norwegian faith—but with true confidence—IT CANT HAPPEN HERE—then—only then—the brave Norse, the stolid Dutch shall NOT have died in vain."

Bud thought "my delivery had been the best of my life."[43] Indeed, the judges selected him as one of the finalists in the entire regiment. "I'll give em a fight," Bud assured his father. The newspapers validated Bud's debate themes. "Well it's good bye France and Great Britain by todays papers," Bud wrote the morning following his speech. "I can't yet conceive of a world without the British Empire. It's going to be terrible. . . . The way England and France are failing, I may have to include them in my list of victims by the time I speak."

Bud gave them one hell of a speech in the finals, taking first place from the Quarterdeck Society. One of the judges, Senator David Walsh of Massachusetts, chairman of the Naval Affairs Committee, told him, "Young man, if you can do this well when you are a plebe, I am wondering what you will do when you are a mighty First Classman." Invited to attend the academy's award ceremony for all academic winners, Bud was the only plebe in attendance and was awarded a gold watch. "They presented me with the watch for the speaking contest. I felt as big as life—marching up before 2300 middies and some 500 spectators. The gold Lonjeen [*sic*] watch was very expensive and engraved on the back with date, occasion."

With final exams looming on the horizon, Bud buckled down. Winning the Quarterdeck award seems to have provided a reality check with respect to what hard work could accomplish. He ended up with "the most brilliant set of exam marks I have built up since the entrance exams a year ago." His 3.44 earned him status as a "star" man, allowing Bud to wear a gold star on his uniform. This was a terrific recovery after his midyear tailspin in math and steam. "I'm really happy," Bud acknowledged. "I must have, unconsciously, given up the ghost a little, as far as drive goes. Second, women must have had something to do with it. From November until March,. I was infatuated with a dizzy dame. When I got over it, I buckled down—the difference—with these grades I have regained a measure of self-confidence. Next year, just watch my smoke, I'll give no quarter."[44]

A plebe year that started with second thoughts about remaining at the Naval Academy followed by his mother's death was ending on a high note, what Bud described as "the salvaging of the glory of previous years." The "crest is riding high. I have never known a two weeks more invigorating to my spirits and exhilarating to a dropping one. First there was my sudden awakening academically . . . starring 48 out of 700. This standing convinces me that I can cut that in half next year— we'll see."[45] He was in the midst of a glorious self-transformation. "I have found myself growing happier each day here dad. More and more I am learning the importance of friends and acquaintances. I believe this plebe year has eliminated more dissatisfaction and restlessness from my system, than I thought possible."

The completion of plebe year brought the midshipmen's summer practice cruise. Between plebe and youngster year, all midshipmen are given the opportunity to discover firsthand what it is like to be in the fleet by going on their first cruise. Aboard the battleship *New York*, Bud was feeling especially buoyant as "we left the Reina Mercedes dock with the band playing, some fifty motor launches took us to the cruisers while parents and sweethearts waved goodbye from the docks."[46]

In a letter home, Bud described his first days at sea as "the same as during those terrible days at the beginning of plebe summer, dazed, confused—but that will all clear up, in two or three days."[47] Indeed, by the time they were off the Florida coast, his letters describe a love for sleeping on deck under the stars and talking with buddies until late hours. Bud delighted in seeing "schools of porpoises stretching miles in single file, leaping in and out of water." He bragged "my luck is holding out. I am the only midshipman on the ship who has not had to stand watch." He joked that he and his friend Zeigler had been able to hide with sundaes to avoid polishing brass and cleaning up paint chips.

He had settled into a regular routine, beginning with 5:00 a.m. morning coffee, followed by rolling and stowing hammocks, getting washed and shaved and turned to, meaning cleaning the ship from 5:30 to 6:30, scrubbing decks and locker rooms. This was followed by a breakfast of beans, corn bread, and more coffee, then more cleaning of the ship until 9:00 a.m., then orientation tours, and lunch. The afternoons were usually spent on lifeboat drills and a range of exercises, including general quarters (i.e., battle stations), man overboard, fire drills, and a heavy emphasis upon damage control drills and gunnery. He stood a variety of watches that included engineering, gunnery, and navigation. Following dinner, the crew enjoyed movies each night. "Where else can a man escape the wiles of women and the worries of civilization with such smothering completeness, as on board a naval ship," Bud wrote to his father.

Looking forward to returning home for the first time since his mother's death, Bud wrote, "Now that the old bean is free of mental exertion, my thots [sic] turn ever homeward. I can see the old homestead clear as a bell. The beautiful sycamore trees lining the sides of the streets brats playing football, etc., in the tar pitted road-neighbors-gossips-dogs and all of you (excuse the dogs first)."

Events in Europe were on everyone's mind during the cruise. "It seems hard to believe that Italy has just added her bit to the Hell in Europe. . . . To me, this era is as significant as the Fall of the Roman

Empire. . . . When Rome fell, the genius of her civilization, the knowledge known only to her intelligentsia—was destroyed with them. Today, the prospect is the same."[48] The British Empire "stands at the brink . . . the last ally in Europe is down—France, rent asunder by the barbaric genius of the Hun. . . . Her last hope is the United States."

Echoing themes from his Quarterdeck speech, Bud lamented the consequences of America's isolationism. His country was demonstrating the "same British conservatism" and "has hesitated too long." England was falling. "That island, source of the greatest empire in history, will be gutted—bombed from the skies—starved at the seas—finally invaded from the shores of her dead ally. . . . Dictatorial jealousy and fear will wipe out what remains when the inferno of bombs and shells have ceased. And WE STILL STAND ALONE having let our friends die— unaided. We must face their murderers—a coalition of Germany, Italy, Japan and their satellites." In the absence of such a coordinated response, "we will go down—just as surely as we maintain our course." Bud endorsed maximum preparedness. "If we can co-ordinate our industry to turn out the most gigantic fleet of planes and navy in the world, we MAY escape the horde that approaches. But having seen the inadequacy of Democracy when facing Monarchy, and having seen the diabolical resourcefulness of the latter, and knowing that the day of the impossible does not happen, one can only WONDER, HOPE AND PRAY."

The coming war also meant changing academic plans at the academy. Due to the emergency created by war and the shortage of officers, the Naval Academy education was being reduced from four to three years. The shortage of officers was so acute that the academy sacrificed a year of training to get them into war.[49] "All of this is very heartbreaking to me, but I can't feel cheated, knowing that all of the young men of this country are going to have to sacrifice to stop that beast in Europe." Bud closed his letter from sea with a promise: "I haven't forgotten you [Dad and Jim]. I think the family has been on my mind more than anything else since the cruise began—and, dad, don't worry about Sept. leave—I'll be home if I have to 'go over the wall' [academy slang for taking 'French leave']."[50] "By the way, when we hit New York July 12, I will be broke—'No mon [*sic*], no fun, your son.' "[51]

Bud received the money from his dad and did not have to go over the wall to reach California. With the midshipman cruise scheduled to end on August 15, leave was shortened to two weeks. The family reunion included a fishing trip with his dad and brother Jim. It was good to be home, although he found "mother's memory haunts every picture, every household object."[52] As soon as he had returned to the academy, Bud learned he had received the highest marks in aptitude for the summer cruise.[53] "The biggest news is that the marks in aptitude for the service for the cruise were posted. Out of 719 'youngsters,' I stood No. 1 with a 3.90. I was darned proud of this,"[54] wrote Bud, who also received a medal for cruise short-range battle practice accuracy, one of two men in the gun crew to get a medal—the pointer and trainer. "I now look like a general with two medals and two stars on my full dress—whoopee. Had them all on at the hop last night."

Bud loved transitioning from plebe to youngster: "I must admit that Youngster year is 100% more fun than Plebe year. Plebe year this means no imperfection—the plebe is the lowest form of life. . . . Youngster year regain self-respect, partake in all the pranks known—water fights, hosing, etc." Looking to the future, Bud advised his father that by third year, known as Second Class, he would aspire to "more poise—more serious attributes—desire to create good impression with i.c—get good grades." And finally, First Class—"conceit—one doesn't desire to speak to inferiors, responsibility."[55] Bud relished everything about being a youngster. "The business of being a 'youngster' is wonderful. Do you remember how I swore I'd never 'haze' plebes? Well I'm already doing it—in a friendly manner, however. It is a THRILL to go down to a meal, relax, be called 'Sir'—ask questions to plebes, etc."

The new incoming plebe class was so large that Bud and his roommates were living five to a room. "The five men in a room is great fun—we raise the devil—but we don't study like we should—its next to impossible. I am trying to concentrate but I am slipping." It may have been twice the fun, but it was also going to be twice the work with an accelerated academic program.

Meanwhile, back in Tulare, Elmo could no longer hide his grief. Life seemed devoid of meaning since Frances's death. He needed to fill the emptiness of her loss, and the shortage of doctors in the military gave Elmo his chance to again serve his country. He made the decision to volunteer for active duty in the Medical Corps, but chose not to tell Bud about his decision while they were together during the summer.

Upon reading the news in a letter, Bud was aghast that his father would consider leaving Jim, who was still a teenager living at home. Marshaling all his debating and persuasive skills, Bud sought to alter his father's decision.[56] "I am going to tell you of my own selfish reasons for dreading your army commitment—then show you two good reasons why you should dread it." He thought his father's action was the "breaking of a vow made to mother—the vow to superintend the advent of Jim's progress into manhood—with the very best you had." Having lost a mother—and for all practical purposes, a brother and a sister—Jim's life would have the "barrenness of a desert," if he lost either a Dad or a hometown environment. Jim would have no father for the "nucleus years" of character formation in high school, a crucial era in a man's lifetime, "during which the long struggle of environment and education with youthful curiosity and temptation, produces the first picture of what he is to be. Suddenly one begins to think seriously about life—he plans his future, judges his qualifications, and dreams about what he will be."

Bud argued that nothing would ever again be the same. "With you gone from Tulare, the shell collapses. The street of the Sycamores has become a dead forest—petrified by the coldness of a relentless Destiny—and the last tie which held me to the 'nucleus' is rent asunder. The thread which bonded me to keep the faith with my 'Tulare self'—is gone." Bud could not understand why someone so devoted to humanity, to curing the sick, a community man, a father to a young son, was trading it all in for literally nothing. Describing his father as a "great man endowed with the ability to care for the sick at heart as well as the physically afflicted—a man possessed of the energy and love of humanity to serve his community in a thousand ways—a remarkable father to his youngest child," that man was now to be "taken from his lifelong service—forced from his chosen role—into the job of a high grade handy boy." By going ahead with this plan, Elmo was trading "his gifted surgical hands for doing things interns do. . . . You should dread with sincerity the reduction in rank from that of a genuine servant of mankind, to that of a general handyman of Uncle Sam's Puny Forces."

Bud closed with a flurry: "I love and admire you, dad. I have known that your motives are always honorable. But I sincerely believe that you have made and are making a serious mistake." Elmo needed to "weigh the scales." On one hand, he had a vow to fulfill and a service to continue; on the other hand, a duty, but not a duty that required a man of his caliber to rejoin the military as a doctor. "I write with tears in my eyes, when I think of Jim, the big, overgrown bear, coming home to a barren house—chuck full of all the little High School triumphs— eager to unload his joys and adolescent problems within the sanctity of his home to an understanding parent. And I shiver when I think of the ghastly emptiness that will be his—when his friends enjoy the family life that I held so dear—the dreadful silence of a lonely supper—the haunting friendlessness of an empty house. Reconsider, Dad, and work fast."[57]

This was one of the few debates Bud lost. Elmo gave quite a bit of thought and consideration when composing his rebuttal, admitting to reading Bud's letter of "criticism" twice, finding it "remarkable."[58] Undecided whether to "rebut or acquiesce or do both," Elmo began his reply with the concept of duty and sacrifice to country taking priority over being a father to a son. Bud needed to understand that the issue was not what one would prefer to do but rather what one might have to do. The previous evening, Elmo had been listening to the radio when isolationist congressman Hamilton Fish was running at the mouth that we must fight "only a defensive war at any time—the Japs do not want war—the Japs will allow us continued trade—we could still buy tin and rubber."[59] Elmo believed that putting one's head in the sand was a ticket to defeat. Of course with only one parent left, ideally it would be best for Jim if his father remained at home, and "there may be a danger that the home on the Street of the Sycamores may be a thing of the past. I recognize that in leaving my community, I may be losing certain position and influence—that my 'career' questionable as it may be even to me, may be decidedly changed. I think all these things were weighed in my own mind, but unlike you, I choose to turn the cloud inside out and gaze at its lining."

Proffering advice that defines the Zumwalt lineage, Elmo had a duty to both his son and country.[60] The storms of adversity had diverted him from the preferred course of remaining a country doctor. "All my life I have had a rather fierce love of Country," wrote Elmo. He enlisted in 1917, a year ahead of the required time because he was a medical intern; he returned as a reserve officer in 1927, thinking that one day his country might need him. For fifteen years, he had been a reserve officer and could not now "show a white feather and say 'let George do it.' That would be a 'bit' yellow. Too many folks in this old USA have felt for many years that someone else should take the burden, someone else's boy should go to war. That attitude is a danger and represents a bit of dry rot in any man's land. So it seems to me that we must look at the map for a moment and realize that our ideal charting has and must be revamped. The storm of war clouds have blown us away from that ideal spot in the sun and we have to modify our plans. And please remember, that all thru your life you too will have to reef those sails and modify your course, that it would be much nicer to follow with always fair weather and no lost time."

No one, of course, understood Bud better than his father, which is why Bud had no chance of winning the debate. "Now your treatise on Jim—was not on Jim—it was upon yourself . . . your temperament was such that you had to have those avenues of 'talking it over' and your estimable Mother filled that niche to perfection. I never could have done that same thing and you know it . . . in the main to her with your temperamental tailspins, you went and had solace." Jim was different. "In the main, he is a self-decided type, seeking but little from the family . . . he is as different from you as day is to nite. You are the hare—he is the tortoise. . . . To me a home is the spirit of those who have made it, and even tho Mother has passed to 'the land where whence no traveler returneth' and you and Saralee have left for good, still, there will always be to me that spirit of camaraderie and I shall feel the touch of your hands and the kiss from your lips, wherever I may be. Because when your mother died—our home passed into that state of change that only a few years could hold together. We both recognize it, except we express it a bit differently." Elmo ended by saying, "You may still not agree, but then I haven't always agreed with your girls. You are a swell son, Love, Dad."

After digesting his father's response, Bud replied, "the answer to my plea was what I expected. I knew you would say what you did and I expect you are right—but then a devoted son cannot be blamed for trying to delay your being called. I want to ask you to give yourself an even break (be sure) when that call comes. All this butting on my part must be expected, because I always tell you how I feel."[61]

On Election Day 1940, Franklin Delano Roosevelt was ushered into an unprecedented third term. Months before the election, Bud had foreshadowed the result in a letter home, saying he was "certainly happy about Willkie's nomination on the Republican ticket. He is a natural for the organizing genius-type of executive we now need. I only hope that he can escape the stigma of 'Big Business' that FDR's group will try to wrap around him." Still, he forecast that "the WPA class will ride with our friend, Franklin, back into a 3rd term however."[62] A week before the election, Bud saw the handwriting on the wall. "Willkie is losing ground at a frightful pace. God help us. Japan joins the Victory Parade. Oh, how I'd like to blast those yellow dogs out of the sea."[63]

In the aftermath of the election, Bud demonstrated his abiding respect for civilian authority over the military. "The election is over—the will of the people has been represented and so, we must buckle down to create our niche under the system we have selected."[64] Looking at events in Europe, he wrote, "I am more determined than ever to get over there and do my bit as soon as I get out of this hole. . . . Greece will go down for sure. But she will have taken a bite out of Italian power that isn't easily sparred against England—all of which contributes to an ally—Time, time, time. Yes, I feel like I am going to do something for them someday." In a letter to Saralee, Bud signed off with a doodle of Ensign ERZ, Jr., and the phrase "Z—Bring on yer G-D-Japs."[65]

It was only a matter of time before Bud fell head over heels in love again. This time the object of his affection was Ann Austin, who worked for the chaplain and was the daughter of Commander Charles L. Austin, who was the officer in charge of the Midshipman store. Bud first met Ann when he went "stag" to the hop. "To show you how beautiful she is, the first date I could make with her is *April*—four months away. However second class summer is coming up. Wait to see if you don't hear more about Ann Austin after four months are over."

True enough, Bud soon found himself invited to dinner at the commander's home. He proudly informed his dad that "all of mother's preparations at those family dinners paid off." He had demonstrated perfect manners and etiquette. He "also got approval of the colored servants." The bond between Bud and his father is revealed in the details Bud shares about his date with Ann. "The mood was right, and I bent over to kiss her." She stopped him, saying that a Second Classman considered her his girlfriend. Ann promised to break it off. "Dad, I have never in my life known a girl who reminded me so much as she did then of Mother as you described her in college days. I had thought such sweetness, such absolute 100% chastity had vanished with your generation. She wouldn't let me kiss her until she considered herself free. And then I made a move which if I told anyone but you would win me the name sucker. I told her to wait. Under her influence I played absolutely square as I have never done before in this business of dating. I told her how suddenly I had changed on other occasions and that we must wait until I was sure—although I felt completely sure now. And I left that house 'walking on air.' I am still in the clouds. I tell you this and no one else. Please do not pass on this letter."[66]

Bud was smitten with Ann. "I still find myself utterly unable to have a good time with any but the one girl—Ann Austin. . . . All this melancholy writing must seem very childish, Dad, and yet, it's hard to rationalize the turmoil that can be created by falling in love. I want to be with her every minute, I hate to see her with anyone else, the whole future seems black, etc. . . . Here I am, at the peak of a rise in grades, I suddenly plunged head over heels in love. Believe me, Dad, this time it's even worse than it was in my senior year at high school. If you thought I was dopey then, you should see me now. I can't eat, sleep, and worse yet—I can't study."[67]

Bud caught one big break in pursuing Ann when he was selected to

march in the inaugural parade, which he called the President's Corona-
tion. He and Ann arranged to meet afterward, taking advantage of a 10
p.m. liberty. That day, Bud was feeling quite ill with flulike symptoms
but knew that if he went to the hospital, he would miss both the parade
and a date with "the most beautiful of girls. I decided to beat the hospital out of the victim." After marching for three hours in bitter cold,
he went directly to the Mayflower Hotel. "I wish you could have been
there. She is by far the most beautiful girl I have ever seen in my life—I
am not exaggerating. She has black hair, dark brown eyes, gorgeous
complexion and the sweetest smile you ever saw." They ate, danced, and
had a wonderful time until the party broke up and Bud headed back to
the bus for Annapolis.[68]

In January 1941 Bud published a fascinating and foreboding article in the
Log, a midshipman magazine. "The Greek Horiatis: A Timely Story of
Modern Greece with a Smashing Surprise Element" was the fictional
story of a Greek man, Themistocles, who during the invasion of Greece
by the Italians, made the decision to blow up a wagon carrying munitions in order to stop the advancing Italian forces. The subtext for the
story was that Themistocles' son, Demetrios, was riding on the wagon
that his father destroyed with his own weapon.[69]

The story is narrated in the voice of the author watching the events
unfold. "Fascists had crossed the border and were steamrolling their
way through everything. I saw Demetrios and his father . . . the great-
ness of bond between them." The narrator heard the father saying, "I
have reared my son to be a man. Then I heard a shot. The cart blew up
in a bloody flame. The bridge, the car and the leading tank rent the air
with a million pieces. The tanks were halted. The hole was plugged.
The line remained intact. But Demetrios was dead. The thought struck
me like a barb—Demetrios, the magnificent, the shepherd who talked
and lived like an immortal. Gone were all his father's hopes and dreams.
Lost to him forever was the peacefulness of this valley and his dreams.
Lost—the pride of Themistocles's life."

The wagon had been destroyed to keep the valley safe. Themistocles
"was staring strangely down at the chaos below." In his hands was a

smoking rifle. "My son was a brave man," Themistocles said quietly. And that is how the story written by Midshipman Bud Zumwalt ended. Years later Paul Stillwell accurately remarked in conversation with Admiral Zumwalt that the story had "an eerie, ironic parallel to the situation" involving Bud and his son Elmo.[70] Bud agreed, but an equally compelling analogy could be made between Bud and his father Elmo, as they both prepared to go off to war.

Tularians gave Elmo a grand send-off before his departure for Fort Lewis, near Tacoma. "As a doctor, as a servant of your community, and as a dad, you have given us an ideal to cherish as long as we live," wrote Bud. "When-if this mess clears up, get back to your community, your passing from the community is a loss that cannot be replaced and you owe it to them to return as soon as duty permits. God Bless you, Dad."[71]

While in Washington, Elmo met Doris Streeter. In a letter dated September 21, 1941, to "those most dear to me," Elmo wrote on the subject of "Family Affairs," admitting that for the past two years he had been a restless man, what Bud had accurately described as "a derelict on the Ocean of Life." Elmo's army orders had provided relief "from a mental state which few saw and even I did not admit." He was even considering remaining in the army in order to "escape a void I saw facing me if and when this service is over." With his children embarking on their own journeys in life, "what was left for 'the old man'? Little but memories and the empty house that had been home." Elmo wanted those most dear to him to understand that since September 1939, when "Requiescat in Pace was pronounced in that place we had called home," the street of the sycamores had been barren.

After these lonely days and longer nights, Doris had entered the picture. He described her as a petite, slender woman of forty-two, a divorcee with a seventeen-year-old child. She was quite poised and, like Elmo, "young" in spirit. They had been introduced through mutual friends, the Dugans, who described her as "pure gold fine loyal—a real friend." Elmo reminded his children that as Frances was dying, she had said, "very calmly, very sanely (as I now see it) she urged me at some future time to re-establish a home. At the moment the thot [sic] was abhorrent. But as time has rolled on I find that her perception was no doubt far greater than mine. She could see the emptiness that life would bring—which I could not. Now, with all the reverence a man can have for the Memory of Her, I feel I can make another decision."

Elmo did not want to remain alone for the next twenty or thirty years. He and Doris were going to marry. "So—with Doris and myself, we have had two people seared by Destiny's hand, who may bridge a Chasm of Emptiness together, and perhaps find a Shang-ri-la, when we have crossed over."

The news evoked mixed emotions from Bud. There was the "strange melancholy" in seeing his father with another woman because of "the child's reluctance to let go of the haunting melody of the past. It is his innate classification of his mother and father as one rather than two." But who was Bud to oppose this union if it offered his father a chance at happiness? "I say sincerely, dad, that I do not oppose this change. I do not think it unwise. Per the contrary, in spite of that inner feeling of maladjustment, I realize that it is best for you. I do not feel it traitorous to the memory of Mother . . . before she died, she told me more than once that you would someday remarry. . . . She knew, she saw, in this matter as in all things, the true and inevitable. . . . She hoped it would be a sweet woman, seeking the same companionship, the same escape from loneliness, from the persecution of the poignant memories. Thank God it was. She knew then and she knows now that you can never love as you loved her. She approves and condones your choice."

Showing insight and compassion, Bud wanted his father to know that he "must never feel that this choice has alienated me in any way, dad. Necessarily we will be unable to have quite the same closeness as before because the woman who is the bond that links us must be kept in the background of memories. But the love of a son for a father who has been his inspiration for 20 years, can never lessen. I wish you all the happiness you so richly deserve. Your life has been hard and you have followed a courageously stronger road. I hope with all my heart I may be like you."

Elmo's medical partner, L. E. Watke, wrote on October 7, 1941: "If you picked the gal she must be OK. . . . Your letter seemed to be in the vein of apology and justifying yourself for your decision, but why? It is your business, and besides, it doesn't need justifying. It is your own happiness which is concerned and which you are entitled to. Jim is the only one that your anticipated adventure might have any effect upon that is to be particularly considered. When he first read your letter, I think it rather stunned him for a while, but he soon recovered. He apparently had not anticipated any such move, and it rather took him by surprise! I believe he thinks it OK but can't quite envision anyone taking his mother's place. Which of course is only a natural reaction, but I don't feel he has any antagonistic feelings towards the matter and will probably like the idea once he gets used to it."[72]

Scheduled to graduate in June 1942, Bud was concerned about his own— as well as his father's—imminent deployment. "It has been written that everyone needs a faith. I have never developed a strong religious faith. Ever since I was old enough to toddle after you in the backyard of our H St. home, you have been my back and my faith. Now that I have become old enough to think seriously, that faith has only increased until, I think, were I to lose that anchor, my life would be as aimless as a derelict's wanderings."[73]

Bud and his classmates understood that the odds favored that they would be in the thick of things, because regular midshipmen were assigned to destroyers, and the reserves went to battleships, which Bud saw as "the less responsible jobs." All of which meant that "we get the worst not if, but when. This next leave coming up may well be the last reunion for us warriors. I hope like hell I have the guts to take my medicine. I certainly don't enjoy the thought of dying—but then again I wouldn't want to be on the sidelines."

Summer cruise had been canceled so that midshipmen could attend classes to make up the extra year. "The whole regiment of midshipmen talk and act like a battalion of doomed men. Not that there is any great dissatisfaction. It's just that they all feel we will be at war soon—that they may well bear the brunt—the regular ensigns will get the death details. . . . As a result there is a great tendency to 'live' right now. . . . There is a growth in drinking, lovemaking, etc.—all the soldier boy pursuits are fast growing up. It's hard to find a good logical reason for keeping yourself in line when you may be dead in a year. So far I've behaved—perhaps because of my affection for a decent girl—but very few of us do. All of this probably explains some of the melancholy you have observed in my letters."

Elmo was so concerned about Bud's mental state that he wrote the senior chaplain of the Naval Academy, William N. Thomas.[74] In reply, the chaplain soothed an anxious father: "By this time I feel sure that you realize that the worry of your fine son was momentary, and that he is now on solid ground. . . . His letter to you concerning conditions in the regiment was no doubt true to his impression at the time, but I am quite sure that as a whole the spirit and the faith of the regiment are very high. You and I know what shocks young people are receiving in these uncertain times when even those of us who have gone through such times are not free from confusion."[75]

Everything changed on December 7, 1941. "I had a date that day, and we were in violation of regulations drinking beer in the place where this young lady stayed. As I recall, there was another fellow and his date with us when we heard somebody shouting, 'Turn on the radio! Turn on the radio!' The broadcast was being repeated over and over again—just the very bare-bone details. So we all went trooping back to the Naval Academy promptly, knowing that was the thing to do."[76]

With the attack on the homeland, attitudes changed. "We, of course, had already been speeded up and were on the fast track, so I don't think there were many changes made in the academics. But suddenly everything took on a much more serious nature. There was no longer any fooling around when you were in a gunnery class. Perhaps far less motivation to study Spanish or mechanical drawing. But everybody really concentrating on the professional subjects, because you knew it could well end up meaning a question of life or death. So it was kind of 'Eat, drink, and be merry, for tomorrow we die.'"[77]

On the same day that the academy debate team lost to Harvard, Bud wrote his father that "Ann Austin pulled a fast one. I broke all my dates with her. We are through." Bud did not remain alone very long, next setting his sights on Jane Carey, whom he met about a year before graduation at a dance, while she was dating another midshipman.

Longtime friend Dave Bagley, who attended Rutherford Prep in Long Beach and the Naval Academy with Bud, recalled, "I often tell people that all his actions before the academy were deeply embedded in an interest in individuals—and often resulted in stealing other guy's girls, including mine."[78]

Bud and Jane dated regularly, and a few months before graduation, Bud proposed marriage. Bud was hoping that his father, brother, and sister would be able to attend June Week graduation festivities as well as his wedding, but Elmo was unable to secure military leave. Elmo requested that Chaplain Thomas speak on his behalf at the wedding. Bud needed money for his graduation ring as well as for a miniature or replica that traditionally went to the girl a midshipman expected to marry. "I remember that Mother left one ring for me and I am wondering what you would suggest. Should I have the stone from that ring set in my miniature or should I keep that intact and buy a new stone for the miniature. If I do buy a new one where will the money come from, etc.? Can you help me out?"[79]

Elmo agreed to Bud's request by sending him Frances's ring, which he gave to Jane. Family and friends soon received from Mr. and Mrs. Joseph Lambert Carey of Philadelphia an announcement of the engagement of their daughter Jane to Elmo Russell Zumwalt, Jr., midshipman, United States Navy. The wedding was scheduled for Saint Andrews Chapel in Annapolis on Friday afternoon, June 19, 1942, at 3:30 p.m., immediately after Bud's graduation.

Elmo Russell Zumwalt, Jr., graduated summa cum laude with one of the most demanding extracurricular loads of any midshipman in his class. He twice won the gold watch awarded at the public speaking contest[80] and distinguished himself as a member of the four-person debating team representing Annapolis in intercollegiate competition during his sophomore year. His athletic interests were small-boat sailing, long-distance running, and marksmanship. During senior year, Bud had been promoted to regimental adjutant, giving him responsibility for training the entire regiment in competitive athletics and professional exercises. He supervised military security and, during parades of the regiment, led the midshipmen in preliminary maneuvers on Worden Field before presenting them to the regimental commander. Bud relished the opportunity to lead as both company commander and later a regimental "Three Striper."[81]

Professor of thermodynamics Leon S. Kintberger considered Bud an "outstanding student. His final standing in Thermodynamics was about eighteen in a class of six hundred and nineteen. I consider this record remarkable in view of the fact that his extracurricular activities at that time were not only extensive but also of a literary nature, indicating that his greatest interests were along that line. My analysis of his intellectual capacity is as follows: Outstanding powers of perception and logic coupled with excellent retentive capacity."[82]

When Parson Thomas learned that Elmo would be unable to attend the graduation ceremonies, he wrote directly to Elmo, "It is not necessary to tell you that Bud has made a fine record at the Naval Academy. The confidence of the Academy has been shown by the responsibilities it has placed upon him. He will make an excellent naval officer, and if I live long enough I expect to see him at the top of his profession."[83]

The graduation day speaker on June 19 was commander in chief, United States Fleet, and chief of naval operations, Admiral Ernest J. King. Still recoiling from December 7, 1941, King addressed the challenges that lay ahead for the navy. He shared Churchill's promise of a future marked by "blood, sweat and tears" and cautioned against interservice rivalries' "narrow-minded jealousy" that might jeopardize the joint effort needed to defeat the enemy. "Machines are as nothing without the men who man them and give them life," and "men are as nothing without morale." King went on to define true military discipline as the "intelligent obedience of each for the effectiveness of all" and pointed out that "traditions, of themselves, are no more than testimonials to the successes of our predecessors" and that it was not enough for us merely to boast of tradition, but that we must "make some traditions ourselves." King concluded that when the war was over, with victory, "We Americans—under the leadership of the President—will take steps to see to it that the ability of any person or any people to enslave others, physically or mentally or spiritually, shall be forever destroyed."

With those words, Admiral King commissioned the class of 1943.

Bud was one of about eighty classmates who were married on graduation day in Saint Andrews Chapel with all the glamour of a military wedding. Resplendent in their new ensign uniforms, the naval officers were permitted to marry as soon as they were sworn in, and an unprecedented number of marriages occurred on that graduation day.[84] It was big news back home, the local Tulare newspaper headline, accompanied with a photo, reading, TULARE ENSIGN WEDS AT ANNAPOLIS. The photo caption read, "Ensign Elmo Russell Zumwalt, Jr., and his bride, the former Jane Carey, of Philadelphia, walk arm-in-arm out of the academy chapel under an arch of swords held by classmates of Ensign Zumwalt."[85]

In a final letter to his father before both men headed to war, Bud noted, "Our family of six scattered to the winds, two united in the final abode, four separated by the necessities of life, I feel more strongly than ever the inextricability of the spiritual tie that reaches out across the boundaries of time and death to unite us all—the tie of our association will be the bravest, noblest soul we will ever know."[86]

CHAPTER 4

WAR YEARS

The success of the torpedo attack by Attack section Two is attributed in a large degree to Lieutenant Zumwalt's skill and courage. . . . His outstanding skill and judgment as well as exemplary conduct under fire were in keeping with the highest traditions of the naval service.

—COMMANDER ELONZO BOWDEN GRANTHAM, JR[1]

In preparing for deployment overseas, one of Bud's classmates at the academy sold his car to a buyer on the West Coast. Bud was due to ship out from San Francisco on July 7 and offered to drive cross-country to make the delivery. The drive from Annapolis to San Francisco provided the newlyweds free transportation and time alone. They made one detour—to Seattle for a family reunion, where father and son introduced their respective brides. Bud and Jane were then joined by new stepsister Irene for the drive down the coast to San Francisco, where Ensign Zumwalt reported to the commandant of the Twelfth Naval District. Jane flew with Irene to Seattle, where she lived until Bud returned from the Solomons campaigns in the fall of 1942.[2]

Joined by a score of fellow classmates, Bud boarded the *Calamares*, a United Fruit Company merchant ship that had been pressed into logistic service by the Navy. The *Calamares* was a transport vessel that was responsible for getting sailors to Pearl Harbor. From there, Bud transferred to the *Dixie* and a few days later to the troop transport *Zeilin*.[3] A number of classmates were still with him, including longtime prep school friend and classmate Bryan Pickett.

The *Zeilin's* logs show it departing Pearl Harbor on July 21 to join a convoy en route to Suva in the Fiji Islands. By August 7, they sighted the west end of Guadalcanal Island, and the marines commenced landing operations as part of the first U.S. counteroffensive in the Pacific War, aimed at gaining control of an airstrip on Guadalcanal and preventing the Japanese from controlling the surrounding air and sea regions.[4] The Battle of Guadalcanal lasted for six months.[5] The *Zeilin*, under Captain Pat Buchanan, was one of three cargo vessels attached to the Espiritu Santo Group, Section Two, under Rear Admiral Norman Scott. "The Espiritu Santo Group carried the First Marine Aviation Engineer Battalion, Marine replacements, ground personnel of Marine Air Wing ONE, aviation engineering and operating material, ammunition and food," wrote Colin Jameson in *The Battle of Guadalcanal*.[6]

The operations ran with precision until the evening of the Battle of Savo Island in the early hours of August 9. It would be Bud's first experience witnessing combat, in this case the worst defeat ever inflicted on the U.S. Navy, described by Admiral Ernest J. King as "the blackest day" in naval history. A fleet of Japanese heavy cruisers under Vice Admiral Gunichi Mikawa engaged a patrol group led by the heavy cruiser HMAS *Canberra*. Shortly after 1:43 a.m., the *Canberra* was struck and disabled by a barrage of gunfire.[7] "More than thirty Japanese shells struck the Australian heavy cruiser, killing her commander, Captain Frank E. Getting, and other senior officers," wrote James Hornfischer in *Neptune's Inferno*. "Almost at once both of her boiler rooms were destroyed, and with them died all power and light throughout the ship. She was a floating nest of flame."[8]

Mikawa's force then attacked the U.S. cruiser force guarding the transports that had been landing the First Marine Division at Guadalcanal and Tulagi.[9] In the course of some hours, the heavy cruisers *Quincy*, *Vincennes*, and *Astoria* sank without inflicting any significant damage on the Japanese fleet. "In thirty-seven minutes, more than a thousand American and Australian sailors had been killed, fatally injured, or forced into the sea, where they would drown," wrote Denis Warner. "Another seven hundred were wounded. Four invaluable cruisers had been lost, and the U.S. Navy had suffered the worst open-sea defeat in its history."[10]

"I was able to witness this whole thing," recalled Bud, who had been asleep below deck when the battle began. Bryan Pickett rushed down to report on a tremendous naval battle going on.[11] "We watched and cheered along with the rest of the crew on the *Zeilin* as we saw these four huge fires, because we were all convinced that it was the U.S. Navy sinking Japanese ships. It was only in the wee hours of the morning that the terrible news came in that the ships that we had cheered for being sunk were, in fact, the *Canberra*, *Quincy*, *Vincennes*, and *Astoria*, and that there were many losses, among whom were some classmates who had just recently been transferred to those ships in the week before, having traveled that same voyage with us from San Francisco and down from Pearl Harbor."[12]

Nothing had prepared Bud for this first experience in battle and its carnage. Bud was especially saddened by the loss of classmates Salty Eversole and Pete Hamner. "It was a time of great sadness and personal reinforcement of the tragedy of the war . . . these were individuals that I'd known well during my Naval Academy days. Indeed, they were battalion mates of me. . . . Seeing death reinforced . . . and sobered us all."[13] Midshipmen often glamorized the idea of dying in battle for the glory of their country. He and his friends had talked repeatedly about it. They now wondered aloud about their own odds of surviving the war.

In August 1941, future admiral Lord Louis Mountbatten made a war-rallying, goodwill trip to the United States. One of his stops included the U.S. Naval Academy, where he gave a rousing speech. Bud was in his final year at the academy and was selected to be part of a small cadet detail to conduct then captain Mountbatten around the campus. When he had a moment, with thoughts of looming graduation and a ship assignment, and Mountbatten's recent experience in the Mediterranean, Bud asked Mountbatten what it was like to be in battle. Mountbatten thought for a moment and said, "Well, there you are on your ship. High in the sky you see an enemy plane turn and dive to begin its strafing run on you. Intellectually, you know that the plane's equivalent of .50 cal. or 20mm bullets can cut through the 3/8" steel skirting on the bridge wing like a knife through butter . . . , but at the appropriate moment you duck down behind it anyway."[14]

It must have given Bud comfort to understand that, regardless of one's station in life, everyone is the same—scared—in battle, and that it

is OK to be scared. When he was finally tested, Bud found the best medicine was immersing himself "into what you know you've got to do with
intense concentration, in part because it's your duty to do so, and partly
because that helps you to forget any concern you might have."[15] Bud was
detached from the battle raging at Savo Island, but Lord Mountbatten's
advice would be handy in later encounters during the kamikaze aircraft
attacks and torpedo attacks in the Surigao Strait.

A few days following the defeat at Savo Island, Bud was transferred
to the ship his orders had assigned him to—the USS *Phelps*. "I was transferred to the *Phelps* by high-line. In those days, we didn't have the sophis-
ticated boatswain's chair. Everything was done in the cargo sack that
food went over in. So I sat absolutely out of sight, as one was ordered to
do for safety, while being brought aboard."[16] At this time Bud weighed
about 195 pounds, the heaviest he had ever been. Emerging from the
cargo bag in the dark of night, he overheard the chief storekeeper say in
a loud voice, " 'Huh! A new ensign; there ain't nobody going to give him
no shit.' That was my official greeting on board the first ship to which I
was officially assigned."[17]

The next morning, Lieutenant Commander E. L. Beck asked Bud for
a debriefing of what happened at Savo Island. Beck was surprised and
appalled by the firsthand report. He took Bud to see the squadron commander, who happened to be on board. Bud was ordered not to say anything about the damage and losses in the battle to anyone on the *Phelps*.
"My lips were officially ordered sealed about the news of the things that
I had seen firsthand."[18]

Something happened to Bud in those first days on the *Phelps* that
ended up plaguing him for the rest of his time at sea—seasickness.
Former high school classmate John Sturgeon recalled that during a
father-son fishing outing in the Pacific, Bud got so seasick that he never
got his line in the water.[19] Somehow this never seemed a deterrent or
even a consideration when Bud was weighing his applications to West
Point and Annapolis. Destroyers have a peculiar high pitch and high roll,
and with the exception of standing watch, Bud remained in his bunk for
almost three months. "I carried my bucket with me and vomited on

watch and retained very little food," said Bud.[20] The one benefit was losing twenty-five pounds, which he never permitted himself to regain.

A new ensign becoming seasick so quickly created credibility problems with shipmates. During the first weeks aboard the *Phelps*, he was an object of "ridicule, initially, because they expected me to get over it, as most people did in a few days, and sympathy as it went on and on. Then I sensed that after a few weeks, it became one of admiration that I was still hanging in there and doing my duties on watch and getting to division parades and issuing the orders in between heaves."[21]

The *Phelps* returned to Pearl Harbor on September 28, 1942. Within days the ship was on its way to San Francisco for overhaul to receive its first surface radar before departing for the Aleutian campaign.[22] Bud was detached for radar school at Treasure Island and sent to Long Beach for three weeks' intensive study to serve as a combat information officer (CIC). In San Francisco, he reunited with Jane and was given ten days' leave so that they could both join Saralee and her husband, Bill Crowe, in Dos Palos.[23] "We spent our ten days on vacation in that little hometown getting caught up with my sister and brother-in-law."[24]

This was a difficult time for Bud and Jane. It was obvious to both of them that their marriage was not working. True, they had had little time together before Bud headed to sea, but Jane's loneliness led her to be unfaithful, and she started to drink heavily. Bud was devastated. "Her reputation was a complete embarrassment to my father and his (Saralee) family who promised they would look after her while he was away. The community talked about it," said daughter Ann. Jane's behavior led Bud's brother Jim to think he could never trust a woman, but Elmo interceded with sound fatherly advice. "You once said after Bud's episode that you thot [sic] you would never trust a woman. DON'T hold to that concept. Most women like most men are GOOD. If it were not so the world would go to pot! But you and I must admit that we will never know women."[25]

Jane and Bud had been married only twenty months. "We didn't think it was a compatible situation, and I persuaded her to agree to let me go into a divorce court in California."[26] An interlocutory decree was issued in 1943 in California and became final in 1944.[27] The dissolution of his marriage was difficult on "the crown prince of the clan Zumwalt," which is how his father often addressed Bud.[28] Writing to his son about the lessons to be drawn from the experience of a failed marriage, Elmo explained that "you have already found that life is not a straight line but rather a curve, when at times we are in the depths and again on the heights—the bitter and sweet of life. That is a typical course. If life was a bed of roses, we should soon hate that beautiful flower and its perfume would become obnoxious. . . . I have always told you that regardless of the ups and downs of life, I was with you every time. I want you to always remember that statement. 'To thine own self be true' and there can never be regrets or remorse. We all have our periods of depression, but if we are fair, we can admit errors and gather up our forces. . . . Never feel sorry for yourself."[29]

Saralee joined the chorus of emotional support. "Hope you are all O.K. emotionally now," she wrote. "I'm sure you will find some wonderful girl someday who will love you like I love Dick. If I were a young girl, I'd be mad about you. In fact, I'm generally in favor of you as a brother."[30] Three months later another letter from his sister reached Bud aboard the *Robinson*: "I hope you will have great happiness in the future. It seems you've had rather a rotten deal so far. Sometimes, I wish you had married Billie because I feel she would have been wonderful for you. Bud, I know things will work out."[31]

No longer married, Bud's mental approach changed. The war became an adventure with "no place I would rather have been than fighting for my country at a time of need. I enjoyed the brotherhood of the wardroom and the ships—the sense of power and the sense of adventure. I had no wife or children to worry about and therefore felt none of the concerns about what 'would happen to those I left behind.' "[32] He and his fellow sailors were doing heavier drinking, thinking that their days were limited. "I was promptly seasick every day and would get over it as soon as we got into port, and would misbehave the next night."[33]

With its overhaul complete, the *Phelps* proceeded on a weeklong shakedown and refresher training.[34] The *Phelps* then deployed from Long Beach to the Aleutians and began the campaign involving the capture of Attu and landings at Kiska. One of the first things Bud did was extraordinarily personal in nature. Somehow Bud had convinced Jane to return his mother's ring. When he shipped out to sea on the *Phelps*, he decided to throw the ring into the ocean because it stood as a symbol of defilement to the memory of his mother. "When he told me of his plan to do that I tried to persuade him to sell the ring and use the proceeds for a worthy charity—one that our mother would have approved of," recalled Jim. "When I next saw Bud, he told me that he had, indeed, tossed the ring into the Pacific. I think he felt he had to do this in order to erase the guilt that was tormenting him."[35]

Japan began operations in the Aleutian Islands in June with the bombing of Dutch Harbor. By June 11–12, substantial Japanese forces had landed and established themselves on Attu and Kiska.[36] The *Phelps* was part of Task Force Roger, under the command of Rear Admiral Francis W. Rockwell. The *Phelps* served as the control vessel for the northern landing group. Bud's battle station was officer of the deck, always beside the skipper during actions. He found the invasion of Attu fascinating because the ship patrolled very close to the beach, where Bud was able to observe the army closing in on the Japanese stronghold. The close proximity provided a sense of being "almost a unit of artillery in that action."[37]

The six months in the Aleutians proved to be the most unpleasant part of the war. The weather was usually rough and foggy, so bad during the Attu campaign that Bud was often unable to see the battleships they were escorting just a few thousand yards away. "The feeling of oppression was so great that numbers of sailors really developed the kinds of reactions that led them to be emotionally insecure," recalled Bud.[38] The one benefit of so much fog and drudgery was that Bud was able to do a lot of reading and began studying the Russian language.

The *Phelps* was next involved in the August 2 bombardment of Kiska. Between August 2 and 15, the *Phelps* played a prominent role in ten bombardments conducted by destroyers. On August 12, Task Force Baker, with the *Phelps* taking the place of the *Farragut*, carried out another predawn bombardment. Landing operations started on August 15, but U.S. troops did not encounter enemy forces, as the Japanese had already evacuated the island.[39]

The invasion of Kiska provided Bud with a particularly memorable experience. The *Phelps* had been detached to an area south of the Aleutians in order to rendezvous with a group of transports bringing the invasion force to Kiska. It was a nighttime rendezvous, and the next morning a blinker message came in to the *Phelps*, "To Ensign Zumwalt, I am on the transport behind you. See you in Adak. Love, Dad." Unbeknownst to Bud, his father's field hospital had been assigned to the Kiska invasion forces. Elmo was commanding officer of the combat field hospital. Bud was jubilant about the prospect of being reunited in Adak.

As the *Phelps* was about to enter Adak Harbor, a Japanese submarine was sighted near Dutch Harbor, several hundred miles eastward. The *Phelps* was ordered to break off and investigate. Bud's commanding officer was considerate enough to call Bud to the bridge to explain there would be no family reunion in Adak. He offered to maneuver the *Phelps* close enough to his father's ship so perhaps the two men could get a look at one another. The skipper sent a signal alerting the other ship of this maneuver. As the *Phelps* positioned itself, Elmo was standing on the wing of one bridge and Bud on the wing of the *Phelps*'s bridge. "We were waving at each other as the skipper came relatively close. I was very much afraid that that would be the last time that I would see him."[40]

The *Phelps* did not return to Kiska until it was time to lead the invasion force. All the while, Bud studied the operation plan with extra care, intent on locating his dad at the first opportunity. Doc Elmo was in the third wave to go ashore in his combat field hospital. The *Phelps* was at general quarters, and Bud served as officer of the deck for general quarters, monitoring events on the beach as well as reports of casualties. Almost two weeks after the invasion was initiated, Bud was able to get ashore, "After trotting around through the tundra and mush, to find my dad, who was in the headquarters pup tent for his hospital, where he had been subsisting on K-rations for the period since the landing, and without shower and with a heavy growth of beard." It was an emotional reunion. They had not seen each other since Bud's honeymoon visit in Seattle, and both men were now in harm's way. There was so much to talk about that Bud persuaded his dad to come back to the *Phelps* for a shower, shave, and hot meal. "He did," recalled Bud, "and with a clean

set of underwear and a shave, he came into the wardroom and sat down
to a full meal of steak and vegetables and ice cream for dessert. When
he got through, he turned to the skipper, Commander Edwards, and
said, 'Captain, I'd rather be an ensign in the Navy than a colonel in the
Army any day.'"[41]

Doc Elmo saw extensive action during the war and was one of the
first Allied doctors to enter the death camps, where he agonized over
the fate of an orphaned Jewish boy about the age of five. Each day the
emaciated child would appear at Elmo's tent, never speaking, just looking at him with his piteous and pained eyes. Elmo offered him choco-
late and bread each day. Then one day he found the boy dead, with
the bread under his bed, where he'd been hoarding it as the means for
survival. "Dad shed tears for his young protégé to whom he had become
so attached."[42] In a letter to the local Tulare newspaper from Germany,
Lieutenant Colonel E. R. Zumwalt described feeling "dazed" by what
he had seen, but "it is the truth."[43] In a letter to Bud, Elmo observed,
"we thot [sic] the Nazi better than the Jap. He seemed a cleaner fighter,
but our experiences with the prisoners repatriated and certain atrocities recently viewed, has shaken the thot [sic] of 'civilization' to its very
core. I stand amazed that men can have done these things to others. If
the Russians wipe out their share of the Germans I can feel no regret for
they as a group have been worked as slaves and died thru starvation by
the thousands."[44]

The *Phelps* had been ordered to leave the Aleutians and proceed to
Pearl Harbor, where Bud received orders to join the new crew of
the USS *Robinson*, which was in the final stages of construction. Bud
welcomed the detachment, believing that destroyers were where the
action was for a junior officer. The *Robinson* was superior in damage
control, electronics, and gunnery to the older *Phelps*. The *Robinson*
was placed in commission on January 31, 1944, under the command
of E. B. Grantham.[45] While the *Robinson* undertook its shakedown
cruise in San Diego, Bud returned to Pearl Harbor for Fleet Gunnery
and Torpedo School, the requisite skill set for a Combat Information
Center officer and evaluator.[46]

Bud's service on the *Robinson* encompassed the invasions of the Palau Islands, Guam, Saipan, Tinian, the invasion of Leyte Gulf, and the battle of Leyte Gulf, "the biggest and most multifaceted naval battle in all of history."[47]

A personal diary kept in violation of orders by *Robinson* sonarman James Heinecke provides a window into the ship's operations leading up to the Battle of Leyte Gulf.[48] On June 15, the *Robinson* steamed between Tinian and Saipan islands, taking a position off Yellow Beach. Their first target was Saipan, with orders to shell the town and a sugar refinery.[49] "We came in sight of Saipan; star shells being fired by bombarding cans, wagons and carriers. The island was the same except there's no more town and blown up quite a bit more. Saipan was about ours—next comes Tinian. Two miles from here a fierce action, and we're not even at general quarters. Crazy war!" wrote Heinecke.[50]

On July 7, 1944, the Japanese initiated a final banzai attack on Saipan, resulting in the death of 406 Americans and over 4,300 Japanese. The *Robinson* lent fire support and then dispatched motor whaleboats to rescue survivors from the north reef of Tanapag harbor.[51] The *Robinson* was next sent to Guam to support operations there with shore bombardment and screening duties through August. In early September, she joined Task Group 32.5 for the invasion of the Palaus, providing close fire support to the underwater demolition teams and providing harassing and illuminating fire.

After a hurried overhaul at Manus Island, in the Admiralty Group, *Robinson* sortied on October 12, 1944, for the invasion of Leyte, Philippine Islands. Three days before the landings, the *Robinson* supported the minesweepers, until October 20, when the *Robinson* began shelling the eastern beach area of Leyte near the town of Rizal, where the assault troop landings were to take place. The weather was so rough that depth charges were thrown from their racks. On October 24, word came that the Japanese fleet was steaming up the Surigao Strait from the south, bound for Leyte Gulf.

The Battle of Surigao Strait began when the southern force was sighted before noon. Admiral Thomas Cassin Kinkaid correctly estimated that the Japanese would try to penetrate Leyte Gulf via Surigao

Strait and had deployed Admiral Jesse Bartlett "Oley" Oldendorf with six battleships, four heavy and four light cruisers, and four destroyer divisions to where Surigao Strait enters Leyte Gulf. By late evening of October 24, 1944, a Japanese naval force under the command of Vice Admiral S. Nishimura sought to make its way through the Surigao Strait into the Leyte Gulf to attack elements of the U.S. Seventh Fleet. This force consisted of two battleships, two cruisers, and seven destroyers.[52] The three-pronged attack plan began with a carrier group, coming down from Japan, hoping to lure the big carrier Task Force Thirty-eight away from the landing areas. This would be followed by two surface forces converging on the defenseless transports, one coming from the north through San Bernardino Strait and the second coming from the southern entrance to Leyte Gulf, Surigao Strait. "The *Robinson* was in the group assigned the job of wiping out the Southern Force."[53]

The *Robinson* was part of the squadron organized into three attack sections, leading the second section attack down the east side of the strait. Bud was the evaluator in the Combat Information Center, the ship's below-deck nerve center. Bud's eyes were glued to the large, circular, tablelike radar screen brightly lit from the underside of its clear plastic top. "Ever-changing crayon marks offered a kaleidoscopic overview of their position in relation to other ships in the formation." Bud's job was to send a steady stream of vital information up to the ship's bridge. Exact courses were plotted, and radar information was interpreted. "At that moment, sonar antisubmarine information punctuated the tense air with its rhythmic and penetrating ping, mixing with staccato instructions and questions from the bridge."[54]

The U.S. battleships and their screens were arranged across the mouth of the Surigao Strait. By capping the course the enemy would take, the U.S. ships would be able to bring the full weight of their broadsides to bear on the enemy who, "forced into a column formation by the narrow channel, would only be able to fire a few forward guns. Destroyers and motor torpedo boats were lined up on either side, ready to make torpedo attacks."[55]

Jason Hammer, a young radioman aboard the *Robinson*, described the evening as "black velvet"—total darkness, "except for the eerie phosphorescent glow of the ship's wake. The only stars in the sky were man-made bursts of heavy-caliber gunfire. If there was a moon, it was obscured by a heavy pall of thick, black smoke." The *Robinson* was ordered to proceed down the strait from the enemy's right flank and engage the battleships. It was a challenging task, requiring getting within nine thousand yards for accuracy. General quarters was sounded, and the force formed a battle formation into three rows. The *Robinson* led the left flank. Radar served as the eyes of the ships on an evening when lookouts were helpless. Hammer was at his battle station, operating a radio when he noticed Lieutenant Zumwalt displaying a "seemingly unflappable aura which seemed to surround him even under the most nerve-jangling circumstances." It left an indelible impression. "Anyone dependent on another human being for leadership, and in desperate need of some degree of assurance under hazardous conditions, immediately will recognize the feeling. Here was a man from whom I repeatedly gained some measure of peace of mind. His quiet strength and obvious calm, whether during torpedo run, kamikaze attacks or retaliatory fire from hostile shore batteries, never failed to reassure me with his always observable control of any situation."[56]

The *Robinson*'s objective was an enemy battleship. "It looked like suicide, but we still kept going closer to our objective."[57] At 2:30 a.m., the Japanese battleship came about—left full rudder, all-ahead flank. "The wagons and cruisers hung back to furnish surface fire to cover the destroyers, whose job was to steam right under the Japs' noses to fire their deadly torpedoes, then fall back and retire." The *Robinson* initiated its torpedo run, closing rapidly on the Japanese battleship while facing large- and small-caliber shells being exchanged by the opposing fleets.[58]

"God first made His presence known to me in the battle of Surigao Straits when the Japanese battleships had opened up on my destroyer with their main battery guns," wrote Bud. "The first salvo was over. The second salvo was short. The third salvo would have hit but, at that point, our torpedoes hit them and our U.S. battleships opened up on them. The Japanese shifted their fire and we were not hit. Afterwards, we discovered that every man on our ship was praying at that moment. I have not always been constant in my prayers, but I have found that, in adversity, He is there and quite forgiving of my lapses."[59] The *Robinson* was one of only a few destroyers to ever make a torpedo attack on a battleship and survive. Miraculously, the *Robinson* was not hit. Bud felt he had cheated death. With their torpedoes deposited and their target in flames, obscured by a thick pall of smoke, the *Robinson* was ordered to retire. Bud noticed "a suspicious-looking blip on the radar screen" and immediately notified the bridge.[60] "Radar contact. Unidentified object in the water dead ahead!" No reply came from the bridge. Bud saw the ship heading directly on a collision course with Little Hibuson Island. The skipper had come left in the retreat and was much too close to the beach. "Bridge! On collision course with Little Hibuson Island dead ahead! Acknowledge! Acknowledge!" blared Bud. The ship was going at a speed of 30 knots plus, approaching the island at the rate of one-half mile every minute. They were heading for disaster unless evasive action was taken immediately. "So we swung to the left, laying out a smoke screen. The enemy was within sight. While making this wide turn, land loomed up dead ahead," recalled Hammer. Crew members feared they would not survive the impact. "A collision at this speed, with an accompanying boiler explosion, would demonstrate very spectacularly why these ships were called tin cans," wrote Hammer.

Bud increased the volume in his voice and finally, with a great degree of unfamiliar urgency bordering on total disbelief and frustration, shouted: "Bridge! Back all engines emergency full immediately! You are going aground!"

Finally, a response: "Bridge aye."

"Providence" finally had responded. The order reached the engine room. The skipper yelled out, "Port engines, emergency stop, left full rudder." There was the expected loud whine from the engines as the ship began to shudder violently. "I grabbed onto the bulkhead," recalled Hammer. "I would have sworn we were going aground. We could easily make out the features of the island in the dark. We were so close I could see the trees, rocks, and sand. I ran into the sound room to take a fathometer reading, zero." People on the bridge reported that they could see mud being churned up.[61]

Coming so close to death forced Bud to grapple with fundamental questions. By his own admission, "I was still an embittered man that October morning in 1944 as the Japanese battleships were engaged in the Surigao Straits."[62] Embittered by the loss of a brother and mother and by a divorce, Bud wondered why "one life has been spared while another has not. . . . I did not attempt to rationalize why I was to be given more time to achieve my purposes in life, for such rationalization defies logic. I did come to accept the fact that we are only in this world for a limited period of time to achieve our individual goals and, there-fore, must devote our fullest efforts to reaching them within that little time we have been given to do so. . . . In the days following the Battle of Surigao Strait, I came to focus upon the fundamental beliefs which had taken root during the first quarter of my life and upon which I have attempted to build during my subsequent years. Simply stated, those principles are threefold—compassion for one's fellow man, service to one's country, and peace through strength—all of which, I believe, are inter-related. It took the horrors of a world war to finally cause those principles to blossom into career-oriented goals."[63]

Bud was awarded the Bronze Star for meritorious conduct in action during the torpedo attacks.[64] The *Robinson*'s skipper, Elonzo Bowden Grantham, Jr., gave Bud full credit and actually recommended Bud for the Navy Cross, but Fleet Command downgraded it to Bronze Star. In-stead, Grantham earned the Navy Cross for heroism.[65]

As the war was coming to an end, Bud began thinking about leaving the navy and attending law school at the University of California. In a letter to his father from the *Robinson*, Bud said he was "fed up" with the navy's "thou must" attitude. He dreamed of being free from the limita-tions that regulations placed on individual freedom. He wondered if he could be happy never having roots in one community. He thought that life in a small town with old friends, a home and family, and nearness to people would be Shangri-La.

As usual, Bud sought his father's advice and guidance. In a flurry of letters, Elmo tried dissuading Bud from studying law. "Hoping that the shock of cold water will not be too great," Elmo wanted his son to consider that "you must take four or five years out of your life to be retrained and then BEGIN to establish a business that will build slowly! IF you can weather that TEN years wherein you are almost penniless, then your step is okeh [sic]! If not, then that step seems illogical to me! In contrast, you are on the threshold of becoming a top bracket officer! With the still large navy, there will come much more of shore duty and you should stand well enough with superiors to determine the type of service desired to an extent wherein you may be happy! You will rate good quarters. You will live with a loyal group! You are well liked. You can offer more now, than you can ten years from this moment. These are all thots [sic] to be considered and I would have you give mature consideration to them before you make any radical step."[66]

In a follow-up letter, Elmo pointed out that "your navy career is definitely earmarked" and "I feel sure you are above the common herd and soon will graduate into what the Army calls 'field grade rank.'"[67] Elmo also noted that "in three very short years you have lived a lifetime as compared to many of your topside officers. You have seen more fighting and ability demonstrated and ingenuity used than most anyone will in a lifetime. YOU have been there and made good!" In closing, Elmo played his cards openly: "I want you to know that should you decide otherwise, I am still with you but I just cannot see a change. Your niche is MADE. KEEP IT!"[68]

Elmo had one more card to play, having learned from Doris that Bud was corresponding with his former girlfriend, Billie Nelson.[69] Billie was looking for a way out of an unhappy marriage. Bud's letters professed the mistake he had made marrying Jane rather than Billie. "I will guess, you both made an error! We know YOU did," wrote Elmo, "and I presume hers followed from the disappointment! I still think her tops (as does Doris). She has a problem! If she loves you and you love her (I am presuming) she still has a legal hurdle to take! Until that is done YOU must be circumspect, otherwise three of you can be seriously hurt! En-garde, mes ami [sic]!!!! Until that is accomplished you haven't even a right to THINK."[70] Elmo spoke bluntly: In the wake of "the last semester at Annapolis and your marital fiasco, I trust injecting Billie into the picture will not disturb you too much but I have felt inasmuch as you asked for advice and DIDN'T lay all the cards on the table, I had a right to let you know that I knew of the extra pair you held up (in this poker game of life)! I have given this in all humbleness, for I like the girl immensely and if you do too, and if circumstances allows you to finally marry her (as I presume is a desire) I should be very happy. So don't mistake this expose. All of this is said with the love the Old Man has for the Crown Prince whom he wants to see established—happy and a credit to posterity some of these days."[71] Two weeks later, Bud's younger brother Jim wrote, "What are your post-war plans? I think if you don't stay in, you are crazy. Or does your desire to return to Tulare for a certain reason outweigh all else?!?"[72]

On August 15, 1945, in the Far East and August 14 in the United States, Japan capitulated, ending World War II. While allied ships sailed into Tokyo Bay some forty-four months after Pearl Harbor, the *Robinson*, accompanied by twelve minesweepers, was sent to Shanghai. Arriving near the mouth of the Yangtze River, thirty miles from land and fifty miles from Shanghai, the minesweepers created a safe passage to Shanghai. Meanwhile, several Japanese vessels were trying to make their getaway to Tokyo. The *Robinson* intercepted the largest of the fleeing vessels, the twelve-hundred-ton gunboat *Ataka*.[73] The *New York Times* and *Washington Post* reported that three times her normal complement were aboard, leading to suspicions that several war criminals were trying to escape. "Their personal belongings appeared far too rich for ordinary navy personnel. One even had a bottle of Johnnie Walker Black Label Scotch," the *Post* reported.

Ataka personnel were placed under guard, and a thirteen-man prize crew from the *Robinson* was put aboard.[74] Bud was made prize captain of the *Ataka*. His mission was to chart and sail up the Yangtze River to Shanghai, essentially to see if they could get through the minefields and deliver the crew to authorities there. It would be "one of the greatest experiences of my naval career," said Bud.[75]

With a heavily armed crew of just fifteen sailors and two ensigns guarding the Japanese, Bud got the *Ataka* under way at 7:00 a.m. on September 13 "with Japanese controlling all functions on ship and prize crew guarding against sabotage and also checking and recording navigational data." As Bud later recalled, "The way the ship was built, the most defensible place for three officers and 15 sailors to be with a crew of a couple of hundred Japanese was to live in the wardroom, because there was only one access to the wardroom. That way you could always defend yourself against people from just one direction. And we kept all our guns and all theirs that we could find there. We had taken away all their swords and samurai daggers and that sort of thing, so we felt we had all kinds of lethal instruments there. We ate and slept in the wardroom, cooked out of a little hibachi kind of a thing. Except in twos and threes, we stayed in the wardroom in order to make sure that we didn't get overrun."[76]

By a stroke of good fortune, Bud located the charts to the minefields in a compartment in the *Ataka*. By 8:49 a.m., they had entered the Yangtze, the first vessel flying the U.S. flag to enter the Yangtze River in four years.[77] As the ship made its way up the river, the Chinese people lined the banks and thronged the river in junks, cheering, applauding, and tooting on whistles to demonstrate their happiness at the return of the American navy to Shanghai. In his official report, Zumwalt noted that "the Japanese on board the *Ataka* grew insolent and restless over this demonstration, making it necessary to put guns in the ribs of certain officers and to line up all of the ship's complement not on watch at their quarters under guard."

Bud's orders read that, once in Shanghai, he was to locate Rear Admiral Milton E. "Mary" Miles, the commander of U.S. naval forces in China. The *Ataka* needed fuel and water, so Bud made the decision to tie up at the Japanese Imperial Naval Headquarters, the only pier he could locate with space available. When Bud noticed a vehicle dropping off a Japanese naval officer at the dock, he commandeered the car. The chauffeur drew a pistol and had to be forcibly disarmed.

Bud was soon at the private residence of Rear Admiral Miles,[78] one of the most inspirational characters Bud encountered in his lifetime of service. "My memories and recollections of him are in the same con-text as those of General George Marshall, with whom I spent a day, or of Paul Nitze."[79] Miles had been sent to China in 1941 to organize the whole of China into guerrilla units, using navy personnel to train and lead them. He and his men carried on in Lawrence of Arabia fashion, be-coming the one American most loved by the Chinese and hated by the Japanese. Miles and his team had arrived in Shanghai just days earlier. Miles offered Bud and his entourage a scotch, the first they had seen in months. When Bud pointed out that they were on duty, Miles said that his eyes were teary and he could not see very well.

Miles was concerned that the *Ataka*'s situation was precarious, be-cause the Japanese might try to overrun the prize crew, so he sent his troops to help guard the ship. The following day, Miles went to the *Ataka* and informed the Japanese commanding officer that the Ameri-can navy was officially assuming command of his ship; the crew was removed and interrogated. Admiral Miles asked the *Robinson* crew why they had not taken any souvenirs. "Then he went over to one of the sea bags and saw one of these samurai daggers and said, 'Well, since you're not taking any, obviously you won't miss this one.' He helped himself to that one." Bud was given a sword that had been presented to him by the commanding officer of a surrendering Japanese vessel. Bud gave the sword to shipmate Mel Knickerbocker, who kept it as souvenir for twenty-six years; when Bud's appointment as CNO was announced, Mel wanted Bud to have the sword. In an August 3, 1970, letter, Bud let Mel know that his brother Jim had delivered the sword prior to the change-of-command ceremony.[80]

Miles then addressed the prize crew, reminding them that there were currently 175,000 armed Japanese and just a few thousand Chinese and a few hundred American troops to oppose them. He reiterated that the presence of the American Navy in Shanghai was very important to the Chinese people. Bud's instincts were not to grant his men liberty, but Miles authorized liberty for one third at a time to show an American presence during the transfer from Japanese to Chinese national control. He warned his men that while on liberty it was essential to impress the Chinese with good behavior and to comport themselves properly.

News spread quickly throughout Shanghai that the war was over and
the proof was in the harbor, where a Japanese ship was flying an American flag. Literally everyone in Shanghai wanted to see the liberators
and the ship. One of those joining the throngs was Mouza Coutelais-du-
Roché. A few days later, one of Mouza's close friends wanted to thank a
few Americans from the ship who had given him a job. He offered them
a home-cooked typical Russian meal and asked Mouza's aunt and uncle
to host the dinner. The next day, Bud was invited to the dinner party.
The hostess was Mouza.

Mouza had been born in Harbin, Manchuria, in 1922, into a community of eighty thousand pro-czarist Russians in the northern province
of China. Both her parents had been born in Siberia. Her mother, Anna
Mikhailovna Habarova, was a White Russian loyal to Czar Nicholas II;
her father, André Coutelais-du-Roché, was a French national and fur
merchant.[81] Her parents escaped Siberia in 1921 and settled, along with
thousands of other White Russian émigrés, in Harbin, where a miniature czarist Russian society emerged. During the Russian Revolution,
roughly two hundred thousand White Russians escaped Siberia into
Manchuria, where they re-created czarist Russian culture in churches,
homes, and schools. "My parents had learned the transitory nature of
success in Asia and made a conscious decision that, although they had
very little, they could raise me as nearly as possible in the same way as
they would have done had they remained in czarist Russia. They gave
me piano lessons. They taught me to love good music and art," recalled
Mouza.[82]

In 1931 Manchuria was occupied by the Japanese, and the family was
forced to house Japanese civil servants.[83] While living under the Japanese occupation, Mouza's mother was diagnosed with stomach cancer.
There were no adequate medical facilities in Harbin, so André took
Anna to the Rockefeller Hospital in Peking for treatment. Mouza was
left in the care of her gradmother, Elena. "Hello my dear girl Muza,"
wrote Anna from Peking. "Yesterday we saw two doctors; they cannot
understand where my tumor comes from. Today I will go again. Today
I am going to see the doctor of a very great standing. I do not know
what he will say, but I think I will have to have surgery. Muzochka,

don't worry, be strong, God willing everything will be fine. . . . Poor Dad, everything is so expensive."[84]

Following surgery, the doctors said there was little else that could be done. While Anna was recuperating from surgery, André returned to Harbin to take care of his business matters. He made the decision that Mouza would accompany her mother to Shanghai so that Anna could recover at the home of her sister Christina in the French Concession. André promised to join them shortly thereafter. Before going to the railway station to take her mother to Peking, Mouza's father kissed her and said, "I'll see you in three days." It was the summer of 1940. Anna Mikhailovna Cotelais-Du-Roche (as written in her death announcement) passed away on April 28, 1942. Mouza spent the next four years with her aunt and uncle in Shanghai because, with the war's termination, Stalin sought to increase his empire by sending Soviet troops to occupy Manchuria. André was unable to leave Harbin, his business was seized by the occupiers, and he died in Harbin in 1946 from pneumonia brought on by malnutrition.[85]

Bud brought a bottle of hard-to-find scotch whisky to dinner. "I had never tasted Scotch so it didn't impress me one bit," Mouza remembered. "What did impress me was him."[86] In a letter to his father, Bud described what happened when he entered the home. Four girls soon entered the room, the first "a gorgeous blonde, lithe and well-formed with a lovely soft complexion and the same air of regality."[87] When the second woman entered the room "my heart stood still. Here was a girl I shall never be able to describe completely. Tall and well-poised, she was smiling a smile of such radiance that the very room seemed suddenly transformed as though a fairy waving a brilliant wand had just entered the room. I never saw the remaining two girls."

Bud could not speak; in fact, he could not stop looking at Mouza until finally she motioned everyone to sit and he got the seat next to her for one of the most memorable meals of his life. "Beyond being pretty, she had a radiant air that was enchanting," wrote Bud. Mouza spoke fluent Russian, French, Chinese, and Japanese but not a word of English. This allowed Bud to hatch his plan. He had been studying Russian for career enhancement but not necessarily to marry a Russian.[88] Bud promised to teach Mouza English if she helped him with Russian, telling her that the next war would be with the Russians. For the next ten mornings, he went for tutoring and started spending more time with her. "This was the beginning of a new period in my life," wrote Bud to his father. Language tutorials quickly turned into long picnics in the park, dancing at the Russian Club, and suppers at Mouza's home. For the picnics Bud brought along army rations as well as caviar and champagne from the black market. Mouza was intrigued by food from a sealed can. "And he was very surprised but he ate with me and I enjoyed it very much."[89] On their seventh day together, while riding on the second level of a bus, Bud proposed marriage, quoting two lines from the poet Christopher Marlowe, "Come live with me and be my love / And we will all the pleasures prove." At first Mouza was mad, thinking that Bud was asking her to live with him. She thought of punching him until Bud explained it was a poem and that he was proposing marriage. Mouza asked if this meant going to a church. He said, "Yes, of course."

Mouza knew she would accept the proposal, but there were details to be sorted out, beginning with the fact that she was already engaged to an Italian marine. Bruno had been stationed in the British concession of Shanghai doing some sort of work with the Italian embassy. The two dated for years until the Japanese imprisoned all Italian military in Shanghai. Bruno disappeared, first to a prison camp outside Shanghai and then to one in Japan. At the end of the war, the Americans freed these prisoners and sent them home. Bruno tried explaining to authorities that he wanted to join his fiancée in Shanghai, but he was sent home to Italy.

Mouza had no idea if Bruno was alive. She had received no word from him. Two days after receiving Bud's proposal in what she described as the most difficult decision of her life, knowing that she would have to leave her homeland and might never again see her father, Mouza agreed, although she also later joked, "I would have accepted his proposal at the end of the third date."[90] When Mouza told Bud about Bruno, he made every attempt to ascertain Bruno's whereabouts. Years later, when deployed in the Mediterranean, Bud went to Venice and located Bruno. The two men took an immediate liking to each other. Bruno explained to Bud that he had tried reaching Mouza, but it was hopeless at the time. Bruno had by now married and expressed approval for Mouza's choice of a husband. The two men remained in touch throughout their lives, exchanging Christmas cards and gifts. When Bud received the Presidential Medal of Freedom in 1998, he invited Bruno to join his family for the White House ceremony, an offer Bruno declined. When Bruno's son Stephan, a cardiac physician, visited the United States, he stayed with the Zumwalt Coppola family in Boston. Mouza chose never to write Bruno or have any communication because her heart belonged only to Bud. She would never look back.

Mouza was a member of the Russian Orthodox Church, which required Bud to be interrogated by the bishop, who asked that Bud furnish two witnesses who could attest to knowing him for five years and could vouch for his family. Two shipmates "stretched the truth" a bit. Another complication arose when Bud and Mouza were required by the church to wait two weeks before the actual wedding could occur. With the *Ataka* turned over to the Chinese, Bud had been reassigned to the *Robinson* and was no longer his own boss as prize-crew captain. Meanwhile, the American consul informed Bud that in order for Mouza to get a visa, there would also have to be an American wedding ceremony prior to the Russian service. But the *Robinson* had just received orders to sail, presumably for the United States. Their wedding date for the church had been set for November 1, 1945, yet the *Robinson* was scheduled to depart on October 23. Using all of his persuasive powers, Bud successfully lobbied for a waiver of the two-week waiting period so that the wedding could be scheduled for October 22.

On the morning of the 22nd, Bud and Mouza both visited Mouza's mother's grave and then went to the American embassy for a brief civil ceremony. Mouza then went with friends to prepare for a traditional Russian ceremony while the *Robinson* crew spent hours trying to get Bud drunk at the Palace Hotel. The wedding ceremony was held in a Russian Orthodox church at noon on a beautiful fall day with over 150 men from the *Robinson* present in their dress whites. The church was small and had no seats, so guests stood in a semicircle facing the priest and the bride and groom. Bud entered the church portals alone with a bouquet in his hand. Throughout the ceremony, a crown was held over the groom's head by the best man, Ensign Lawler. Before the wedding ceremony, the *Robinson* crew made a guard of honor for the newlyweds. Two shipmates recalled the moment: "When we saw them we had the idea that it would be a nice touch if we were lined up facing each other for a salute as each party went from their cars to the entrance of the church. . . . We all enjoyed the surprised and amazed look on Lt. Zumwalt's face."[91]

A reception followed at the home of the bride's aunt and uncle, the same home where Bud had first met Mouza a few weeks earlier. Once in the home, the men were directed to gather in front of the fire; glasses were filled with vodka, emptied, and then smashed into the fireplace. A grand party ensued. Members of the radar division purchased an Irish linen tablecloth with matching napkins from the Palace Hotel gift shop in Shanghai as a wedding gift.[92] A shipmate made the following handwritten entry in his diary: "Liberty today. Attended a wedding. One of our officers was married to a Russian girl. It was very impressive. She was a beautiful bride, dressed in a Russian style wedding gown. The ceremony was conducted in the Russian Orthodox rite. Lots of ritual and pomp. Lt. Zumwalt looked very happy with his new bride. He sure worked fast, we've only been here since September 9th. But I'm sure this marriage will endure a long time."[93]

One wonders what Bud must have been thinking. His first marriage had failed because of war and separation; he had known Jane for less than a year before proposing. He was also corresponding with an old flame, Billie Nelson. Bud was now marrying someone he had known for less than two weeks, who had been engaged at the time they met, who spoke no English, and the *Robinson* was scheduled to depart the next morning from Shanghai to the United States. Bud was also under the impression that it would take a few years before his war bride's visa would be approved. Yet he rolled the dice. Luck was on the side of these two people whose life experiences had placed them at this point. The morning after being married, Bud said good-bye to Mouza and went to the *Robinson*. About three hours later, he returned to Mouza, because when he arrived at the *Robinson*, Bud was told that new orders had arrived from Washington detaching him for transfer to serve as executive officer of the USS *Saufley*, which was not scheduled to arrive in Shanghai for another several days. "I thought how nice of them to do this because they found out probably he just got married so they do something to him," thought Mouza in her innocence.[94]

In his fitness report detaching Bud from the *Robinson*, Commander Ray Malpass wrote that "the officer has been studying the Russian language in his spare time. An opportunity should be given the officer to improve his knowledge of Russian, and thereby increase his value to the Naval Service."[95] "His performance of duty is the best of any officer of his rank in my experience in the naval service. . . . He is highly recommended to promotion when due."

Bud and his new bride would now be able to honeymoon in Shanghai. The music never stopped.

CHAPTER 5

CROSSROADS

*Our new commanding officer was also a hard charger with
a marvelous sense of humor. His "can do" spirit quickly
earned him a nickname—The Road Runner: Beep, Beep.
There goes Zooomwalt.*

—REAR ADMIRAL ROBERT J. HANKS[1]

The newly married couple took full advantage of their week alone in
Shanghai. They were able to spend additional time together once the
Saufley arrived, because the ship operated from the mouth of the Yangtze transporting pilots to and from merchant ships.[2] After a few weeks
in Shanghai, the *Saufley* was ready to depart on a five-month deployment. This was the first of many times in his naval career that Bud left
Mouza alone, but things were especially complicated this time because
Mouza had just learned she was pregnant. The city of Shanghai arranged a farewell party for the crew of the *Robinson* on December 11,
1945, the night before the *Robinson* was set to cast off its moorings and
depart for home. It was at the farewell party that Mouza told Bud she
was expecting a child. "I remember the night of the *Robinson* party in
Shanghai when you had to try five times to tell me Elmo was to be
born."[3] Bud's number one priority was getting his wife to the United
States so that his father could deliver their baby. All this would need to
be sorted out while Bud was on deployment. In the interim, Bud took
consolation that for now his wife could remain in familiar surroundings
with the support of friends and family.

"My darling Mouzatchka, we have only been apart 24 hours and already it seems like torture," wrote Bud on their first morning apart.[4] It
was a sentiment that virtually every newlywed sailor experienced. Describing their five weeks together as "like a dream," Bud found himself

constantly looking at his wedding band and Mouza's picture. "It will be terribly hard to live away from you . . . I love you, dearest. More than you will ever know, I love you."

Bud advised his pregnant wife to get plenty of rest, eat healthy food, and "don't worry about anything." He closed by writing, "Wherever you are, no matter what, I am always with you sweetheart, loving you with all my heart and soul." Hours later on the same day, Bud wrote again, this time from the mouth of the Yangtze River. "My darling bride, I am terribly lonely for you! I never knew that it could be possible to miss a person so much. It seems like 1000 years since I last held you in my arms."[5]

Bud had left clear instructions for Mouza to go with his friend Ed Martin to the U.S. embassy in Shanghai in order to begin the visa application process. The GI bill had already passed Congress, but the embassy insisted that it would not be processing visas for war brides. Bud was determined to find a way of making the bureaucracy responsive. In a December 20, 1945, letter written three days before arriving in Pearl Harbor, Bud confided how difficult it was to be halfway across the ocean, knowing that he would not be seeing his wife for at least four months. "It is very hard to be without you," wrote Bud. He promised to take care of everything. "When I go to bed each night I kiss your picture and then say a prayer for you and for our baby."

By December 23, Bud was in Pearl Harbor and about to depart for San Francisco, where he planned on reaching out to everyone and anybody for help. "Everywhere I go I will ask people to help us and maybe someday I can find the right person. I will try very hard. . . . Please try very hard to come to America."[6] In another letter Bud explained, "I have been very busy trying to get you into the U.S. . . . Hurry to me sweetheart."[7]

Worried about the health of both Mouza and the baby, Bud sounded as though he was at the breaking point: "I hate the Navy life. I love you and want to be near you . . . I have missed you terribly. Darling, I feel so bad that I have not been able to make you so happy or do very much for you. But I will try harder next time."[8] The big break came in early February 1946, when the *Saufley* made a stop in New York en route to

Charleston for decommissioning. Bud had been planning this day for weeks, securing a day's leave so that he could catch a train bound for the nation's capital. In a February 3, 1946, letter to "my darling," Bud informed Mouza that "all my papers came. I am going to take them to Washington, D.C. right away. It has been a terrible long time for you. I pray that soon you will have visa and everything will be all right."[9]

Bud had no contacts in the State Department; in fact, he possessed nothing except a desperate single-minded focus on getting his wife into the States. Arriving at Union Station, he asked the taxi driver to take him to the State Department, which was then located in the old Executive Office Building next to the White House. "I went up the steps and turned right, and the first sign I saw on a door was Assistant Secretary of State for Far Eastern Affairs." Bud thought that title was close enough and opened the door. The secretary said, 'What can I do for you?' I said, 'I need to see F. Everett Drumright.' His name was on the nametag."

The secretary ushered Bud into the office. He was a lieutenant in uniform and mistakenly assumed he was one of many navy messengers who came to the office daily. Here is Bud's account of what happened next:

> He looked at me and said, "What can I do for you?"
> "I'm here to get my wife liberated from China."
> "Well, tell me about it."
> "Well, I am married to this woman who is a French citizen, and who isn't able to be admitted."
> "You know, we've got a law that we've got to obey, and she can't come in for a couple of years."
> I pointed out to him, and said, "No, that group right over there, the U.S. Congress, passed a law December 28 authorizing the wives of servicemen to come in immediately."
> He said, "We have no information of that."
> I handed him a copy of the *New York Times* article.
> He said, "Lieutenant, I'll of course be happy to at least check, but if this story is accurate, I guarantee you we'll put a message in a pouch going out to Shanghai."[10]

Less than a week later, Drumright cabled Bud, who was by then back aboard the *Saufley*: "With reference to our recent conversation, I am happy to notify you that the following telegram dated February 17, 1946 has been received from Consul General Josselyn in Shanghai for transmittal to you. 'Not necessary to file affidavits. Your wife sailing on Navy transport SS *General Scott* leaving approximately March 1.' "[11]

As the *General Scott* pulled away from Shanghai harbor, the band started playing American songs, and the soldiers on board were happy to be going home. Mouza was happy for them, but inside she was torn with regret and sadness. "And so little by little, very, very slowly the ship started moving away from the land and it is getting further and further and further. And suddenly they start to play the song 'America.' And of course they try to throw their hats, 'We're going home! We're going home!' "[12] But for Mouza, Harbin was home and with that came the realization that her father was alone and she might never see him again.

In Charleston, Bud was ecstatic. *Robinson* shipmate J. M. Reid, Jr., who had attended Bud and Mouza's wedding in Shanghai, wrote, "And as for Mousa [*sic*]—please be *honest* and should there be *anything* that I might be able to do to help her on her arrival, please feel free to say so— because that is the way I would like it. Maybe there is something we could do or someone to see to help speed things up—I would be elated to help in any way."[13]

On March 11, 1946, Saralee received a cablegram from Shanghai. "Arriving Seattle March 21st on SS *General Scott*—Mouza." Saralee was frantic because the ship was diverted from San Francisco to Seattle. Bud, who was in Charleston with his destroyer, had orchestrated arrangements so that Saralee would meet Mouza in San Francisco, but neglected to tell his wife about these arrangements. "Please notify me what to do," wrote Saralee. "I can't possibly be there to meet her but could get the Red Cross or Navy League to meet her. But what then? Do you want her sent east—if so where? Does she have money?"[14]

The actual voyage was a harrowing experience for Mouza. The ship encountered several typhoons, and she was placed in sick bay, where nurses could provide support. When the ship was diverted to Seattle, several of the well-meaning soldiers on board the *General Scott* worried how Mouza would fare once she was off the ship. She had no idea where San Francisco was in relation to Seattle but assured the soldiers that they need not be worried because Bud would be there for her. One beneficial aspect of the trip was that Mouza befriended a German woman who had married a marine. She spoke English quite well and tutored Mouza throughout the trip.

Fortunately for Bud, he had received news of the ship's diversion and arranged for his stepsister Irene and a few family members to greet Mouza. The only problem was that Bud had never told Mouza that he had family in Seattle. Irene was a new stepsister with a different last name than Zumwalt. Mouza did not believe they were family, even suspecting they were the police, until Irene was able to show family photos from Bud's earlier visit, omitting the photos with Jane.

In a heartfelt letter to Mouza written just a few days before her arrival, Doc Elmo sought to reassure his new daughter-in-law. "We know that you are due to land in America within the next few days and will be met by Irene as the first family contact in this foreign land. Knowing Irene well, we are sure your welcome will be a happy one and that she will initiate the first steps needed to get you on your way to Bud, now on the far east coast. Please accept this letter as a greeting and as our rather long distance manner of saying 'welcome to these United States.' We sincerely hope you will find this, your new home, cordial, friendly and happy. We trust that the customs, mannerisms and language, will not prove difficult. We are sure that any of us would feel a bit lonely, going to another country and no doubt at the moment, you too, feel just a bit lonely."[15]

The next day, Saralee wrote that Mouza "must feel lost in this strange new country and its likewise people. I should be petrified if I were her." Saralee wanted her brother to know that "I do want this marriage to turn out right for you after all your unhappiness."[16]

Mouza's first days in Seattle were spent getting oriented to the customs of her new country.[17] She then flew to Charleston, where Bud was stationed for mothballing the *Saufley*. Mouza's arrival could not have been timed better. Bud essentially had three months' shore duty, which enabled him to show Mouza how to do many of the things needed to run a household in preparation for his next extended deployment.[18] With the decommissioning of the *Saufley* complete, Bud's next orders were sending him to serve as executive officer on the USS *Zellars*.

One of the first family members to meet Mouza was Bud's brother Jim. Still in uniform, Jim hitchhiked from the U.S. Marine Corps base in Cherry Point, North Carolina, to their modest rental in Charleston. When Jim got there, Bud was still on his destroyer and not expected back until dinnertime. Mouza was still quite reluctant to speak English, so whenever Jim asked a question, she would offer a timid *da* or *nyet* in reply. The icebreaker came when Mouza offered to make tea. Jim discovered that no one on earth made tea like Mouza, a secret learned from her Russian ancestors.[19]

Just before the baby was to be born, Mouza flew to Tulare, where she would be under Doc Elmo's care. On July 14, Bud wrote his "darling Mouzatchka by the time you receive this maybe we will have the baby. I hope so darling. I know everything will be all right."[20] The next day, Doris wrote Bud that "Mouza is fine but also getting very tired which is to be expected. Everything is ready for the event—even to a bathinette [sic] which is being loaned to her by a girl in Officers' Wives Club. . . . We will let you know when things happen here."[21] Three days later, Elmo joined the choir, writing Bud that "we are glad to be serving Mouza and shall try to see her safe and courageously threw [sic] her delivery. . . . She is a lovely girl and we shall inform you by telegram of the delivery."[22] A week later an impatient Bud wanted to know "when will the baby come, Mouzatchka? How many more days must we wait to know? I am very nervous and very excited . . . I love you so very much. Sometimes it seems that I will die without you. It is very hard to be so far away from you."[23] The next evening, Bud called Mouza to celebrate their nine-month anniversary.[24]

Elmo was born on July 30, 1946, at a hospital just three blocks from the family's Sycamore Avenue home. A telegram was sent immediately to Bud, who was now aboard the *Zellars*: "Mouza presents your son. Seven PM tonight. Nine and Half. All OK. Doris and Dad." The arrival resulted in an unusual and certainly unique birth certificate: child,

Elmo Russell Zumwalt III; father, Elmo Russell Zumwalt, Jr.; delivering physician, Elmo Russell Zumwalt.

In a letter to his brother, Jim noted that "my first impression of him was that he was rather ugly. He had a long funny shaped head, and his cheeks were so fat that it was difficult to see his mouth or chin. This made him look something on the order of a mole. Don't be alarmed tho [sic] Bud, as his looks have improved remarkably already."[25]

Bud had secured a three-day pass to be with Mouza and the baby in Tulare. The delivery was normal and the plan was for Bud to return to duty in Charleston while Mouza rested for the next month in Tulare. Coincidentally, in June of that year, Bud's brother Jim returned to Tulare, having been discharged from his military obligations. Jim was planning to "live it up" with old high school friends until the beginning of semester at Berkeley. A day after Bud returned to his ship, Mouza became seriously ill, with a fever of 106 degrees. Both Doc Elmo and his medical partner worried that she might not survive, suspecting some type of virus contracted in Shanghai that had been triggered by Elmo's birth.

Jim volunteered to take over the baby duties, because by this time Doris was showing signs of instability. "Abandoning my carefree life of socializing, I assumed the responsibility. I became adept in preparing formula—altering the mix as Elmo slowly gained weight; washing and sterilizing bottles; laundering, ironing and folding diapers; and pinning diapers on a wiggling body that seemed intent on thwarting my best efforts. At night I kept Elmo in a crib close to my bed." Jim was baby Elmo's primary caregiver for several weeks. As Mouza slowly regained her strength, she and Jim would walk the baby stroller along the avenue of the sycamores.

By late August, Mouza had regained some of her strength, but Jim needed to be back on campus for the start of classes. Jim would not leave Mouza and the new baby with Doris, whose drinking and temper tantrums were already wreaking havoc in the once tranquil Zumwalt home. Doris had banned Saralee from the Sycamore home, and Jim "was almost completely cut off from dad."[26] Elmo sided with his wife, making things difficult for the siblings. Bud would soon be referring to Doris as the "wicked stepmother."[27]

Bud's new quarters in Charleston were not yet ready, so some type of arrangement needed to be made. Jim drove Mouza and baby Elmo to Dos Palos, placing them in the care of Saralee for the final two weeks of Mouza's recovery.[28] Saying good-bye to Elmo was traumatic for Jim. "Having been his sole parent during his first month of life, I regarded him more than just a nephew and Godson. He had become 'my baby' and for him I felt an affinity that few uncles have ever known."[29] Elmo felt the same way. A day prior to his own death in 1987, Elmo dictated a note to his Uncle Jim and Aunt Gretli, "who between them, nurtured me at birth and in terminal circumstances, with love and appreciation for their lifelong support."[30]

Mouza and Elmo then flew to Charleston, where they would have only a brief time together before Bud deployed. When Bud left Mouza for the first time in Shanghai, his new bride was pregnant. This time he was leaving her with a new baby in a new country with few friends. Moreover, he worried incessantly about a relapse of the mysterious infection. In the first two years of their marriage, they would move twelve times. "We always lived in furnished places. Mouza had one suitcase for the baby's things. I put the baby carriage on the ship when we moved. She'd fly with two suitcases, one for her things, and we'd rent a place wherever the ship had been ordered, and we'd pull a drawer out of the chest of drawers for the baby to sleep in, and that's the way we lived for two years."[31]

While at sea, Bud received a series of letters from Jim observing that "Mouza seems very lonely." Jim asked, "What are your plans? Are you getting out of the navy or not? Dad is very worried as he fears that you will experience serious financial difficulties if you get out. . . . Whatever you do, make Mouza happy, Bud. If you can do that you shall be successful from now on."[32] Several months later, Jim wrote, "I feel terribly sorry for Mouza as her letters sound lonely again."[33]

All of this led Bud to again give serious thought to leaving the navy and pursuing his dream of a career in medicine. "I found myself longing to go to medical school and follow in the footsteps of my parents and perhaps enter into my father's medical firm."[34] Bud had been accepted by the University of California Medical School, but Doc Elmo was still pushing Bud hard toward a career in the navy. Former *Robinson* shipmate J. M. Reid, Jr., had recently visited Elmo and Doris in Tulare. In a letter to Bud about his "unsettled future," he explained that Doc Elmo "still felt your best bet was the Navy. That if you took a fling at law or medicine that you might find it rather tough financially with the wife and baby, while going to school. Also, that you could be retired at age of 41 with a fair pension. Altho' I know how you feel I am rather inclined to agree with him in certain respects, especially at the present time. . . . I know how tough it is being away from Mousa [*sic*] and the boy, but if you could get shore duty for a time and could wait until things straighten out, I really feel that you would be far better off."[35]

Service aboard the *Zellars* was the most difficult in Bud's naval career. "We were in the midst of Secretary of Defense Louis Johnson's economy era when supplies and spare parts were so badly curtailed that it was extremely difficult to keep the ships operating."[36] Navy personnel were being released as a cost-saving measure, and there was a great gap in experience at all levels. Ships would often steam for eight hours and then drift or anchor so that people on board could sleep. Bud's skipper on the *Zellars* was his former instructor at the academy, Commander L. S. Kintberger, who let him run the ship, so Bud was able to gain valuable experience in readiness, maintenance, and personnel, but his thoughts were on exploring new career opportunities.[37] Just prior to a lengthy deployment in the Mediterranean, Bud submitted his letter of resignation to Captain Kintberger, who wisely kept it in his desk drawer for several weeks.

One intriguing possibility for Bud was the Honour School of Modern History at Oxford as a Rhodes Scholar. In August 1946, Bud learned that a special selection board of officers had deemed him qualified to compete for a Rhodes Scholarship. He was ordered to appear before both state and district committees and to complete the application forthwith.[38] By

September 1946, Naval Academy Superintendent Vice Admiral Aubrey W. Fitch nominated Bud for the prestigious scholarship.[39] Writing his application aboard the *Zellars*, Bud addressed the need for "international cooperation, and amity must be based on understanding." In his statement of intellectual interests, the candidate emphasized "my experience in South America, Malaya, and with a cross-section of troubled China have convinced me that the well-prepared American can do more for the pursuit of peace by a life's work in international affairs than in any other field of endeavor."[40]

Captain Kintberger's letter of support for the Rhodes Scholarship lauded Bud's leadership on the *Zellars* during the period of demobilization. "His task has been the difficult one of maintaining a maximum of organization, training and material readiness during the period of demobilization. His performance has been far above the average for his experience. His personnel policies have been wisely and impartially carried out; morale has improved; an esprit-de-corps has been developed. In my estimation he has a full measure of the 'qualities of manhood, truth, courage, devotion to duty, sympathy for and protection of the weak, kindliness, unselfishness and fellowship.' The candidate's academic and war record indicate to me that he has promise of outstanding achievement in later life."

If Bud was going to remain in the navy, he needed postgraduate training in law. He had been accepted for law school, but his detailer reported that he would not receive orders because there were no suitable replacements to serve as executive officer. During the Louis Johnson era, the navy was not sending anyone to law school. Saralee could not understand what was going on. "Neither Dick nor I can understand why you have become so important to the Navy. To us and to many ex-officers with whom we have talked it is inconceivable that they wouldn't release you when you have two such great opportunities to attend law school. What's the dope? Certainly hope you can get out of navy."[41] A month later, Saralee wrote Bud that "I personally don't think Mouza can stand the present situation too long."[42] She urged Bud to get out of the navy. Jim wrote soon thereafter: "I certainly hope that you can obtain shore duty for a spell. Mouza is very tired of being alone—nor can I blame her . . . she has no reason to be happy here therefore—till her husband is with her."[43]

Mouza's December 1947 letters to Bud detail financial hardship, with only $33.84 in the checking account and $27 in cash. "I hope we'll have enough money for everything," she wrote.[44] From aboard the *Zellars*, Bud began a 1947 New Year's Eve letter. "I am still so sick about the Law School that I can hardly talk about it. It just seems when our bad luck will never leave us. I am worried to hear how you will feel about it."[45]

Bud's detailer knew how upset he was about law school and offered to send him to any Naval Reserve Officers Training Corps (NROTC) program. By this time, Bud had served six years in the navy, all of it at sea. He yearned for a shore assignment so that he could spend time with his family. He was offered an assignment at the Naval Academy but felt that it would be better for Mouza's cultural and social adaptation if she lived in a civilian environment. Bud told his detailer, "Send me where it's warm." In February 1948, Bud was detached from the *Zellars*. He got his wish with an assignment at the University of North Carolina as assistant professor of naval science in NROTC.

As they drove up over the hill from Durham into Chapel Hill, "we both agreed it was like suddenly dying and going to heaven, to enter this very beautiful spot where we were really going to be away from the Navy stress and trauma for a while." They secured university housing at Victory Village, a grouping of little clapboard houses built for students during the navy flight-program era, which after the war had been turned over to the university for junior faculty housing. Many of the young professors there had married foreign women, "so that we had a remarkable situation where on Johnson Avenue, a long sloping curved avenue that we lived on, there were probably 15 or 20 families, and probably ten of them were GI brides."[46]

The Zumwalts made lifelong friendships in Chapel Hill, growing especially fond of their neighbors Jim and Caroline Caldwell. "In the spirit of Victory Village we got to know each other quickly and became fast friends," said Caldwell. "We had much in common and we just liked each other." Jim and Bud were both instructors at the university, Bud in naval science and Jim in history. "Bud and I were friendly adversaries in politics, and we spent much time disagreeing about Truman policies and politics while Caroline was helping Mouza make the transition from the ways of the Orient to the ways of this country. Young Elmo did much to seal the bonds between the families. The most friendly and outgoing of youngsters, he wandered in and out of our house at will. Caroline sometimes just read to him, and sometimes he just sat in our living room and played."[47]

North Carolina was also the first time Mouza came face to face with a distinct type of discrimination. "A wives' club was established, that is, a naval officers' college graduate wives' club. The only member who didn't qualify, as non–college graduate, was my wife. This was her [the organizer's] southern way of discriminating against a foreign wife, so that second year was kind of less happy, although the other wives were great about making sure that Mouza was included in all other things. They weren't able to deal with that particular form of exclusion, another reason why I became convinced that prejudice in any form can be a very harmful thing," Bud recalled.[48] Until Bud was promoted to the rank of admiral, Mouza was haunted by the fear that she would be the reason he could not reach the pinnacle of his profession.

The Caldwells' neighbor on the other side was Christina Wright, whom everyone called Aunt Tina. Tina was a lifelong friend of Mrs. Katherine Tupper Brown Marshall, wife of the legendary General George C. Marshall, who lived in Pinehurst, North Carolina. Aunt Tina and Mouza became close friends, and Tina was always speaking of wanting to visit Mrs. Marshall. Bud volunteered to drive Tina down because it presented an opportunity to meet Truman's former secretary of state and soon to be secretary of defense. The Marshalls lived in Liscombe Lodge on Linden Road, a modest clapboard bungalow set in a smallish garden and so well screened by trees that it was barely visible from the road.[49] Mouza and young Elmo were also invited to Pinehurst. Bud and General Marshall spent several hours talking together. Their conversation erased any doubt Bud had concerning his career path. "I spoke of my concern for what was going on in the world and the great

pessimism I felt about the political situation in this country, and its reluctance to do the necessary things to maintain a readiness to resist the
kind of aggression that we saw going on elsewhere." Bud thought that
his country was literally liquidating its power overnight.[50]

General Marshall spoke about how the American people had "girthed
its loins" and come together in World War II in building tanks, ships,
and aircraft and training a million men to fight. Looking at Bud, he
said, "Young man, don't ever sell the American people short. They have
vast reserves of hidden strength ready to be marshaled when the crisis
is clear." Peering over his glasses and looking directly at Bud, Marshall
added, "And when that time comes your country will need dedicated
career men like you."

If there was a single point when Bud chose a military career, "in my
case this was it, under the benevolent stare of that magnificent man."[51]
Bud believed it was just a matter of time until the United States would
have to get involved in another war.[52] Years later in testimony before a
congressional committee, Bud recalled the conversation with Marshall:
"I recall commenting to him that as a high school student in 1938 I had
gained the impression from public statements of government officials
that in the event of a war with Japan, the U.S. would prevent the invasion of the Philippines." Marshall said that was indeed declaratory
policy, but, "Lieutenant, you need to understand the difference between
your country's declaratory policy and reality. The reality was that we
knew we lacked the military capability to carry out our declaratory
policy. We always assumed privately that we would have to surrender
the Philippines and fight our way back after mobilization."

General Marshall went on to say that "in peacetime in democracies
the dynamics of media criticism, public interest in and preference for
social welfare programs, and party politics, all tend to overwhelm national security requirements. He pointed out that prior to both World
War I and World War II the conventional wisdom in the US, England,
and France was that the military was never satisfied, want too much,
and have too much. It was his judgment that the millions who died
in these wars need not have died had the democracies been prepared

for war. Presciently, before the Korean War, he predicted that the then ongoing disbarment [*sic*] by the west would soon lead to another war."[53]

Bud understood that he was forgoing a career that could have brought professional satisfaction and intellectual stimulation, a career that both his parents had chosen and one that he admired greatly. His decision was between two service-oriented careers. "I saw the opportunity, at sacrifice of financial and family opportunities, to bring my own personal commitment to bear in the service of my country. I felt patriotic; I was patriotic; and feeling that way I saw in my own case with seven years of experience, having been Executive Officer of a destroyer and knowing that I was now considered ready to command one, that I had more to contribute to my country by continuing in the profession to which I had already dedicated seven years of my life than to start a new profession where I had nothing significant to offer in competition with the thousands of others who would also make that choice."[54]

Bud's first command was the *Tills*, a destroyer escort operating out of Charleston, South Carolina.[55] His time aboard the *Tills* was brief, primarily involving reserve training exercises. The *Tills* would typically travel to Miami and then to Key West for antisubmarine warfare training and then return to Charleston. He and Mouza were settling into a wonderful family life, having purchased a new home in Charleston and welcomed a second son, James, into the family in November 1948. Then, after a few months on the *Tills*, Bud received orders detaching him immediately to serve as navigator on the battleship *Wisconsin*. On a week's notice, the family sold their home and moved back to Chapel Hill, because the *Wisconsin* was clearly going into the Korean War. Chapel Hill was a place where Mouza had friends and a support system.

As navigator of the *Wisconsin*, Bud distinguished himself in the Korean theater, receiving a commendation medal with combat distinguishing device for meritorious service during combat operations against North Korean and Chinese Communist forces in the Korean theater from November 21, 1951, to March 30, 1952. The *Wisconsin* alternated between serving as an escort for carrier Task Force Seventy-seven and providing gunfire support to the marine division located on the east side of the Korean peninsula at the bomb line. The navigator's job during the escort operations involved making sure that the officer of the deck was trained to deal with the maneuvering board solutions and screen rotations and prepared for battle stations in case of attack. During gunfire support operations, the navigator's job was to locate targets precisely during twenty-hour days on and off the bomb line. Bud's fitness report noted that as navigator, "his competence and untiring vigilance in ensuring safe navigation of the ship" allowed the commander to focus on planning and gunfire operations. Lieutenant Commander Zumwalt's "performance of duty was consistently superior in bringing the ship through dangerously mined and restricted waters, frequently under adverse weather conditions and in poor visibility. He assisted in the planning of combat operations consisting of numerous gunstrike missions along the coast of North Korea."[56]

The *Wisconsin* tour provided Bud with the chance to help out his younger brother. Upon completing his college work, Jim went off for a year's junket by motorbike around Europe. He then enrolled at the University of Paris and met a Swiss girl named Gretli, a schoolteacher who was doing graduate work. They fell in love and wanted to get married, but her father would have no part of this. Jim asked if Bud could possibly get to Switzerland and meet the family, thinking it might have a positive impact. Bud and Jim arranged to meet in Paris, taking the train to Zurich and then to the little suburb of Wallisellen. The patriarch of the family could not understand how Jim could marry when he had no job and was still in school. Bud explained the GI Bill and his own situation with Mouza. By chance, while they were in the train station, Bud and Jim noticed a magazine counter with the *Coronet* featuring the *Wisconsin* in the pictorial section, a picture of the skipper and Bud poring over a chart, and a picture of Bud lecturing the officers in the wardroom. Jim gave a copy to the family, embellishing the story by saying his father was mayor of Tulare and a physician and his brother navigator of the *Wisconsin*. Gretli soon called Jim and reported that they had received permission to marry. ·

Next, Bud was detached from the *Wisconsin* in order to attend the Naval War College in Newport, Rhode Island, a career-enhancing assignment. The family moved into navy housing, joining a host of other people getting ready to enter the command and staff course there.[57] However, their idyllic life was tossed into disarray when the discovery was made that Elmo had been born with a hole in his heart. The news was devastating to Bud and Mouza, who had already seen their son win a battle with polio years earlier.

Doc Elmo's judgment was that there had to be some underlying cause for his grandson's frequent bronchial problems. Bud and Mouza took Elmo first to the naval hospital, where "they did the usual superficial job that Navy medicine does for junior people, and found shadows in his lungs, and said he probably has TB." Bud consulted with his dad, who said, "That is absolutely out of the question. . . . He can't have tuberculosis. Get him somewhere else." Doc Elmo advised them to go to Children's Hospital in Boston. "The shifting diagnoses is confusing and but depicts the fact that medicine is an inexact science," wrote Elmo to his family. "To you and Mouza, Bud, my prayer goes that the last appointment in Boston will point the way to a clearing future. I can hope that the Ductus Arteriosis may be the answer for it would seem to offer the best chance for an outstanding result. On the other hand, if it is decided that Elmo has a damaged heart, you still have the opportunity of careful guarding and living that may allow him to grow into a competent citizen whose life work must be limited to non-active things."[58]

Doc Elmo's hopes were confirmed by doctors in Boston, who discovered that the shadows in the lungs were the result of distended capillaries, distended under pressure from the blood that was supposed to go throughout the body but was shunting back through the hole in his heart to the other side and putting pressure on the lungs.[59] Much of the year was spent undergoing the workup and learning about a problem that would be surgically corrected when Elmo was old enough.

The year at the War College in Newport (June 1952–June 1953) was followed by the first of two tours in the Bureau of Naval Personnel (known as BuPers). The assignment was wonderful for the family. Bud was home at a reasonably decent hour and free on Saturdays and Sundays. "We were therefore able to do the picnicking, we attended church, and I was a Cub Scout leader for a while . . . I did some Little League work with at least one of the two sons."[60]

Juxtaposed to the idyllic home life was the job itself, which initially looked like a career killer. "I don't know who did it or how it happened, but I'm dead," Bud said to Mouza after his first day on the job. "I'm assigned to the complements and allowances branch and to the shore establishment division. All they do is write the documents that show what the wartime complements and the peacetime allowances of shore activities are. I can't possibly have been given that job as a challenging job. It's got to be a place for a person who's going to seed." Bud was certain he had been dead-ended. "My God, I've really been killed."[61]

Bud came to understand that the people who wrote the individual allowances held quite a bit of influence. They were the ones who collected the data that determined who needed ten thousand boatswain's mates and not six thousand or ten thousand machinist's mates. It also provided him with a broader way of looking at problems and new ways of thinking about how many aviators and how many submariners were needed. The job became an opportunity rather than a dead end. "I began to think about ways in which that might be exploited."[62] From his years at sea, Bud knew the navy was losing boatswain's mates and machinist's mates, radarmen and electronics technicians by the thousands. The single most important reason was years of sea duty with no prospect of extended shore duty. He remembered how the long deployments had affected his outlook on remaining in the navy. "When I looked at the totals that were coming out of the machine, from these allowances that we were writing, it was clear that we had written billets in the ratio, in the case of radarmen, which were the worst, there were 26 seagoing billets for every two shore billets—13 years at sea for one year ashore. So these guys did have an expectancy of more than 20 years before they got their two years of shore duty."[63]

Bud and his staff looked at literally every job in the shore establishment for petty officers and came up with approximately nine thousand different jobs that could be assigned to anybody. He put together a presentation and went up the line with it to show that "if I could be authorized to shift these billets, I could fix in a single sweep, just as soon as we had trained the people to fill them, the ratio of the sea and shore duty for the four or five most rigorously seagoing ratings."[64]

Bud eventually made the presentation to Vice Admiral James Holloway, Jr.—the future Admiral Holloway and father of Bud's successor as CNO, James L. Holloway III—who at that time was serving as chief of naval personnel. Also at the briefing was Albert Pratt, assistant secretary of the navy. They were both impressed enough with Bud's presentation to have him next brief the assistant secretary of defense, Carter Burgess. Using data to show that the entire division of Complements and Allowances, not just the shore side, but the seagoing side as well, was really a Requirements division, Bud was able to get approval for a name change to Requirements. This one presentation "kind of led to my being discovered. We made a big fix there. Admiral Holloway sort of adopted me as a personal assistant and was constantly sending for me on this or that," recalled Bud.[65]

What Bud thought was going to be a disaster turned out to be one of the most exciting and stimulating of jobs, enabling him to be innovative and get things done. Not everyone would have been able to seize this institutional opportunity. Throughout his career, most notably in Vietnam and as CNO, Bud saw the playing field differently than others. He also learned that the people who often had the best solutions to problems were the ones in the field, not the bureaucrats in an office. In this case, the ex-enlisted men who were allowance writers, the lieutenant mustangs (enlisted men who rose in rank), usually knew what had to be done. "It was amazing how much you could learn about how to fix things from these fellows who really knew how the system worked, and yet they hadn't really done anything to fix it."[66]

Perhaps the ultimate benefit of working in BuPers was getting to personally know the detailers, because there was a tradition in BuPers that you took care of your own. When it came time for his next assignment as a lieutenant commander, Bud was able to secure orders to command the *Arnold J. Isbell*, homeported in San Diego. On Bud's detachment report, Holloway handwrote, "On our list to come back to the Bureau."[67] Bud had found an early mentor in Holloway, whose later support would be instrumental. Bud thought that Holloway possessed the largest span of control of anybody he had ever worked for, and he learned much from watching him operate. The two men formed an unusually close relationship. When Admiral Holloway died, his son Jim, a former classmate of Bud's at the academy, wrote to Bud that "Dad always loved you."[68]

The *Isbell* was in poor shape, as were most of the World War II destroyers that had operated hard during the Korean War. *Isbell*'s performance level was dismal—she had stood eighth in a squadron of eight for battle efficiency the previous year. She had one of the lowest reenlistment rates in the force, and no reserve officer had requested regular navy in over a year. "I got the ship that had stood the lowest in the squadron, which was a great place to start . . . I've said that many times. When you've got nowhere to go but up, you just can't lose," said Bud of his new assignment.[69]

San Diego harbor was covered by fog on the day *Isbell*'s new captain was set to get under way to accompany an aircraft carrier to sea for operations.[70] An engineering problem delayed departure, but eventually the new captain backed out of the slip at the naval station. He conned the ship into the channel and headed out, but the surface search radar quit working. Future rear admiral Robert Hanks was aboard that day. "Using lookouts in the very eyes of the ship, what little information we could get from the air search radar, the fathometer, and radar estimates of our position from the carrier, we fell in astern of her and eventually made our hairy way out to sea. Every man aboard immediately recognized that we were now in the hands of an experienced and thoroughly competent destroyer sailor."[71]

Bud knew that to get *Isbell* back on her feet, the crew needed a real shot in the arm. A number of actions were necessary, running the spectrum from leadership to management. Bud started with the ship's voice radio call, Sapworth. A ship at sea is known by its voice radio call, which also carries over to the shore environment, officers' clubs, enlisted men's clubs, and in liberty areas. Sailors love to be proud of their ship, but the Sapworth was not an easy name to be proud of. The local joke was that a previous commander had sought a name change from Saphead, and the office of the chief of naval operations had approved a change to Sapworth.

Sapworth was undignified, and Bud was determined to get it changed. On November 20, 1955, he wrote Rear Admiral Henry Bruton, director of naval communications, "Since recently assuming command of the *Isbell* this Commanding Officer has been concerned over the anemic connotation of the present voice call." Other ships in the company had calls like Fireball and Viper. "It is somewhat embarrassing and completely out of keeping with the quality of the sailormen aboard to be identified by the relatively ignominious *Sapworth*." Bud wanted the name changed to Hellcat, "so that *Isbell* may carry on the '31-knot Burke' tradition and proudly identify herself to all and sundry consorts." Bud believed that "approval of the requested voice call, in my opinion, will lend a great deal of impetus to the surging team spirit . . . it is important to me and my ship."

It helped that Admiral Bruton had been Bud's commanding officer on the battleship *Wisconsin* during the Korean War. At the bottom of Bud's memo, Bruton scribbled, "Let's do this if possible. If not, let's give him something better than SAPWORTH."

The request was approved! The first thing Bud did was to arrange for new patches to be made depicting a black cat with the flames of hell coming up around a devil's tail. *Isbell*'s newfound identity was exemplified the day she was returning from a WestPac (Western Pacific) deployment and suffered a breakdown of one her chain-driven lube-oil pumps during a line-abreast formation. *Isbell* slowed to about fifteen knots in order to make repairs, while the rest of the division headed for the horizon.

A message came from the USS *Frank Knox*: "Hellcat. [This is] Viper, sorry about that. We'll say hello to your families when we get to San Diego." "Commander Zumwalt was furious," recalled Hanks, who credited Bud with teaching him seamanship, leadership, and compassion for those under him."[72] "Repairs made and he ordered full power, but there was no chance to catch sister ships. Then, over the radio came word that *Knox* had an engineering casualty and it could not be fixed at sea." Soon, boiling along at 30-plus knots, *Isbell* overhauled the crippled *Knox*. A smiling Zumwalt eased up to the officer of the deck and said, "Lay a course to leave the *Knox* about a hundred yards to port." As *Isbell* came abeam of the Knox, Bud picked up the radio handset: "Viper, this is Hellcat. I'd like to speak to Viper himself. Over."

"Hellcat, this is Viper himself. Over."

"Viper, this is Hellcat. *Beep, Beep. Out.*"

And away *Isbell* went.

"*Arnold Isbell* was a very close-knit wardroom. I really regretted having to give up the command," recalled Zumwalt.[73] In Bud's opinion, "the combination of hard work to get the training done, melded with the esprit that comes from a kind of a team approach to things, just made us a really highly successful ship. We won, as I recall, every one of the E's, including the battle efficiency pennant in my second year and stood, of course, number one."[74]

Bud sent a copy of *Isbell*'s battle efficiency pennant to Admiral Bruton, along with a note of thanks for assigning the Hellcat voice call. With the help of a fine group of officers, he succeeded in raising the re-enlistment rate 250 percent in just four months. Moreover, two reserve officers had requested augmentation and retention and had specifically asked to remain aboard. "We are doing everything humanly possible to provide motivation, incentive, and esprit de corps. We are making progress."[75] Another key part of the assignment was the contribution made by Mouza in the big-sister and mentoring role she played for the wives of the officers who were trying to decide whether to make a career out of the navy. Bud always credited Mouza for the fact that *Isbell* had a much higher retention rate of reserve officers than did any neighboring destroyer.

Bruton forwarded the letter to Holloway with a note saying, "Here's a guy that's really getting results." The next thing Bud received were new orders to return to BuPers to be a lieutenant detailer. He was relieved of command on July 13, 1957, ending his destroyer tour. During his final weeks on the *Isbell*, Bud received a personal "Dear Zig" letter from Bruton. "Of course it is no surprise to me that Zig Zumwalt continues his clearly outstanding and exemplary performance in every assignment. I am so certain that such will continue in the future that, without your permission, I have intervened in connection with your next assignment."

Without discussing it with Bud, Bruton had gone to Holloway and recommended that Bud replace Commander Tom Weschler as aide to CNO Arleigh Burke. Holloway embraced the idea and sent a memorandum to Burke registering his strongest possible support. Both men felt that it was time Bud had a "front corridor" job. As Burke's aide, he would become known to every flag officer in the navy and to many officers of all services. "Whether this works out or not, you know my feeling and that of ADM Holloway with respect to you," wrote Bruton.

Bruton was not discounting the lieutenant detailer job; rather, he believed being Burke's aide would be far better for Bud's career advancement while also keeping him in Washington. Admiral Burke might prefer someone more junior than Bud, and "if ADM Burke decides not to disturb your present orders, it will be for the great benefit of all the lieutenants in the Navy. . . . I hope this meets with your approval, and I think it will. The very few people of your high caliber must pay the penalty of having senior officers interfere (possibly) with your own plans, in the greater interest of the Naval Service."

It was a flattering gesture, but Admiral Burke selected a classmate of Bud's, Ray Peet, as Weschler's replacement. Peet had also commanded a destroyer, so he and Bud were at the same stage of their careers. By default, Bud received his preferred assignment, a second tour in the Bureau of Naval Personnel. Returning to Washington, the family moved into the home they had bought during Bud's first tour with BuPers. The new job involved working only with surface officers, providing Bud his first encounter with the navy's appalling racist system. One of the first briefings Bud received involved how to detail lieutenants. When it came to the question of detailing the few blacks in the pool, Bud learned that "if you got a black officer, you sent him to recruiting duty, which was then the least popular and the least important in our priority of tours. The dregs were being sent there. . . . You then, at the end of his two-year tour, extend him a year. If he hasn't left the Navy by then, send him to the worst broken-down auxiliary or tanker you can find. That's bound to give him a record that will get him passed over, and you'll be rid of him."[76]

Bud had no idea how high up in the chain of command this type of unwritten directive emanated from, but it was evident that "we were practicing racism of the highest order." There were Filipino stewards who did not want to be stewards but wanted to be, in one case, a gunner's mate and, in another case, an electrician's mate. "To get that approved at a time when the Navy's policy was racist, I had to be on that telephone 25 times per case to get it done. That's the kind of thing that convinced me that the system was insensitive and racist and needed shaking up. I knew firsthand from the experiences I had in command."[77]

After Bud had been on his detailer job for a few months, Holloway broadened his portfolio by assigning him to solve the problem of what to do about doctors, who were leaving the navy in droves, getting Bud heavily involved in the Doctor Incentive Pay Bill as special assistant to Richard Jackson, the assistant secretary of the navy for personnel. "Bud was now the officer in charge of active duty military personnel, the billet most intimately related to the Bureau of Naval Personnel. Holloway counted on that officer to keep the assistant secretary placated and satisfied with BuPers, as well as keeping him informed if there were problems."[78]

"His performance was outstanding," wrote Holloway in recommending Zumwalt. "He handled many special projects for Mr. Pratt and for me, notably the medical officer Incentive Bill. He is extremely well seasoned administratively and operationally. He is an outstanding leader. I am confident Zumwalt is absolute tops for your staff. As I previously remarked, it is in your interest, in the Navy's, and particularly in mine to have outstanding people in your office, even though I may make some sacrifices over here."

In essence, Bud was now an action officer in the Bureau of Naval Personnel. He believed that if naval personnel were to have adequate medical care, increased benefits would be necessary for those choosing a medical career in the navy. He advocated increased pay for doctors in order to compete with the private sector, but the position was not very popular with line officers, who believed that doctors should not receive a higher salary than those who had seen extensive sea duty, especially those with much greater responsibilities during war. Bud's friends began referring to him as "Doc Zumwalt," a title that he had once aspired to have in civilian life.

Another part of his increased responsibilities involved being the action officer for the navy's equivalent of medicare legislation, making it possible for naval personnel and their families assigned to places where naval medical facilities and doctors were not available to be cared for by doctors within the community at the government's expense. Bud received his first experience of the congressional process by personally lobbying as many members and staffers as possible. The experience served him well as CNO. "I learned the slow, deliberate and painstaking methods necessary to get programs through in the face of many other lobbies that impact on Congress."[79]

Another lesson Bud learned involved staffing. "To get any one of those projects, whether it was a change in sea-shore ratio or the Doctors Incentive Pay Bill, or the Medicare Bill, or any of the others through, there were at least 50 people in the Bureau of Naval Personnel that had to initial. I learned very early that in each office there was likely to be someone who, just to cover his own ass, would always say, 'No, bad idea,' or ask a question that had to be researched." In each office there was also "someone who was a doer," and at this early stage in his career, Bud "learned and compiled an actual list of whom to see and whom not to see in each of the bureau offices. And by the time I left there, I could get something staffed in a day that in the first effort might have taken three or four months to get staffed. That came in to be of great use in later years too."[80]

In the preceding six years, Bud had had the opportunity to learn the personnel business from three vantage points: establishment requirements, distribution, and policy management. "These tours served to drive home to me, as I never understood it before, the extent to which bureaucracy on the part of those preoccupied with plans and operations were impacting on the ability of the Navy to solve its personnel problems. And I resolved that if I ever reached a position of authority I would do my best to solve these bureaucratic constraints."[81]

Bud Zumwalt was now seen as a comer. In April 1959, he was one of a select group of commanders chosen to be interviewed by Admiral Hyman Rickover for the posts of commanding officer of a nuclear frigate and executive officer of a nuclear cruiser.[82]

It was a Friday when Bud arrived at 10:00 a.m. for the interview at Rickover's office in the former State, War, and Navy Building. Rickover and his staff occupied perhaps the most austere offices in Washington. There was no carpeting, only torn and worn linoleum, and perhaps most distinctive, all the chairs had front legs shorter than the back legs. Bud was taken to a waiting room, where he sat alone for ninety minutes before being escorted to the first of three interviewers whose job it was to screen candidates before they ever got to Rickover. Passing these initial tests, he was taken to a room and told to sit next to a sign that admonished, AFTER HAVING BEEN INTERVIEWED BY ADMIRAL RICKOVER NO ONE IS TO RETURN TO THIS ROOM EXCEPT TO PICK UP HIS PERSONAL POSSESSIONS AND LEAVE IMMEDIATELY WITHOUT TALKING OF INTERVIEWS.

The three screening interviews took up most of the day. It was already past five p.m. when Bud learned that his interview with Rickover would soon begin. By seven thirty, someone came into the room to say that Rickover had left at five and that Bud should stand the next day, which was a Saturday. After waiting almost all the next day, Bud was informed that Rickover could not meet until the following week. For almost a month, he heard nothing. It was all part of the Rickover ritual of playing mind games and testing potential commanders in the nuclear fleet. It was also just the beginning of an elaborate process bordering on the insane.

A month passed, and out of the blue, Bud received a call asking if he could see Rickover that day. Looking over his appointments, Bud said that he was free any time after 10:00 a.m., but he would have to first notify his boss. A few minutes later, the call came that Rickover wanted to see him immediately, at 8:30 a.m. As a final preparation, knowing that Rickover was especially hard on aides, Bud removed his aiguillettes before coming over. Walking down the long corridor, he felt prepared. "I well knew that one must neither rise up and smite him nor be accommodating and obsequious." Bud was first told by Rickover's assistant that the admiral had already determined that Zumwalt lacked sufficient engineering experience and, if selected, would be assigned as an assistant engineer on a carrier, a position usually held by lieutenants. Zumwalt considered this akin to the hazing process from plebe year at the academy, although he did agree that he could use the experience.[83]

The moment finally arrived to meet Rickover. Bud entered through a swinging door into what looked like an interrogation room in a police station. There was a desk, in front of which were two chairs. Behind one was a "gnome-like figure" who motioned Zumwalt to sit in one of the chairs. Behind the chairs was another small desk where a witness sat—a Rickover ritual. Physically, the two men could not have been more different—Bud in his navy uniform, Rickover in his dark, baggy suit, white shirt with oversize collar, and nondescript tie.[84]

Rickover reviewed the notes from the three screening interviews, saying nothing at first. "Everyone who interviewed you tells me you are extremely conservative and have no initiative or imagination," said Rickover to break the silence. "And what do you have to say about that?" Bud was taken aback, saying he needed a few seconds to think, since "it is the first time I have received a charge like that about me." Rickover shot back, "This is no charge, God damn it. You're not being accused of anything. You are being interviewed and don't you dare start trying to conduct the interview yourself. You are one of those wise Goddamn aides. You've been working for your boss for so long you think you are wearing his stars. You are so accustomed to seeing people come in and grovel at your boss's feet and kiss his tail that you think I'm going to do it to you."

Rickover told Zumwalt to get out and go sit in the other office until he was ready to be interviewed properly. "And when you come back in here you better be able to maintain the proper respect."

Bud was escorted to a barren room referred to as the tank; it held a table and chair and no reading material. He was there to think and reflect on proper behavior. The chair faced a blank wall. There was a small window that faced out into the corridor so that people passing by could look inside to see who was getting the treatment this time. People would peer in, making Bud feel like an animal in the zoo. Thirty

minutes had gone by when an assistant opened the door and told Bud
he should return to Rickover. But when he arrived, Rickover was on
the phone, haranguing the caller about a construction program. Even
though Zumwalt already had a Top Secret clearance, Rickover yelled to
his aide, "Get him out of here. I don't want him listening to this."

Ten minutes later Bud was back in the hot seat. Rickover started off
by asking, "Now what is your answer to my question?"

"I have initiative, imagination, and I am not conservative," replied
Zumwalt.

Rickover moved quickly through questions about Tulare, Bud's
parents, summer employment, and the Naval Academy. "What were
these special extra-curricular affairs you are so proud of?" asked Rickover. Bud replied that he was a debater, an orator. Rickover sneered,
"In other words you learned to speak equally forcefully on either side
of the question. Doesn't make a damn bit of difference what you believe is right—just argue the way someone tells you to—good training
for an aide."

Bud tried defending himself. "No sir, I consider that debating taught
me logical and orderly processes of thinking." Rickover's voice began
to rise. "Name one famous person who was able to argue on either
side of a question." Sensing he could win this point, Bud said, "Clarence Darrow," and began listing reasons when Rickover cut him off.
"You're wrong. Absolutely wrong. . . . I warn you here and now you
better not try talking to me about anything you don't know about.
I know more about almost anything than you do and I know one
helluva lot more about Darrow than you do. I warn you, you better
stop trying to snow me."

Rickover turned to his assistant. "Get him out of here. I'm sick of
talking to an aide that tries to pretend he knows everything." Back he
went to the zoo cage. He set a record with four trips to the tank. Another hour went by before a kind soul came in with a sandwich. As
Bud was taking his first bite, the aide returned to say Rickover wanted
to see him *now*! Still chewing, he walked into the room, and Rickover
shouted, "Don't you even have enough sense not to chew gum when
being interviewed?"

"All right, now are you ready to talk sensibly about Clarence Darrow?" Bud continued to insist that Darrow could take both sides of a question because everyone deserved counsel. Rickover could take no more. "I give up," he said and changed the subject. If Zumwalt were superintendent of the Naval Academy, what would he do with the curriculum? Bud replied that he'd eliminate some English and history to provide more math and science. "Thank God you are not the superintendent," replied Rickover. "It's just the kind of stupid jerk like you that becomes superintendent." In Rickover's mind, all he was going to graduate were illiterate technicians.

Rickover next wanted to know the name of one book of philosophy that Bud had read since graduation. "Plato," replied Zumwalt.

"Now I warned you not to try to impress me. I told you I was sick of having an aide trying to impress me. I proved how stupid you are on Darrow," said Rickover. Rickover wanted to know if Zumwalt had ever read the *Republic*? When Bud replied in the affirmative, Rickover asked, "What's it about?" Anticipating that Rickover wanted an answer in six words or less, Bud said, "The ideal man or the ideal democratic state." Rickover turned to his aide. "You see what kind of stupid jerk this guy is?" The *Republic*, Rickover said, was about justice, and he wanted to know whether Plato would have advocated eliminating history and English from the curriculum. "No sir, but Plato was postulating a perfect world and we don't have one." Rickover could take no more. Seething with anger and shouting that Zumwalt was trying to conduct the interview, Rickover told his aide, "I am sick of this guy. He is trying to act like an aide again. Get him out of here."

Bud was so angry he could hardly speak. The aide tried consoling Bud, saying, "Don't let it get under your skin. We all have to go through this." He remained alone in the tank for another forty-five minutes. Re-entering the room, Rickover dressed him down again and asked if there was any reason to continue the interview. When Bud replied affirmatively, Rickover asked, "How long have you been interested in nuclear power?" "Five years," replied Bud. Rickover wanted to know what he had done to prepare for a job in the area of nuclear power. "Very little," admitted Bud, leading Rickover to ask his secretary to come in and take a letter to the president of Chase National Bank in New York: "Dear Mr. President. For five years I have wanted a million dollars. Please send me a check today. Yours very truly, H. G. Rickover. P.S.: I have done nothing whatsoever in the last five years to earn this money, but send it anyway."

Looking at Zumwalt, Rickover asked, "Get the idea?" Rickover went for the kill: "Why haven't you done anything?" Zumwalt replied that his modus operandi was to master areas based on current assignments and he was currently studying four hours a night on personnel matters. Rickover cut him off again. "God help us, that's what is wrong with our personnel when we have guys like you working on it."

Rickover wanted to know how long Zumwalt had known about this interview. When he responded, "Four to six weeks," Rickover multiplied 4 hours by 28. "That's 112 hours at least since you've known that you had available for study by your own admission. Now why haven't you studied anything about nuclear power in this period?" Bud tried defending himself, saying he had studied the BuPers booklet on nuclear physics. Rickover scoffed at this, asking if he was prepared to take a test right then. Bud said that he was not, but that he could pass it anyway. "I doubt it," said Rickover.

"What is leadership?" asked Rickover. Zumwalt responded, "It is knowing your own job well, knowing the job of your subordinates, and inspiring them to do better." Rickover sneered and asked if Bud's father was a great leader. Bud said yes, explaining that his father had been a country doctor for forty years, active in community and civic affairs, and was an outstanding citizen. Rickover showed a trace of a smile, changing the subject again, this time to Bud's grades at the academy, his assignments during the war, and his wife and children.

Rickover asked if Zumwalt planned on running up and down the Pentagon's E ring telling everybody about the interview. "Admiral, I'm going to say it was the most fascinating experience of my life." Rickover threw him out again, saying, "Now you're being greasy."

The two men would not meet again until 1963.

Bud went directly to his detailer to report that he had completed the interview and was certain he would not be selected. Bud requested orders to the USS *Dewey*. He returned to the Pentagon and in direct disobedience of Rickover, typed up a complete report on the interview, routing it directly to the CNO, the secretary of the navy, and the assistant secretary of the navy for manpower. This infuriated former classmate Ray Peet, who knew Bud had violated his pledge to Rickover.

Bud was shocked when the call came from BuPers to tell him that Rickover had selected him for one of the two nuclear jobs, either as commanding officer of the guided-missile frigate *Bainbridge* or as executive officer of the cruiser *Long Beach*. Fate ordained that the other person selected was Ray Peet. Rickover didn't care who took which assignment, believing that each man needed this type of command at this stage of his career. Bud and Ray had been in competition since their years at the academy, when each was a three-striper in command of a company as midshipman lieutenants. They had both come up in the destroyer force. Bud had won all E's on the *Isbell*, and Peet had done likewise in command of the *Barton*. Ray Peet had been selected for captain a year before Bud and two years early.[85] In September 1960, Bud would be promoted to captain, a year ahead of his class. He was the first in his class to be promoted to rear admiral, a year ahead of Peet.[86] Their rivalry reached its boiling point during Bud's CNO years.

Bud spurned a tour with Rickover on the *Bainbridge*, choosing instead to command the world's first guided-missile frigate, the first naval vessel constructed from the keel up to be a guided missile ship, the USS *Dewey*. Wanting to avoid the Sapworth experience, Bud went to see the person responsible for voice calls to select Sea Rogue as the *Dewey*'s call.

On December 7, 1959, which was the ninetieth birthday of George Goodwin Dewey, Admiral George Dewey's only son, the *Dewey* was commissioned. After the ceremony, George Goodwin Dewey said to his son Charles, "I am going to keep an eye on that young fellow because he is going right to the top!"[87] During sea trials, the *Dewey* performed very well, with the exception of its weapons system. "The Bureau of Weapons had not, in my judgment, done a very good job of preparing operability tests, tests for checking out every element of the system before you went into a tracking situation," said Bud.[88] This was the first BT-3 weapons system, a wingless, tail-controlled model of the Terrier surface-to-air missile, and the system had significant bugs. The components were built by different manufacturers. The fire-control computer was different from the weapons console and built by a different company than the radar, and the system designed to tie all the components together was abysmal.

During tests, the missile would come out of its little hopper and plop into the ocean uncontrolled, or it would start out on a controlled flight and then depart from it before it ever got to the target drone. "We must have fired eight or ten missiles over a period of a number of months in heartbreaking efforts to get the thing working . . . the great day came when we got a hit," recalled Joe Roedel, who worked in the radio shack. "At that time we were improving the effectiveness of the Terrier missiles and at times I know he became somewhat discouraged due to the fact that sometimes the booster malfunctioned and sometimes went further than the missile. On several occasions while delivering messages on the bridge during a firing he would jokingly bet which was going to go further, the missile or the booster. Another time when the testing had progressed to a better degree, the missile was supposed to fly alongside the drone, but on this one occasion the missile struck the drone and blew it up. But during these mishaps he always had something funny to say. I don't think I ever saw him get angry."[89]

Russian ships in the Baltic routinely harassed U.S. ships. On their first day in the Baltic, a Russian destroyer bore down on the *Dewey*, making a complete 180-degree turn as she passed to approach on the *Dewey*'s port bow and rapidly pulled abreast in a move obviously designed to embarrass the U.S. ship. Having anticipated some such encounter, the *Dewey* crew was ready and accelerated from 17 to 35 knots, leaving the Russian ship wallowing in their wake. The *Dewey* continued at the accelerated speed for thirty minutes before reverting to 17 knots, the Russian ship still following three miles in the rear. The next day, they drew even, some 1,800 yards abreast. For the next hour, the Russian ship played cat and mouse, eventually closing to a mere 750 yards, way too close for comfort and safety, akin to autos traveling hubcap to hubcap.

"When Capt. Zumwalt saw the Destroyer pull in front of our

designated course we were doing the cruising speed of 16 knots," recalled Joe Roedel. "I was on the Bridge with the message board and he said 'I think he wants to play chicken,' he then ordered the engine room to go full speed ahead. We went to 27 knots and headed for the Destroyer's mid-ship. We were within a half mile from the Destroyer when you could see the screws of the Russian destroyer churn the water and the stern of their ship dip down and she pulled out of our path. Then Capt. Zumwalt stated 'I guess we won that one' and we continued at full speed for a while. This is when he got the nickname of 27-knot Zumie. 27 knots was always the speed of which he liked to run at when he wanted to get his adrenaline moving."

Bud consulted with the admiral on board and raised international signal flags, translatable into any language from the International Code of Signals, a system entirely different from the system used by the United States for naval operational communications. "We have the right of way" was hoisted by the *Dewey*. After a timely interval, the Russians raised a reply: "My present course . . . by chance . . . accidental," and the Russian ship dropped astern. "The S.O.B.," said Bud in amazement. All night the Russian destroyer remained astern. The next day, Bud saw the Russian ship speeding to overtake the *Dewey*. "She approached— her men were lined up on deck, she piped honors (a salute) and hoisted signal flags that translated to 'A Pleasant Voyage.' "

It was with "great regret" that Bud left the *Dewey*. "There is a picture that I have somewhere," Bud told Paul Stillwell, "that was taken of me in England just as I got halfway across the gangway, looking back. The look on my face is the look of someone who's just lost his lifelong mate." Bud received a Bravo Zulu report for his command of the *Dewey*. "Commander Zumwalt was observed to be an outstanding officer in all respects. He was enthusiastic about his ship, the first guided missile destroyer leader, and about the officers and men assigned. His exceptional leadership rapidly developed excellent ship spirit and high morale. He is courteous, friendly, intelligent, thoughtful, a gentleman in the best tradition. He is unquestionably competent and qualified for his command the finest to which a Commander might aspire, but also for positions of even greater authority and responsibility. Commander Zumwalt, in short, is a superior officer who should be given every consideration for accelerated promotion to Captain and flag work."[90]

The fitness report, dated July 22, 1960, "will stand with the best of them and should give you a good boost toward captain and beyond. I certainly hope so anyway, because I know you have earned every word," wrote Lieutenant Commander W. L. Read.[91] Sure enough, in September Bud was promoted to captain a year ahead of his class. "We are all very excited out here," Bud wrote Saralee in sharing the news of his accelerated promotion.[92]

Bud Zumwalt would next enter Paul Nitze's orbit.

CHAPTER 6

PLATO AND SOCRATES

It is difficult to talk about Paul Nitze without getting lyrical.
—BUD ZUMWALT[1]

Bud Zumwalt had completed two tours in Washington that provided him with a tool kit for looking at and thinking about navy personnel issues comprehensively. He was now eligible to be considered for one of the plum assignments for those on the fast track—the National War College in Washington, D.C. The applicant pool came from the top cohort in each of the military services, the foreign services, and civilian agencies such as the Foreign Service, the U.S. Information Agency, and the Central Intelligence Agency.

There were approximately 130 members in Bud's War College class.[2] Assignments and grades were geared toward developing logical solutions, seemingly tailored to Bud's intellectual orientation. "I liked very much the fact that there was no school solution, that any one of eight or ten different imaginative ways of dealing with a geopolitical problem or a military problem was considered acceptable."[3] Everyone was assigned to a set of committees that regularly rotated membership so that during the year each member served on a committee with almost every member of the class. As he got to know people, Bud began compiling a Rolodex of names. "Out of those associations there came one of the most valuable aspects of the National War College, and that was the opportunity to call up a buddy, not just a name, in the State Department, or in the Air Force, or in the Marine Corps and find out what was really going on when you had a problem that interfaced with that institution. The old school tie was just very, very valuable. In one telephone call you could find out whom to go to solve the problem, what kind of a guy or gal that person was, how best to approach them. Very often your classmate would also know enough about the problem in general to advise you whether or not it was something that could be solved easily or with difficulty."

All students at the college were required to work on a major thesis. With his interests in the Soviet Union, Bud focused on the problem of succession in the USSR as a way of understanding the dynamics of the Soviet system. He was able to interview a number of visitors to the War College campus, including Ambassador Llewellyn Thompson, Chip Bohlen, and several people in the Office of Naval Intelligence.[4]

Paul Nitze, assistant secretary of defense for international security affairs (the ISA division was often referred to as the "little State Department") was one of the distinguished visitors invited to speak at the college that year. Nitze had just returned to government service after eight years as president of the Foreign Service Educational Foundation. A 1928 cum laude graduate of Harvard University, Nitze joined a prestigious New York investment banking firm, Dillon, Read and Company. Then in 1941 he left his position as vice president of that firm to become financial director of the Office of the Coordinator of Inter-American Affairs. From 1942 to 1943, he was chief of the Metals and Minerals Branch of the Board of Economic Warfare, until being named director of foreign procurement and development for the Foreign Economic Administration. From 1943 to 1944, Nitze was a special consultant to the War Department, and from 1944 to 1946, he was vice chairman of the United States Strategic Bombing Survey. He was awarded the Medal of Merit by President Truman for service to the nation in this capacity. For the next seven years, he served with the Department of State, becoming deputy director of the State Department's Policy Planning Staff in 1948, and director the following year. He left the federal government in 1953 to become president of the Foreign Service Educational Foundation in Washington, D.C., until January 1961.[5]

Nitze had served as a national security advisor to the 1960 presidential campaign of Senator John F. Kennedy and anticipated receiving a high-level appointment in the new administration. Kennedy offered Nitze three choices—undersecretary of state for economic affairs, national security advisor, or deputy secretary of defense. Unfortunately for

Nitze, Kennedy gave him thirty seconds to make up his mind. "I choose the post of Deputy Secretary of Defense," Nitze told the president-elect. "Fine," said Kennedy.[6]

Days went by, and Nitze heard nothing more about his appointment. It soon became clear that Kennedy would be unable to honor his commitment to Nitze because secretary of defense nominee Robert McNamara had insisted on being given a free hand in appointing his immediate staff. When selecting Roswell Gilpatric, McNamara "told Kennedy that he would prefer a deputy who would be his alter ego and carry out his programs without argument or confrontation."[7] McNamara then offered Nitze the position of assistant secretary of defense for international security affairs.

Nitze gave a magnificent talk that afternoon at the War College. As was customary, a group of faculty, along with the commandant, joined Nitze afterward for coffee and discussion. Instead of being asked about his own remarks, Nitze was startled to hear the commandant and professors only wanting to discuss a lecture someone by the name of Captain Zumwalt had given the day before to the entire student body for Alumni Day on the problem of succession in the USSR. "The Commandant and his staff, rather than discuss my subject with me, had words only for the superb individual research project, and the lecture based upon it."[8]

Nitze recalled being "irritated about being overshadowed by some student."[9] He began asking questions and taking notes about this Captain Zumwalt, learning from the commandant that Bud was a top student at the War College with a wide range of intellectual interests. The next day, upon returning to Washington, Nitze placed a call to Vice Admiral William Smedberg III, chief of the Bureau of Naval Personnel, asking that Zumwalt be diverted upon graduation from the War College to his staff in Washington as a naval aide. Smedberg objected to this type of interference. Nitze told Smedberg that he was willing to take the issue up with the new secretary of defense, Robert McNamara. Smedberg replied that that would not be necessary.

A few weeks later, Bud's detailer informed him of the request from Nitze. The detailer thought it was a terrible idea and recommended J-5

(Joint Staff) as the more career-enhancing opportunity for someone on the fast track. Bud was not particularly turned on by the idea of going to the Joint Staff, with its interservice rivalries, but understood that it was the logical move for getting an important ticket punched in the selection process for rear admiral.[10] Taking a job with Nitze was risky for someone viewed as a comer, selected a year early for captain. Bud had already said no to Rickover, and the detailer was adamant that going to ISA would derail any chance Bud had for early selection for rear admiral.[11] After thinking about it for days, Bud rolled the dice, requesting that his orders be cut to ISA. "More in sorrow than in anger," the detailer agreed to do it.[12]

So Nitze brought Bud Zumwalt with him to ISA, assigning him to the Division of European Affairs as desk officer for France, Spain, and Portugal. According to Nitze's account, Bud "rapidly won the respect of his superiors as well as his opposite numbers in the State Department, operating at the policy level on certain matters."[13] Nitze later often joked that he "hijacked" Bud for his own purposes. Bud worked on negotiating military base rights with Spain and Portugal and on studies involving nuclear policy. Nitze quickly realized that the quality of staff work he was receiving "confirmed his advance reputation" from the War College. Bud was quickly promoted to director of the Arms Control Division, a position that ordinarily would have been given to someone more senior, and his pay grade was increased from a 0-6 slot to a 0-8 slot equivalent to the slot of a rear admiral.[14]

The new assignment thrust Bud into issues of nuclear proliferation and foreign policy.[15] It was apparent to Nitze that some type of arms-control negotiations was going to be undertaken, and he needed someone to ensure that all proposals received the most searching evaluation from the perspective of sound military judgment. Bud developed a series of memoranda analyzing the strategic implications of several versions of nuclear test-ban constraints on weapons development, the prospects for proliferation of nuclear weapons with and without the test ban, and mandatory safeguards under a test ban. "His work was instrumental in helping this government to achieve an intelligent consensus and to avoid militarily unsound clauses in the treaty," explained Nitze. "He was insistent that the views of the JCS [Joint Chiefs of Staff] be obtained and considered by civilian authority." As director of Arms Control, Bud also served as Nitze's immediate staff person on the Committee of Principals, which had been set up by McNamara and Secretary of State Dean Rusk to deal with arms-control issues.[16]

Immersed in test-ban treaty negotiations, Bud was now in a position of greater visibility and opportunity. As a regular staff participant at meetings of the principals, Bud drafted most of the papers used by Nitze. "I had the intellectual excitement of getting all that done in advance of the meeting and then of sitting in on the meeting and taking notes so that I knew firsthand exactly how to deal with the subsequent revisions and so forth."[17] Bud additionally served as an advisor on Sino-Soviet affairs. Indeed, he had evolved into a specialist, "particularly effective in analyzing the implications for U.S. policy of various actions of the Soviet leadership."[18]

One of the early papers written by Zumwalt for Nitze took the position that the national interests of the United States did not require getting involved in Vietnam, that it would be a strategic error to do so in a land war.[19] Bud's own thinking was heavily influenced by Nitze, who believed that U.S. forces should be brought to bear only by air and naval-surface-ship bombardment and by a blockade of North Vietnam. If this did not cause Hanoi to cease its infiltration into South Vietnam, amphibious landings to seize Haiphong and Hanoi would have to follow. This strategy amounted to cutting off all logistical lines to North Vietnam by land and sea. Bud estimated that if this option came to pass, five thousand American casualties would result. U.S. Army forces in South Vietnam would be limited to advisory efforts to equip and train the Vietnamese to fight their own war against indigenous communist forces as the United States contained the threat in the north. He did not think the People's Republic of China (PRC) would challenge these actions.

Alternatively, if the United States was going to get more heavily involved, Bud believed the best way to do so would be with massive air power applied at the source and with a small number of advisory personnel to help bolster a rapidly deteriorating in-country situation.

The worst possible course of action would be to introduce forces on the ground. The record shows that Zumwalt and Nitze believed Vietnam was not the place to enforce the U.S. policy of containment because neither man saw South Vietnam as a viable national entity, believing as well that the conditions for nation building did not exist. There were better places for building economies and military capabilities in the region—Thailand, Malaysia, Singapore, Indonesia, and the Philippines. Under the Nitze-Zumwalt concept, U.S. efforts in Vietnam would have been limited to modest support and military advisory personnel. "To commit any ground troops at all meant in the end to commit as many as might be required to ensure the security of those already committed. There was no such thing as being a little bit pregnant," Nitze wrote of his early 1961 disagreements with McNamara.[20] Secretary of Defense McNamara, of course, rejected this line of reasoning because his Pentagon computers all demonstrated that the United States could accomplish its objectives by escalating the ground war. McNamara rejected the Zumwalt-Nitze recommendations; he favored a grand strategy of attacks on limited targets followed by pauses and, if necessary, massive U.S. ground forces in the South.

Robert McNamara and Paul Nitze had what can only be described as a love-hate relationship. When Nitze was in the hospital with an incipient ulcer, he told Bud, "You know, this is McNamara's ulcer." Bud thought that "McNamara both admired Paul's vision and depth and wisdom, and deeply resented it, because he recognized that Paul was so much his superior in that regard. McNamara had twice the memory that Nitze did about names and numbers, but he just wasn't in the same league with Nitze on concept. In my judgment, McNamara wanted in a deputy someone who would just carry out his bidding. He was perfectly content to have Paul be down there as number three, where he could tap him for his wisdom and yet not be compared with him in Cabinet meetings and in press conferences and that sort of thing."[21]

Following the discovery of Soviet missiles in Cuba, Nitze joined a small group of the president's advisors, the Executive Committee of the National Security Council, often referred to as ExComm. Nitze was authorized to take notes on the meetings because ISA was the liaison serving the Joint Chiefs of Staff. After each meeting, Nitze would debrief his carefully selected ISA team.

"I was involved in that totally," recalled Zumwalt. The night before the crisis, there had been a USS *Dewey* get-together at the home of the former executive officer. "That was one of those rare occasions when I had far more to drink than I should have had. It was a Saturday night. I was awakened on Sunday morning about 6:00 o'clock, after having had only three hours' sleep, by a call from Mr. Nitze's personal secretary telling me that I must be in the Pentagon within an hour. That's when we began the staffing for the President's speech." For the first twelve hours, "I was suffering so much from the hangover, that it was very hard to be brilliant, and I kept saying to myself, 'How tragic it is that on the one time in a year when I pull this, I get pulled in on the most important work of my life?' "[22]

Nitze had just returned from a meeting of the National Security Council, where it had been confirmed by photos that the Russians were bringing missiles into Cuba. "Paul came back and decided to set up a very small group to work with him. The President authorized people to work with some of their staff but had made it mandatory that there be absolutely no disclosure to any other people, no matter how high their clearances." Nitze's team included Henry Rowen, John Vogt, John Mc-Naughton, Daniel Ellsberg, and Bud.

Nitze's instructions were clear: "You fellows are all peers; there's no hierarchy here. We're going to be under terribly short deadlines, and when I come back from each meeting, I'm going to be giving each of you an assignment, and you're going to have anywhere from 30 minutes to two hours to get it done." Each staff member was given something to analyze. In Bud's case it was how a blockade might work, which at the time was just one of several options being considered. Nitze later wrote that "I adopted the practice of assigning different analytical tasks to each of these advisors. Frequently they had to respond within a few hours or less, timed to give me their written analyses to review, to revise, and to distribute and use at the next ExComm meeting."[23] Bud offered a similar recollection: "I don't ever recall, during the period of the five or six days that this went on, that we had more than an hour and a half to get back to Paul Nitze."[24]

Following days of intense deliberations, during which all possible responses were considered, Kennedy settled on a naval quarantine. With Soviet ships approaching the blockade, McNamara and Nitze met to discuss the type of directive that needed to be written for the Joint Chiefs and the navy as to options available to them. Nitze sent for Bud to write up the orders, but he arrived without paper and pencil. McNamara said he wanted an order sent immediately to the commander in chief, Atlantic Fleet, to stipulate, "and he went through point one, two, three, four. He got up to about point five, and he saw that I didn't have a paper and pencil," recalled Bud. "And I will cheerfully admit that what I should have done was have asked for paper and pencil, but I was busy memorizing these points. He picked up a tablet of paper and a pencil, and instead of saying, 'Here!' he threw them at me. Well, this infuriated me so much that I left them there—a really dumb thing to do—and he went on and finished his summary, and I left."[25]

Bud dashed back to his office and wrote down first words and then sentences, taking about thirty minutes to write the entire message out. "Then I went back in to Paul Nitze and he spent about 15 minutes studying it. And he said, 'Bud, it's all there, take it up to McNamara, but I want you to know that he said, 'If he missed a single point, fire him.' "[26] Bud later mused, "That was probably as close as I came to losing a career."[27]

The wording in Zumwalt's draft, reflecting McNamara's instructions, was to give Soviet prime minister Nikita Khrushchev two days to respond. Nitze changed "within two days" to "within a couple of days," which to the Russians would come across as less definitively a deadline.[28] McNamara then approved the rules and procedures for the quarantine. "The proposed action was aimed solely at blocking further shipments of offensive weapons bound for Cuba and to do so with a minimum degree of force. Careful monitoring of every ship involved in the blockade was essential, and insofar as possible, control of the situation at sea was to remain in the hands of fully responsible officials in Washington."[29]

President Kennedy wanted to be certain that the Soviets were not backed into a corner. The president wanted to give Khrushchev as much time as possible to think through his response to the boarding of the first ship. To provide maximum flexibility, Kennedy decided that the first ship to be boarded should not be one from the Soviet Union, but from a Third World or Soviet-bloc country. The problem was that the navy wanted the go-ahead to intercept one of the Russian ships. Bud was given the responsibility to relay the president's decision to the navy. He hand-carried this new directive to Captain Ike Kidd, then in a staff position as executive assistant to CNO George Anderson. The message was sent to Vice Admiral Ulysses S. Grant Sharp, then in Op-6 (Plans, Policy, and Operations). Within about fifteen minutes, Sharp sent for Zumwalt: "Look, this isn't the way to do this. I want you to go back and tell McNamara that we're closer to a Russian ship, and that we want to go ahead and intercept a Russian ship first."

I said to him, "Admiral, I can tell you that this has been very carefully thought out, and it's come from the highest levels."

Admiral Sharp asked, "What's your rank?"

I said, "Captain, sir."

He said, "Do you understand a direct order when you get it?"

"Yes, sir."

"I order you to go back and tell McNamara what I want to do—what Admiral Anderson wants to do."

Bud first went to see Paul Nitze, who was outraged. Nitze instructed Bud to return to Sharp and tell him to carry out the orders of civilian authority. He used the word "meticulously."[30] Bud tried convincing Nitze to go with him to see McNamara. Nitze replied, "I most certainly will not. You get back to Admiral Sharp and tell him that I have personally directed that he is to carry out McNamara's orders."

Bud returned to tell Admiral Sharp that the CNO was going to be in deep trouble if he chose to ignore the new instructions. That evening McNamara encountered CNO Anderson at the navy's command center for quarantine operations. McNamara confronted the CNO, demanding to know if he intended to follow orders. Anderson told him that there was nothing to discuss. The navy had been running blockades since the days of John Paul Jones. McNamara did not "give a damn" about what

John Paul Jones might do; he wanted to know what Admiral George W. Anderson was going to do.[31] It was a classic example of military versus civilian authority. This was the first of the series of events that led Mc-Namara to conclude that he was going to get rid of Anderson. "And, frankly, there was, in my judgment, absolutely no excuse for them," said Bud. "If ever there was a clear-cut case of doing something in the way of using military power to send a signal, this was it. And a good, rational way of building up the levels of perception had been thought out. And why in the world Admiral Sharp, who is a very wise man, wanted to do it this other way—or maybe Admiral Anderson had directed him, probably had—defied me at the time and defies me now."[32]

After the president's speech, Bud walked back to the Pentagon, telling Paul Stillwell, "I went out. There wasn't a taxi in sight; everything had been scooped up by the ambassadors. And there was no sign of life anywhere. I decided it was a nice night and it'd been a long week, I walked from the executive office—it was then the state department; it's the executive office of the White House now—to the Pentagon . . . a good long hike but not all that bad. I got about three blocks down and had that eerie sensation that there wasn't any sign of life. It suddenly dawned on me, I wonder if those missiles are en route."

Years later, on the occasion of Paul Nitze's ninetieth birthday, Bud wrote, "I wanted to recall with you the magnificent job you did on behalf of our country during the Cuban Missile Crisis. The almost hour-by-hour debriefings that you were able to give your Cuban Crisis mini-staff, are indelibly engrossed [sic] in my mind. . . . The remarkable way in which you made use of the mini-staff, with assignments from you to each individual upon return from each executive committee, with a requirement for short-fuse 'think pieces' during the 30 minute, one hour, or several hour interludes in the meetings of the executive committee, made it possible for you to maintain the 'high ground' at subsequent meetings. Your recognition that he who had a prepared paper had a major control of the thought process was just one more indication of your insightful knowledge of policy making. With the hindsight of 35 years, I still consider those tumultuous days an example of America at its finest. This episode was just one more indication to me that America has never had your superior as a government servant."[33]

In his report recommending Bud for accelerated promotion to rear admiral, Nitze noted that "during the several days of the crisis he produced a series of military-political analyses, on short notice, at all hours of the day and night. These were brilliant pieces of work which added to the process of obtaining rapid government consensus." When he used these reports in his meeting with President Kennedy, they "had their impact on decisions" and the negotiations. "He was specially commended for the quality of his work during the crisis."

Bud continued working on contingency plans for Cuba, developing the concept of a State-DOD plan for each of several likely contingencies. Nitze sent Bud to New York to work with Soviet foreign minister Vasily Kuznetsov, UN secretary general U Thant, UN ambassador Adlai Stevenson, and negotiator John McCloy on procedures for the UN to carry out its mission of monitoring the removal of offensive weapons from Cuba. He also became the point man for a special study requested by Secretary of the Army Cyrus Vance on U.S. policy options toward Cuba. The report became the primary source document in the generation of policy after the crisis. "My personal observation of Captain Zumwalt's professional competence and remarkable grasp of complex political-military affairs has been a source of continuing admiration," Vance wrote to Nitze. "In addition, the habitually heavy workloads, combined with short deadlines, have always found Captain Zumwalt more than equal to the task. Especially noteworthy was the excellent and comprehensive Cuban policy review compiled under his immediate supervision. . . . I have found him to be: a brilliant analyst, an articulate speaker and writer, a prodigious worker, possessed of the soundest judgments, and uniquely effective and calm under pressure"[34]

In June 1963, Secretary McNamara told Nitze that Roswell Gilpatric, deputy secretary of defense, had decided to return to his law firm. He asked Nitze to take the position as deputy secretary, and Nitze agreed to do so. Nitze had already instructed Bud to get the special security clearances required to be his executive assistant at the new job when Nitze learned that there was Senate opposition to the appointment. There were also a spate of stories that Bud believed had been started by Gilpatric, saying that "Nitze was too much a mirror image of McNamara, and too similarly cold and analytical, lacking in charisma to be a logical counterweight to McNamara."[35]

Meanwhile, Secretary of the Navy Fred Korth was under heavy fire for misusing the navy's yacht, *Sequoia*, for old business associates.[36] Korth was an extroverted Texas banker with many friends. He had offered his successor at the bank use of the yacht for entertaining clients. A copy of the letter was leaked to the *Washington Post*. Nitze was called to the White House and asked by President Kennedy to take the job of secretary of the navy. Nitze considered the job a demotion and one that removed him from national security policy formulation. Service secretaries were responsible for recruitment and training of personnel and for the development, acquisition, and maintenance of equipment. They are not in the chain of command with respect to the use of military forces in combat. That command goes through the unified and specified commanders, up through the Joint Chiefs and the secretary of defense to the president.

Nitze insisted on taking the matter directly to the president. Kennedy listened patiently and then made two points: He needed to fill Korth's position that day and wanted Nitze to take it, guaranteeing him that within six to nine months, he would get Nitze out of the navy job and into a central policy job. He laughed at the idea that anyone in Washington would consider being nominated to be secretary of the navy a demotion. "I had to agree. So we shook hands on it," recalled Bud, who urged his boss to take a job where he would have his own command rather than being a deputy to McNamara.[37] "My view is that it's one hell of a lot better than being Deputy Secretary of Defense. It's the difference between having your own ship and being a staff person." Cyrus Vance was then appointed as McNamara's deputy secretary of defense.

Nitze took Bud with him from ISA to serve as executive assistant and naval aide. "I wanted him," said Nitze. "McNamara told me I could get anyone I needed. He could write a speech, contribute to policy discussion, speak well, possessed a certain eloquence and qualities of character and loyalty that have defined him and his career ever since. I needed someone on my staff to keep me alert on what was going on among uniformed Admirals . . . he was my eyes and ears on uniformed Navy."[38]

Nitze assigned Bud to serve as the transition officer responsible for getting him confirmed. Kennedy's nomination of Nitze on October 15, 1963, was a controversial one. A major effort led by Senators Harry Byrd, Strom Thurmond, Richard Russell, and Stuart Symington was already under way to kill the nomination. While it may have looked strange to place a military man in charge of a political process, it reflected Nitze's deep trust in Bud's talents. It's unusual for a military person to be made responsible for winning the political clearance of a presidential appointee, but in this case personal trust trumped politics. Bud quickly organized a team under Robbie Robertson from the navy judge advocate general's staff to help with the upcoming confirmation hearings.[39]

"Paul Nitze made enemies almost as rapidly as he made friends," wrote his grandson Nicholas Thompson.[40] Nitze had written and spoken candidly on controversial issues for over thirty years. Several members of the Senate Armed Services Committee focused on two documents that they believed portrayed Nitze as a dangerous liberal. Donald Rumsfeld, who had just been elected to the House from a district in Illinois, circulated a letter denouncing Nitze to the committee. The basis of the denunciation was that Nitze had once chaired and moderated a panel of the World Council of Churches that had passed a series of extremely liberal resolutions, including one that advocated admitting the People's Republic of China into the United Nations. Other members of the committee focused on a speech Nitze had given at an Asilomar conference in which it sounded as though he was advocating taking nuclear weapons away from the United States and putting them under NATO command.

In the course of Zumwalt's research, a more complete picture of the truth emerged. Nitze had been asked to be the moderator of the World Council of Churches panel by none other than John Foster Dulles, when Dulles was serving as Eisenhower's secretary of state. Dulles hoped that Nitze would be able to get the participants to water down their recommendations, but "he got rolled over on most of them," recalled Zumwalt. "The only thing that he had failed to do, as you don't worry about doing when you're not in public life, he hadn't put in a specific disclaimer. As a matter of fact, he hadn't even been consulted on the report."

With regard to the Asilomar conference. Nitze had been asked by the organizer of the conference to deal with "far-out new ideas." He began the speech by saying he would follow the definition of an economist, "a man who lightly passes over the minor inconsistencies, the better to press on to the grand fallacy." In reality, Nitze did oppose the Eisenhower proposal of putting all nuclear weapons in one basket. Instead, he favored beefing up conventional power so that the United States did not have to rely exclusively on nuclear weapons. At Asilomar, Nitze listed all of the assumptions on which this proposal rested and then said, "Now, assuming that these assumptions are followed, here's what I think would happen. You wouldn't have the conventional forces to withstand an assault; you'd find yourself progressively backing down here, backing down there." Then in his final paragraph, he said, "Now, let me leap to the grand design or the grand fallacy, as the case may be—I suggest that if all these assumptions are followed, and if all these things come to pass, that about the only thing left for us to do would be to place all of our weapons, our nuclear weapons, under the command of NATO and have NATO be responsive to the United Nations charter."

Nitze decided to visit Democratic senator Stuart Symington, whose father had been a good friend of Nitze's father at Amherst College and who had held young Nitze when he was christened. It was rumored that Symington was one of the major opponents of Nitze's appointment. Symington told Nitze that his confirmation would be dead if he tried defending the speech's logic in the context of the time and circumstances it had been given in, three years earlier. Instead, he needed to say that the entire body of the speech was a grand fallacy and not to be taken seriously. With some reluctance, Nitze followed this advice.

Meanwhile, Nitze sent Bud to meet with Senator Henry "Scoop"

Jackson. Nitze had first won Jackson's trust and admiration for his role in writing the 1950 National Security Council reexamination of U.S. grand strategy (National Security Council Report 68, or NSC 68), which established the main lines of American policy for the Cold War.[41] "Jackson agreed completely with NSC 68's conceptual framework of a long-term struggle between American freedom and Soviet totalitarianism, and with the expensive means that it called for to combat the Soviet threat: a vast military buildup in conventional and nuclear arms, the alternative to which was the enslavement of the free world," wrote Robert Kaufman.[42] Jackson had later recommended that Kennedy appoint Nitze as secretary of state, the position filled by Dean Rusk.

Jackson took an immediate liking to Nitze's protégé. He advised Bud to lie low and find "everybody who is a firsthand witness in both cases that can contribute anything, and get affidavits from them. Stay out of the newspapers and come over here for confirmation hearings with Paul Nitze, having given me all those documents, and I'll give a speech both at the hearing and on the Senate floor and put those documents into the record."[43] When the hearings began, most of the attention was on the Asilomar conference speech. Nitze insisted that he had just "thrown out" the idea of a handover to the UN. It was intended for discussion. "That's not the way it reads," said Senator Harry Byrd of Virginia. "I recognize that, Mr. Senator," said Nitze.

Senator Strom Thurmond made the unprecedented request that Chairman Richard Russell put Nitze under oath, after his testimony to Byrd. It was now personal. Thurmond's first question was, "Mr. Nitze, have you ever been to Cane?" Nitze didn't understand the question. Thurmond repeated, "Cane, France." Now Nitze understood the question, recalling he had been to Cannes, France, for a conference of the Bilderberg Group. Thurmond asked if there were any Russians there. Nitze replied no, but there were plenty of Americans, including General Lyman Lemnitzer, George Ball, David Rockefeller, Dean Rusk, Ted Heath, and Senator Bourke Hickenlooper. Thurmond dropped the line of questioning and next asked whether Nitze owned a farm in southern Maryland and whether he had sold part of the farm to the American Telephone and Telegraph Company. Nitze replied that he had reluctantly sold three acres so that the company could put in a microwave tower and switching station. Thurmond charged that Nitze had received more for the land than it was worth because AT&T was a contractor to the Defense Department. Thurmond requested an eight-week adjournment so that the committee could run down all the details on this transaction. Chairman Russell granted the request.

When the facts emerged, it turned out that when AT&T first approached Nitze for an option on the three acres, he refused, because his wife, Phyllis, did not want microwave towers overlooking their fields and woods. However, the location was necessary for the Pentagon's new communications systems, so after consulting with general counsel in Defense and the relevant congressional committees, attorneys on both sides approved the AT&T option.

Thurmond was forced to drop this conflict-of-interest line of inquiry, choosing now to question the conclusions Nitze had reached at the National Council of Churches in 1958.[44] Bud had already found a Quaker preacher who said he could certify that Nitze was against those positions. "As a matter of fact, I did a lot of debating with him," swore Bud's witness. "He was on the opposite side on all of those issues. . . . Paul Nitze was against me on every one of those issues, and he was taking intellectually honest positions, and it's just unconscionable that they're going after him for this."[45] Senator Byrd provided perhaps the most comical moment in the hearings, saying it was incredible that a man would so carefully and logically consider the pros and cons of a proposition he considered to be a grand fallacy. Byrd ultimately voted for confirmation, unlike Senators Strom Thurmond and Barry Goldwater. The *New York Times* headline, NITZE OPPOSED IN NAVY JOB FOR ALLEGED PACIFIST VIEWS, had little effect on the majority of the committee. Scoop Jackson had kept his promise to Bud by working on the more reasonable opponents. Richard Russell was the first one to fall off, but President Kennedy's assassination ultimately diluted the opposition.[46] Lyndon Johnson was an old buddy of the senior members of the Armed Services Committee and was able to call them up and say, "Look, I've got to have this guy confirmed fast." And it went through just like that in the aftermath of a national tragedy.[47]

One of the first visitors to congratulate the new secretary of the navy was Admiral Hyman Rickover, who could not resist popping his head into Bud Zumwalt's cubicle. They had not seen each other since the interview. "Now remember, Captain Zumwalt, I didn't reject you, you rejected me." That was true. And by his own admission, Bud was fascinated by the man, speaking incessantly to others about him and trying to break the code of what made Rickover tick. Bud came to see Rickover as "essentially a broker of power, and only secondarily a Naval Officer. He would have been the same type of personality had he been a Rabbi."[48] Many years and battles with Rickover later, Bud said candidly, "I suppose at bottom we acted like two wary old dogs who kept circling each other because neither dares lunge first. The hair on the backs of our necks always seemed to bristle when we encountered each other."[49]

In some respects, Rickover was just like Bud—one of a kind. Brilliant, sly, tireless, blunt, devious, arrogant, foresighted, vain, angry—attempts to define him will forever fail. Former secretary of the navy Paul Ignatius used many adjectives to describe Rickover—short, wizened, devious, calculating, manipulative, dominant, demanding, focused, unscrupulous, egotistical, and effective.[50] Bud felt it was his responsibility to educate Nitze about what a formidable problem Rickover was going to be for the navy.[51] Rickover was always on the offensive, always attacking and seeking to discredit people who got in the way of his vision of a nuclear navy. Bud thought Rickover "lived on the edge of madness."[52] Rickover's office was technically only a division within the Bureau of Ships (later Sea Systems Command), and he was nominally under the command of the chief of the Bureau of Ships, his junior in rank. Nevertheless, for all intents and purposes, Rickover was autonomous. His power derived from Congress and from his second hat as director of the Naval Reactors section of the Atomic Energy Commission (AEC). Rickover had built strong support on the Hill with the Joint Atomic Energy Committee, the House Armed Services Committee, the Senate Armed Services Committee, and the House Appropriations Committee. He was literally three-hatted, reporting to the AEC on nuclear issues, to the secretary of the navy through the Navy Material Command, and to the Joint Atomic Energy Committee

(JCAE). Nobody really had control over him. Any directives from higher authority that were not to Rickover's liking received a reply either from the AEC or the JCAE. The Atomic Energy Act literally gave JCAE plenary power in all matters dealing with atomic energy.

Nitze and Zumwalt spent much time trying to figure out how to deal with Rickover, even trying to force him into retirement. Nitze knew Rickover would fight back, so he first checked with key members of Congress, including Richard Russell, chairman of the Armed Services Committee. Russell told Nitze that while he would not actively support Nitze's action, he would not oppose him. Rickover reached the mandatory retirement age of sixty-four in 1964 but was recalled to active duty for a two-year term. Nitze asked Zumwalt to prepare the brief, which Bud accepted with great enthusiasm, seeing it as his chance for ultimate payback. Bud spent a great deal of time writing up a bill of particulars that provided legal authority to get rid of Rickover. "I typed the memo myself and it was never run through the Secretary's mailroom," said Bud. No one else knew, except one naval aide. Nitze took the brief to Cyrus Vance, who approved it. Nitze then set up a meeting with Scoop Jackson, but when Nitze arrived at Jackson's office for the meeting, he learned it had been moved to the committee room. Opening the door, Nitze saw Senator Jackson sitting in one seat and Rickover at the witness table. Nitze was asked to sit at the opposite table. Nitze now understood why Rickover, along with J. Edgar Hoover, was considered the most established bureaucrat in Washington. "Rick doffed his admiral's suit whenever he found himself in conflict with Navy policy, and sniped at the Navy in his civvies," recalled Bud.[53] Without the support of Senator Jackson, Nitze's closest ally in Washington, the plan to retire Rickover was abandoned.

When the time again came to extend Rickover's tenure, Nitze took a different strategy, informing him that he was being extended but that he needed a deputy who would be competent enough to carry out his job once he retired in two years. "Rick was furious about this and said there wasn't anyone who could take his place and so forth and so on. Rick would get rid of every Admiral who was competent to take his place. He wouldn't have anybody succeed him or be in a position to succeed him. He once quipped, 'Deputy? Deputy? Did Jesus Christ have a deputy?' "⁵⁴ A joke around the Pentagon was that Rickover didn't buy a grave site; he just rented a tomb for three days.

One of the major issues on which Rickover and Nitze clashed was which ships in the ship-construction program should be nuclear powered. Rickover believed that all combatants destroyer size and larger should be nuclear powered. No one disputed that nuclear-powered ships were more capable, having longer endurance and freedom from the tanker "tail" required by fossil-fueled ships. But the costs were enormous in comparison to conventional-propulsion ships. Nevertheless, Rickover and his staff could always produce a study proving the cost-effectiveness of nuclear-powered ships in comparison to conventional ones. These studies had great resonance in Congress and forced the navy to produce its own studies of equal or greater weight to defend shipbuilding decisions that ran counter to Rickover's ideas and claims.

Nitze recalled one major confrontation when the navy determined that the next class of destroyers should be powered by gas turbines and not steam turbines. With Nitze's encouragement, the navy was attempting to move ahead on the development of propulsion systems other than nuclear and fossil-fueled steam. The Soviet navy in particular had enjoyed success with gas turbine propulsion, either as a primary system or as a boost system for a diesel cruising configuration. Nitze and Zumwalt saw several advantages in such systems—lighter weight, greater simplicity, and lower maintenance demands. The U.S. Navy had yet to install a gas-turbine system on a large surface combatant, however, and Rickover fought this tooth and nail.

Nitze had all of the studies done and received McNamara's support, but Rickover objected because he wanted them to be nuclear. Nitze was scheduled to make the budget presentation to Congress, the details of which the president, the Bureau of the Budget, and McNamara all favored. Nitze was startled to learn that Representative George Mahon had also called Rickover to testify. Nitze reminded Rickover that he was expected to give the administration's position and that if he disagreed with the administration, he would still need to provide the reasoning for both sides.

Rickover told Nitze he did not know the executive branch's reasoning. Nitze spent the weekend working on a memo, and sent it to Rickover. On Monday the secretary called Rickover asking if there was anything technically incorrect in the memo. "Mr. Secretary, no. There is not a word in it that is technically incorrect. It is totally incorrect." Rickover attacked Nitze for basing his logic on cost-effectiveness, an improper test. Rickover favored the best possible ship regardless of cost and "the best propulsion unit for a destroyer is a nuclear propulsion unit. The fact that it costs more, the fact that you can demonstrate that per dollar, the gas turbine unit is more effective, is beside the point. It is not the best. Furthermore, I believe that Congress will support me, that they believe the Navy should have the best. What I am going to do if I am asked about this—and I will be—is first of all present your memorandum verbatim; then I will present my chain of reasoning, and I will beat you."

A vice admiral was telling the secretary of the navy "I will beat you." Rickover introduced the memo with denigration and personal attacks, telling the committee that neither McNamara nor Nitze had ever been elected, that they were self-anointed. "It was the most revolting insubordination you can imagine," recalled Bud. Nitze was incensed but could do nothing.

Admiral Rickover was still trying to pry Bud from Nitze, suggesting him for command of the *Long Beach*, a nuclear-powered ship. This time the plan almost worked because Bud felt the tug of institutional obligation to his fellow black shoes to accept the command. It was important that a destroyer man have this type of command. Nitze hit the ceiling when Bud told him about Rickover's offer, seeing it as a plan to separate Plato from Socrates. "Bud, that's one of the stupidest statements you ever made. I know, from what you've told me, that the *Bainbridge* would have dragged you down into a four- or five-year command. And now you're proposing to go do the same thing to yourself as a four-striper."

"That's right, but I have got a moral problem here," replied Bud. "Every destroyer man is looking to that as being a former destroyer man's command. I've had at least ten or twelve of my fellow destroyer men come around and tell me that I'm the only guy that can head out the submariners, and I think, therefore, I owe it to my group to do that."

"That's so damn stupid," shouted Nitze. "I'm going to call Dave Mc-Donald right now." He picked up the phone. "Dave, I've got Bud in here, and he's just done a very dumb thing, and I wonder if you'd have him over and talk to him."

Bud walked into the CNO's office with a sheepish demeanor. Mc-Donald asked, "What the hell's going on?" Bud repeated what he had told Nitze, that he wanted to command the *Long Beach* and that Nitze said it was the stupidest thing he could ever do.

McDonald asked, "Well, why do you want to command it? You can have another surface ship, any surface ship you want."

Zumwalt again repeated what he had told Nitze.

McDonald said, "I can understand that. That makes sense. Suppose I commit to you that the next skipper of the *Long Beach* will be a destroyer man. Will that satisfy your problem?"

"Yes, sir."

"OK, get the hell out of here and get back to work."[55]

Paul Nitze did not want to lose his protégé to Rickover's nuclear navy. Nitze thought that anyone who had had command of two or three destroyers had sufficient command experience of ships. Paul Nitze had other plans for Bud.[56]

Bill Thompson's office was in the E ring across the corridor from Paul Nitze's office.[57] Thompson's job was in public affairs, charged with making Nitze accessible to the press and promoting his positions. On his first day of work, Thompson, like all new staff, was instructed to see Captain Zumwalt. "He told me that the Navy was fortunate to have the likes of Paul Nitze as its Secretary." Nitze was concentrating on a few select programs, such as positioning himself as the Anti-Submarine Warfare (ASW) czar, and wanted to move the navy into a better position to combat the Soviet Union's submarine threat. He was also trying to develop a Forward Deployed Logistics Ship (FDLS) concept that would reduce the cost of building ships, and he wanted to modernize ship-yard capabilities in the United States so as to reduce the cost of building new ships. The problem was that Nitze did not like claiming credit or being in the spotlight. Thompson's job was to devise a public relations program that got the navy credit and improved its image, even if Nitze chose to be a reluctant player. Thompson needed Bud's help in convincing Nitze to sit for interviews with media representatives.

Thompson was intrigued by Zumwalt, especially watching him operate in his normal daily activities. "It wasn't an exhibition he gave occasionally; it was his normal behavior. One could perceive a growth he favored extending from his ear but on closer inspection, it was a telephone that was omnipresent. It seemed that he was always talking into a phone that was cradled between the side of his head and shoulder, thus freeing his hands to enable him to shuffle the stacks of papers on his desk or to write. Additionally, he was a juggler, keeping more balls in the air than anyone I had ever been with and constantly maintaining a level head. His demeanor was extraordinary—never wasting a minute, not forgetting anything and always conscious of those with whom he was working. . . . I also learned that he was a delegator and didn't harbor many tasks to himself but remembered all the cards he had dealt and didn't forget all those who were involved in his projects and issues. He seemed to get more out of his small staff than even they thought they were capable of performing. But, that is the basis of leadership—getting others to do things, some of which they didn't think they were capable of doing or didn't care about doing. In addition to all the other interesting things going on about me, I was participating in a Leadership 101 Lab Course, and enjoying it!"[58]

At the encouragement of Zumwalt, Nitze initiated a weekly meeting each Friday comprising the secretaries, CNO, Marine Corps commandant, VCNO (vice chief of naval operations), and deputies. This was the leadership of the Navy Department crammed into the SECNAV office, awaiting Nitze's debrief of his weekly meeting with Secretary McNamara. In turn, each was asked to address the major concerns of his shop. Nitze and Zumwalt espoused a duumvirate (two-man power) in the navy, with the secretary of the navy and the chief of naval operations reading from the same text and acting in concert.[59] The system worked because of the relationship between Nitze and CNO Admiral David McDonald, who had replaced the insubordinate George Anderson. Neither Nitze nor McDonald was a grandstander, and they were committed to the process. On matters involving navy weapons systems or strategic political questions involving the Soviets or allies, McDonald tended to defer to Nitze. On matters involving the mix of congressional political considerations and bureaucratic politics within the navy and the Pentagon, Nitze deferred to McDonald. Bud played an absolutely indispensable role in keeping the two men united. "He has helped to foster the closest collaboration between the CNO and me, thus enhancing our mutual effectiveness," wrote Nitze. Bill Thompson could see that "this concept made for better harmony, a feeling of teamwork and accomplishment." Zumwalt used the duumvirate term and philosophy whenever describing the relationship between Nitze and the CNO.[60]

In the normal rotation of senior officers, especially those considered front-runners, it was Bud's time for assignment to a major sea command as commanding officer. Bill Thompson was the first to hear, as he usually did, the rumor that Bud was getting the cruiser USS *Chicago*. Bill started preparing for staging the public relations aspects of both the send-off and the welcome events. "One day in April of 1965, I was getting ready to present a full plan to Bud about the *Chicago* visit. There was the usual organized chaos about the office that morning, but something was a little out of character and I couldn't identify it." It turned out that the Flag Section Board was meeting that morning in the Pentagon. "I approached Bud at his desk. He seemed to be having a difficult time in the usual realm of activity and I started the conversation with an opener, 'Bud, about taking you to the Windy City, we have to come to grips with some firm dates and when we can make the announcement that you will be in Chicago. As you realize, I can't do this thing right by making last minute arrangements. We have to get down to details.'"[61] Bud gave him a sheepish grin, looking nervously around, and said in a low voice, "Looks like we will have to postpone Chicago for a while." Before Thompson could ask why, he realized that "the Son-of-Gun was on the list! I stuck out my congratulatory hand and he grabbed it, and said, 'We can't say anything yet.'"

Bud had been deep selected, designated for promotion before normal time. Bud was selected two years before his class would normally come before a selection board. At forty-three years of age, he would be the youngest officer ever promoted to flag rank in the history of the U.S. Navy.[62] Paul Nitze had punched Bud's ticket, validating all the choices Bud made years earlier. Nitze rated Zumwalt "1 in 100,000" with respect to both action officer roles and leadership at the policy level. "He has achieved a status in Washington circles unique for one of his rank. He has both the singular capabilities and background of unusual experience to mark him for acceleration, now, to flag rank. . . . Perhaps most important, he has been the one whom I have looked to within the Secretariat for overall continuity of the administration and to insure that no management gaps exist. In this capacity, his performance has left nothing to be desired; he has maintained a sure hold but with the finesse required of his relatively junior rank. I credit him with major contributions to whatever success is judged to accrue to my administration. . . . He has the forthrightness to speak up when he thought I was wrong and to provide sound advice on these occasions." Nitze urged that an exception be made to the tradition that an officer have a major sea command and then be selected in due course. "I would suggest that Captain Zumwalt's three previous commands, one of which was a prototype missile ship (DLG), together with his exceptional capability to serve Navy and country in positions of highest responsibility, indicate accelerated promotion to prepare him for rapid movement into highest military positions."

Bud often used the analogy of Plato and Socrates to describe his relationship with Paul Nitze. Bud considered his mentor to be "a man of brilliant intellect" and became almost lyrical when discussing him.[63] "He is a man who has absolutely no personal hang-ups. He doesn't mind reversing his opinion when a better argument for a better alternative position can be given. He doesn't hesitate to have anything that he has written be used by others. He doesn't hesitate to adapt things that others have written to papers on which he is working. More than any man I have ever known, he is a sincere seeker of the truth, and he is willing to get there by a series of the best known approximations, using his own and others' minds to generate these truths. He is deeply patriotic and has chosen to devote himself to public service, whether in or out of Government, ever since coming to Washington as a young man in the early war years. He does not enjoy small talk in the normal sense. However, he will sit down and converse for intense periods with people of all levels of sophistication, age, and background, if they appear to have interesting experiences or insights from which he can learn. He has read all the great philosophers and can converse at length about them."[64]

Bud's apprenticeship under Nitze was nothing less than the equivalent of a PhD in foreign policy. From Nitze, he learned the art of breaking down foreign-policy problems into their several parts, analyzing them in detail with great care, examining and questioning assumptions, integrating subsets of problems into the overall whole, and then examining them again to ensure that the very best available light had been brought to bear on the subject. "From him I learned the importance of increasing the breadth of one's observations by going beyond the conventional intelligence of the day and talking to informed persons at all echelons in and out of Government and the various nations with whom we must deal. From him I learned the importance of checking the original, first material to insure that secondhand judgments had been properly derived. And from him I acquired a depth of historical background of the previous twenty years of close Government service and observation that made it possible for me to get the running start of continuity toward that moment in history when I would be called upon to lead."[65]

By coincidence, on the same evening Bud learned of his selection to flag rank, he and Mouza attended a cocktail party with Paul and Phyllis Nitze, Robbie and Trish Robertson, and Bill and Zum Thompson. Bud's promotion was still a secret, but rumors swirled that Bud was on the list. The host, Rear Admiral Bill Mack, kept at Bud, saying, "Ah, come on, Bud, get off of it. You know and everybody here knows." Bud broke down and admitted he was on the list. After a few more drinks, the close friends gravitated to the lower level of the Mack home, finding a niche under a staircase where they gathered to slap Bud's back and laugh. At some point, Bud said, "Well, when I get to be CNO, Robbie, you are going to be the Judge Advocate General and Bill, you are going to be the Chief of Information." He punctuated the statement by poking each of his friends in the chest. They laughed at the ludicrous statement. Here was a captain, early selected, not even in an admiral's suit and already predicting that he would one day be the CNO. After a few more jokes, Bill Thompson countered with, "Hey, Captain, I can't wait around very long for that to happen. How long do you think it will be before you are knighted as CNO?" Bud scrunched up his face, pursed his lips, tilted his head, and after a few seconds came out with, "Give me five years."[66]

Fate would make a drunken boast come to pass. In an "Ode to Captain Zumwalt," Bud's friends offered the following farewell:

> As a young lad with a scratched knee,
> He was always hopping from tree to tree.
> One day he slipped, and as he fell
> The woods resounded with a Tarzan's yell.
>
> When he joined the black shoe Navy,
> With this I'm sure you'll all agree,
> The dulcet tones of Tarzan's call
> Oft shattered the quiet of Bancroft Hall.
>
> A dashing Navy Captain is he,
> An Aide to the Navy Secretary.
> He is one who is hard to fault,
> Known to us all as Bud Zumwalt.
>
> From ISA in DOD
> He came to us with Mr. Nitze
> To teach that the finest perfume
> Is found on a DDs engine room.
>
> He often says, "Mr. Nitze I wish you'd learn
> That front is bow, rear is stern.
> A better Secretary will you be
> If you'd learn our vocabulary.
> Soon he'll breathe the salty air
> Sitting in the "Admiral's" chair,

But when the winds begin to wail,
We'll see who's leaning over the rail.

A new kind of WAVE soon he'll see,
One he can't bounce on his knee,
But when she's wearing caps of white,
He'll know he's heading for quite a fight.

The greatest fighting machine ever built
Is the greyhound destroyer, regardless of tilt.
He leaves the destroyers, does this young fella,
To take command of a CRUDES Flotilla.

Again he will have his own command,
A staff to rule with a strong hand,
A hand that's firm—but always fair—
A Flotilla he'll have beyond compare.

"Now hear this"—and to all tell:
Again through the fleet comes Tarzan's yell.
CRUDESFLOT 7—take up the slack.
Gangway, you lubbers, Bud Zumwalt's back![67]

CHAPTER 7

PATH TO VIETNAM

Rear Admiral E. R. Zumwalt, Jr., fighter pilot,
270 F-111Bs killed.

—PLAQUE PRESENTED BY F-111B STUDY TEAM[1]

Paul Nitze wanted his new rear admiral to remain with him in Washington, but Bud had other ideas. "I persuaded him that I could do more for him and do more for myself if he let me go promptly and I got a sea tour checked off."[2] The obvious one for an ex–destroyer man like Bud was to go to a cruiser-destroyer flotilla. He requested a flotilla that was scheduled to deploy within a year so that he would have a chance to do the workups through fleet exercises and then take it overseas. "I figured that's as far as I would have a chance to get before Paul would send for me. As it turned out, I only got the 11 months in and out of San Diego, and didn't get the deployment."[3]

Bud took command of Cruiser-Destroyer Flotilla Seven with the cruiser USS *Canberra* as flagship on July 24, 1965.[4] Based in San Diego, Bud found the job "more like running a good Boy Scout troop than running a ship."[5] With the change of command being held in San Diego, Bud's entire family, along with a large entourage of friends from Tulare, attended the ceremony. The job of flotilla commander involved devising ways of improving the training and readiness of twenty individual ships by dealing with their skippers and working to solve problems involving personnel shortages or schooling quotas. Bud described it as being "a kind of a big daddy. . . . I saw my job as a flotilla commander as one of avoiding getting into the commanding officer's business. . . . I saw my job as one of doing everything I could to improve the training and readiness of individual ships by dealing with the skippers, working to solve the problems they thought they had of personnel shortages or of the need for schooling quotas, where they'd been turned down, that sort of thing. So I tried to operate as a kind of big daddy to get a reclama in where they got turned down on something. And then of, with my own staff, perfecting an ability to manage ships that might come under my command, either in fleet exercises or if I deployed."[6]

As in previous assignments, Bud made the most of the tour. His major professional accomplishment was running three major fleet exercises as well as Exercise Baseline, designed to establish a minimum for communications readiness by collecting data to determine the current effectiveness of communications and the relative performance of individual types of equipment. "That took a massive amount of planning; it took a massive number of observers from Washington and from the fleet support activities. We put in about three months planning for it and an intensive week or ten days running the exercise, and then a massive reconstruction report afterwards."[7]

Being based in San Diego meant Bud and his family had a chance to spend time with his brother Jim's family. Bud invited Jim to go with him on a fleet exercise. They were at sea for a week, and Jim was able to observe Bud at his eighteen-hour-a-day job running the operation. Jim came back and wrote a letter to the family saying that, having observed Bud, he could now understand why he had such a conservative persuasion—he was too busy to find time to read. Bud wrote him a little note about the systems analyst, who decided to find out how a flea could hear. He put a match on the table, pulled one leg off the flea, and said, "Jump over the match," and the flea jumped over the match. He pulled another leg off the flea and said, "Jump over the match." The flea jumped over the match. The third leg, the flea couldn't do it. His conclusion was the flea hears out of his right rear leg.[8]

Bud never had the chance to deploy with his flotilla, because Nitze and CNO Dave McDonald wanted him back in Washington. The navy secretary was quite dissatisfied with the navy's performance in the analytical area, especially in studies requested by Secretary of Defense Robert McNamara. There was no real understanding of the importance of systems analysis. Nitze convinced CNO McDonald to start a systems analysis operation with Zumwalt at its head. "Dave McDonald said to me when he brought me in that he was tired of fighting Nitze and now he was going to join him on systems analysis which was largely a response to the planning, programming and budgeting cycle that McNamara had put in."[9]

Bud returned to Washington to head OP-96, the newly created Systems Analysis Division of OPNAV (the Office of the Chief of Naval Operations). His charge was to establish the division and develop it into an effective organization. This required hiring staff as well as organizing functional duties. Bud defined the mission of his new office to be one of generating a better understanding of requirements and problems and a more effective presentation of those requirements in major program areas that were likely to influence the combat capabilities of naval forces. As the first director, Bud would heavily influence the combat capabilities of U.S. naval forces through the next generation. The major analyses completed under his direct supervision included studies on fleet escort, antisubmarine warfare force levels, tactical air power, surface-to-surface missiles, and war at sea. Additionally, Bud contributed to the restructuring of the Center for Naval Analyses to ensure that completed studies reflected greater thoroughness, comprehensiveness, and accuracy.

While in Systems Analysis Bud gave much attention to justifying a naval base on the island of Diego Garcia. His staff undertook a political-military study and were able to demonstrate by a calculation that it took several carrier task forces to support an effort in the Indian Ocean without Diego Garcia and only one with. "I fought for Diego Garcia in ISA; I fought for it again in Systems Analysis; and then I finally got it done when I was CNO, at a time it was considered just a far-out kook idea," recalled Bud.[10]

The major challenge for Systems Analysis involved the F-111B, a project Secretary Nitze had inherited from his predecessor. Secretary Mc-Namara had endorsed the development of one fighter aircraft adaptable to the air force, navy, and marines. The term McNamara used for the airplane was "commonality"—features that would be more *cost-effective*, a key phrase in the lexicon of the McNamara whiz kids.[11] Naval aviators objected to the concept, believing that the air force version was too big and too heavy to operate from aircraft carriers. Zumwalt was tasked with conducting a systems analysis study of the F-111B for the navy.

Secretary Nitze found himself in the middle, between a navy bureaucracy that was at best lukewarm and at worst actively hostile to the plane and a secretary of defense who regarded the slightest expression of doubt toward the program as an attack on him personally and a sign of disloyalty. Because the air force managed the program, there was often interservice friction, because members of the air force management team felt that the navy was dragging its feet or intentionally trying to make the program fail.

Bud needed help, and the one person he knew was best able to handle the task for Systems Analysis was Charles DiBona, a lieutenant commander assigned to the undersecretary of the navy as a special assistant. Charlie had recently completed a tour in Alain Enthoven's Pentagon systems analysis office, and he was a Rhodes Scholar who had stood number two in his Naval Academy class.[12] Bud was told it would be impossible to get DiBona because his current boss, Robert Baldwin, undersecretary of the navy, would never release him. When Bud asked, he was told "not no, but Hell, No!" Bud went to see Paul Nitze to say, "You've given me an impossible job, unless you will pry Lieutenant Commander DiBona away from Bob Baldwin," which he did. "I then, using Admiral McDonald's authority, brought together a team of eight or ten people." "I spent my first weeks giving Bud Zumwalt tutorials in systems analysis," recalled DiBona, who helped structure the studies, which required technical insight into cost-effectiveness analysis.[13] "We worked night and day, weekends, literally 18 hours [a day]," recalled Bud.[14] He and DiBona worked most of Labor Day weekend in 1966, "with me dashing back and forth from painting my house—he remembers me coming in with my glasses covered with paint specks—and he did the more detailed calculations, and I would come in about every three hours to go over and check-sight where he was at."[15]

One briefing with Secretary Nitze on the F-111B captured the essence of the two men's working relationship. As the briefing was ending, Charlie raised his hand to say, "The admiral got a little screwed up on one point." He then corrected the point and everyone left the meeting pleased. Afterward, Charlie asked Bud, "How'd we do, boss?" Bud replied, "Charlie, we did great, but the next time I screw up, raise your hand and say, 'an additional insight into the matter is . . .'"[16]

DiBona took the lead on the F-111B.[17] The most constant problem with the plane was weight growth, which affected all aspects of flight performance, especially navy carrier suitability. The numbers showed that if one assumed that the F-111B could land on and take off from an aircraft carrier, which is what McNamara had directed them to assume, then there was no viable alternative that could beat it in terms of cost and effectiveness. "But we were all convinced that the F-111B would not be able to land and take off of an aircraft carrier," said Bud. It was DiBona who convinced Bud that a footnote needed to be added with some type of explanation noting that "notwithstanding the study assumption, there was considerable doubt as to the cost-effectiveness, that an alternative should be brought along, an alternative which used the same engine and the same avionics as the F-111B and the same missile system, which is what made it such a lethal system."[18]

This decision to search for an alternative airframe created problems within the navy, beginning with the new CNO, Admiral Thomas Moorer, who had replaced Admiral McDonald. Bud needed Moorer's approval on adding the footnote. He found the new CNO on a golf course playing a round with Rear Admiral Jerry Miller. Both men were aviators and suspicious of black shoe admirals like Bud. "When I briefed Tom Moorer on what the outcome showed, his face fell, but when I told him what our footnote was going to say, he said, 'Okay, I buy it.'"[19] Jerry Miller began hitting the grass with his golf club in frustration, vehemently disagreeing with Moorer's decision, believing that Zumwalt should tweak the data to come out with a different conclusion, one that showed the F-111B was not cost-effective. All Bud needed to do was massage the data by making some slight changes in estimated performance here and there. "But we had done all those estimates carefully and, under a kind of umpire system, to make sure that they would not be criticized. Then to go back and change them would have been not only dishonest but stupid," recalled Bud. "Jerry just hadn't been close enough to the problem to see that this was one where there was both a tactical reason for the study and a strategic reason for the sake of OP-96, in the long haul, to do things honestly."[20] When Zumwalt became CNO, Miller thought he would never get a third star or any fleet assignment. "I had obviously not incurred the favor of the civilian hierarchy, much of it because of the F-111B."[21]

With Moorer's acquiescence, Paul Nitze was able to tell McNamara, "Here's the study that proves you're right, but just in case you're not, give us some money for the alternative."[22] McNamara was able to do that: to justify the F-111B while at the same time preparing for its failure. The sequel is that the F-111B was finally killed, and the preferred alternative, the F-14, was put forward. Bud's study team then presented Bud with a plaque, REAR ADMIRAL E. R. ZUMWALT, JR., FIGHTER PILOT, 270 F-111BS KILLED.

As head of Systems Analysis, Bud came face to face with his old nemesis, Admiral Rickover. Tasked with undertaking a major fleet-escort study, Bud had convinced McNamara that escorts should be built until there was greater protection for carriers. Once that agreement was achieved in the major fleet-escort study, it was possible to do detailed, numerical calculations and show that more major fleet escorts were needed.[23] Zumwalt wanted a separate analysis under the major fleet-escort study, one that became known as the endurance supplement, to provide calculations on additional endurance. Forrest Petersen, a nuclear propulsion–trained aviator-captain, was given the job of running the endurance study. He had extensive nuclear training and was scheduled to be commanding officer of the USS *Enterprise.*

When Rickover got wind of the endurance study, he called Petersen to say that if he wanted a future in the nuclear navy, he had better produce the right results—ones that showed the escort needed nuclear propulsion. If the endurance study didn't produce these results, Petersen could forget the *Enterprise* assignment. "He came to me after a week or two and said that Rickover was really leaning on him to come up with dishonest answers. We both agreed that if he didn't in some way accommodate Rickover, that Rickover would ruin his career."[24] Bud knew he could not defeat Rickover; his only chance would be to find some type of middle strategy. Bud sought the counsel of Vice Admiral Bernard Ambrose "Chick" Clarey, director of navy program planning and budgeting. They agreed to let Petersen work the study and see whether or not honest calculation would justify nuclear escorts. If it did, there would be no problem; Rickover would assume that Petersen had carried out his order, and his career would be saved. But if it didn't, then they would have a much tougher problem. As Bud saw the situation, there was "Navy planning and Rickover planning." Rickover planning usually prevailed.[25]

Fortunately, Petersen's study justified a need for a certain number of major fleet escorts to be nuclear propelled, and the endurance supplement to the major fleet study endorsed that number.[26] Rickover's biographer observed that "Zumwalt and his colleagues developed the 'need' for a supplement to the study that would address warship endurance; in the supplement, nuclear propulsion could be addressed, and very favorably."[27] Petersen's career was saved!

Rickover outmaneuvered Zumwalt on two new submarine designs, the *Glenard P. Lipscomb* and the SSN-688 *Los Angeles* class. The *Lipscomb* had an electric drive, meaning that it was quiet but slow. The SSN-688 subs were faster but far noisier. Both of these designs had been initiated by Rickover. Both the navy and the Department of Defense had questions: Was the *Lipscomb* or the SSN-688 worth the additional cost compared to the currently deployed *Sturgeon*? Should both quiet and fast submarines be built? If so, how many of each type? How should they be deployed?[28] Systems Analysis was created for precisely this type of study of new weapons systems. Bud understood this, but so did Rickover, which led him to farm out his own analysis, which unsurprisingly touted data proving unequivocally that both types of submarines were needed. This was the typical Rickover fait accompli. No one had been previously informed of these studies, and Zumwalt knew why. Bud protested but lost.[29]

During the early-morning hours of January 31, 1968, approximately eighty thousand North Vietnamese regulars and guerrillas attacked over a hundred cities throughout South Vietnam. The Tet Offensive involved coordinated attacks on thirty-five of forty-four provincial capitals, thirty-six district towns, and many villages and hamlets. For weeks prior to their offensive, enemy forces had been filtering into Saigon.

The front page of the February 1 *New York Times* showed a picture of the U.S. embassy in Saigon under assault. Guerrillas had blasted their way into the embassy and held part of the grounds for nearly six hours.[30] The goal was to demonstrate that not only could the countryside not be pacified, but now South Vietnam's cities, including Saigon, were not secure. The offensive set in motion a remarkable sequence of events, including discussion as to whether tactical nuclear weapons should be used at Khe Sanh, where a fierce diversionary battle waged. General William Westmoreland, the head of the Military Assistance Command Vietnam (MACV), requested authorization from President Lyndon Baines Johnson for an additional 206,000 troops, which would bring the U.S. troop commitment to almost 750,000.

Tet became the decisive battle of the war because of its effect on American politics and public attitudes. The intensity of the war and the capacity to sustain it were controlled not by America's superior technology, but by the enemy. There was no breaking point, no crossover point in the enemy's will to continue the struggle indefinitely. On the evening of February 27, CBS anchorman Walter Cronkite told the nation that the war was destined to end in stalemate. "We have been too often disappointed by the optimism of American leaders, both in Vietnam and Washington, to have faith any longer in the silver linings they find in the darkest clouds. . . . For it seems now more certain than ever that the bloody experience of Vietnam is to end in stalemate. To say that we are mired in stalemate seems the only realistic, yet unsatisfactory, conclusion."

Addressing the nation on March 31, 1968, President Johnson called for a partial bombing halt and asked Ho Chi Minh to join him in working toward a peace through negotiations. The United States was "ready to send its representatives to any forum, at any time, to discuss the means of bringing this ugly war to an end," said the president. Then, in a dramatic gesture toward national unity, the president renounced his chance for reelection, "I shall not seek, and I will not accept the nomination of my party for another term as your President."

As the reverberations of Tet rippled through the Pentagon, CNO Thomas Moorer decided he wanted Bud Zumwalt out of the systems analysis job. A few factors were in play, some personal and others business. In the wake of the F-111B study, Moorer came to realize that he was unable to control the conclusions of studies coming out of the Systems Analysis Division.[31] First, the output was not supporting the cost-effectiveness of the nuclear-propelled ships that Moorer and Rickover favored. Of more significance was the fact that studies being conducted in Systems Analysis by Burt Shepherd and others were questioning the effectiveness of the bombing campaign. As an aviator, Moorer understood that with the data at his disposal, Bud Zumwalt constituted a threat to aviator dominance.

Other reasons were more personal. Paul Nitze had now replaced Cyrus Vance as deputy secretary of defense, and the Zumwalt family had been spending just about every weekend on Nitze's 2,500-acre farm along Maryland's Chesapeake Bay. When young Elmo first saw the property, he felt that this is how aristocracy must have lived. The Zumwalt family had their own guest house on the property, a home more palatial than anywhere the family had ever lived.[32] Mouza loved "the wonderful hay rides through the snow from our house to your house on the farm many Christmas seasons."[33] Each evening in the main house perched majestically over the Potomac, Bud and Paul would sit on the brick back patio, smoke cigars, and talk.[34] And this was where Moorer's concern really began. Moorer had always been troubled about Zumwalt being too close to Nitze, but with Nitze ensconced as deputy secretary of defense, Moorer realized that Nitze's protégé and Moorer's subordinate was privy to McNamara's thinking well before the CNO. It gave Bud an advantage over Moorer, who resented Bud's access to the inner strategic thinking of the Johnson administration.

A paranoid Moorer decided he needed to move Bud, but Moorer was astute enough to know that it had to be achieved in a way that would not suggest to Paul Nitze that Bud was being fired. Moorer's vice CNO, Bernard "Chick" Clarey, came up with a near-perfect solution. It was nearly time to replace Rear Admiral Kenneth Veth in Vietnam, and Bud Zumwalt could be sent as relief. There was only one problem with the Moorer-Clarey plan: all of the component commanders under MACV had at least three stars, with the exception of the navy component commander. Admiral Veth had two stars, because he was simply not at the same level as his peers. Bud blamed Moorer because, as an aviator, Moorer recognized that air strikes brought glory to the navy, so that's where the best people and resources went. Admiral Moorer was not going to waste resources fighting the war inside Vietnam. The army owned that war. This disparity was duplicated down the chain of command in the navy, whose task-force commanders were o-6 captains (a rank below the counterpart army o-7 brigadier and o-8 major generals), so the navy was downgraded in the eyes of the whole MACV structure. In a structure dominated by the army, rank was very important.

Knowing that Zumwalt would never take a two-star assignment and that the new secretary of the navy, Paul Ignatius, and Paul Nitze wanted to see Bud get three stars, Moorer upgraded the job to a three-star position.[35] "I'm just absolutely confident that they both thought that was the graveyard where I would end up. . . . I believe the reason Moorer wanted me in Vietnam was that no rear admiral had ever left Vietnam and obtained another job that led anywhere. This was Moorer's way of getting rid of me: Promote the son of a bitch and no one will ever hear from him again."[36]

From Moorer's perspective, the plan looked foolproof. The navy had been in the backwater of events and influence in Vietnam, a situation exacerbated by the navy's reluctance to expend resources and manpower in what was seen as the army's war. Admiral Veth was a man of limited vision who had been content with a defensive strategy, limiting the navy's role to interdiction of supplies coming in from the sea.[37] There was a dichotomy between the Brown Water Navy fighting in the Mekong Delta and the Seventh Fleet offshore, with little interplay between the two. The glory and resources went to the aircraft carriers and air strikes in the north. Outside of the Mobile Riverine Force, a combined army and navy team in the delta, the navy was a passive component with little influence and taking fewer casualties.

Less than six months after Tet, on July 25, 1968, President Lyndon Johnson announced the nomination of Rear Admiral Elmo R. Zumwalt, Jr., to the grade of vice admiral with a new assignment as commander of naval forces, Vietnam, and chief of the Naval Advisory Group of the U.S. Military Assistance Command, Vietnam.[38] For his service as director of systems analysis, Bud was awarded the Distinguished Service Medal for displaying "exceptional acumen, integrity, tact and diplomacy as personal representative of the Chief of Naval Operations, not only in dealings within the Department of Defense, but also in testifying before Congressional Committees."

The Brown Water Navy patrolled the bewildering maze of inland waterways of the Mekong Delta, which comprised one fourth of South Vietnam's total area and which was home to over one third of its population. It was also home and shelter for tens of thousands of Vietcong forces. These waters served as a source of infiltration from the north, its dense vegetation allowing limited visibility and providing cover for guerrillas waiting in ambush. The floating vegetation and heavily silted water concealed mines and other hazards.[39]

Tom Moorer spun Bud's promotion to vice admiral and a third star as signaling a new importance to the navy's role in Vietnam. Bud knew otherwise but was rightly elated about the promotion. His mind flashed back to the day he took command of the *Isbell*. Once again Bud Zumwalt was in a leadership position with nowhere else to go but up. In a July 25, 1968, letter, Bud's brother Jim joked, "As I understand the picture there are currently 14 full admirals and 35 vice admirals in the navy— only 49 more notches for you to climb. As I have often said, I have no doubt that you will make # 1." Jim added that "we hope and pray that your new assignment will bring you personal happiness and success and that God will protect you from the possible dangers that lurk in that war torn land."[40]

In a farewell poem to Rear Admiral E. R. Zumwalt, his staff at Systems Analysis offered the following:

A toast to a gentleman of the greatest renown,
A very great leader who is known all around,
Our choice for the man who is upward bound
And sure to reach the top of the mound.

Under his watchful supervision,
The CNO's Systems Analysis Division
Became a reality, instead of a vision,
And provided assistance to every decision.

Its job to analyze for cost/effectiveness,
They have turned every stone and left nothing at rest,
'Cause their very fine leader always demanded the best,
As evidenced by the bouquets on his vest.

Be it at work or be it at play,
The Big "Z" is the surest to say,
"Let's give it a go and turn not away,
And rise to the occasion with a Hip, Hip, Hooray."

But above and beyond his analytical mind
Is his strong sex appeal and easy line.
We hasten to mention that warm winning smile
That flashes a message: "All ladies beguile."

As he departs, a new Command to boost,
We gather here to say Farewell to the King of the Roost.
He is the greatest in effectiveness at a terrific low cost.
Without him to guide us, we'll surely be lost.

CHAPTER 8

BROWN WATER NAVY

*Seldom does a man by intellect and force of character affect
the lives of so many, and seldom does this run so deep as
to give reality to the concept of complete and total loyalty.
Seldom, if ever, has a naval commander suffered so directly
(especially when the casualties were heavy) and personally,
the consequences of his orders in combat.*

—VICE ADMIRAL EARL FRANK RECTANUS[1]

Bud was expected to be in Vietnam by early September 1968, two
months before the presidential election. As he departed Washington on
August 26 for a series of whistle-stop briefings across the United States
and the Pacific, his thoughts turned to assembling a staff and gaining an
appreciation for the challenges ahead. While the Zumwalt family spent
the last days of August on the beach and visiting friends from their earlier assignments in Coronado,[2] Bud was sequestered in all-day briefings.[3]

Bud was soon joined by his new flag lieutenant and aide, Howard
Kerr. Kerr had enlisted in the navy in 1960 and was commissioned after
graduating from Officer Candidate School in Newport. He served suc-
cessive tours on the carrier *Bonhomme Richard* and the destroyer *Walke*
before joining the staff of Cruiser-Destroyer Flotilla Nine. By September 1966, he had enrolled in the Fletcher School of Law and Diplomacy
and was writing his master's thesis, "The U.S. Military in Vietnam from
1954 to 1963."[4] Kerr was next scheduled to command a patrol boat in
Vietnam when Bud offered him the position. Kerr respectfully declined,
preferring a command assignment of his own. "After I'd hung up the
phone, I went back and I was reading the *New York Times*, and I got to
about the middle of the first section of the *Times*, and my eye caught
this article that said that the Navy had just named the youngest three-
star admiral in its history and that he was going to Vietnam to take

command of U.S. naval forces in-country. I put the paper down, and I thought through that a little bit and I told my wife that I thought I had just made a big mistake."⁵

After making a few calls to friends for advice, Kerr realized he should have accepted the job. He decided to call Bud back to see if there was any chance for reconsideration. Bud readily accepted the reversal, insisting only that Kerr be the one to contact BuPers. "I had to sign in blood that I had, in fact, turned down a command tour and violated all the tenets of the religion." Kerr's detailer warned that the assignment could be career suicide because "there are a lot of people in Washington who were delighted to see Zumwalt leave town."⁶

On September 10, 1968, Zumwalt and Kerr, accompanied by their spouses and the Zumwalt daughters, Ann and Mouzetta, departed from San Francisco for Pearl Harbor. Kerr had thoughtfully called ahead to ask United Airlines to have the flight attendants let him know when it was safe to move about the plane. Bud's orders read that upon departure of the continental United States, he was authorized to assume the title of vice admiral, United States Navy, and wear the insignia, which Kerr had purchased because "we intended to have this frocking ceremony on board."⁷ Admiral Moorer had given strict orders that Bud was not to be pinned until the aircraft was not only in the air but halfway over the Pacific. Bud saw it as one final attempt to "jerk him around," recalled Ann. "I remember being very intrigued that Dad literally waited until the wheels were up and just over the Pacific when the pinning occurred. I realized he obeyed the orders but with a slightly different interpretation."⁸

As soon as the seat belt sign was off, Mouza pinned one set of stars, Kerr's wife, Patricia, pinned another, and the final honor went to Ann. "As I pinned I wondered when Dad would receive his fourth star knowing it would be soon. Even though I was told this duty was a career ending, I knew Dad would come out some way, somehow with a fourth."⁹

Kerr had brought aboard a bottle of champagne, but United had already gotten behind the idea and threw a big party with champagne for everyone on the flight and a special cake for the new vice admiral. In a "Dear Clan" letter, Bud described the moment: "We left continental

United States and immediately after take-off, high in the air over the Pacific, amid bubbling champagne glasses and happy cheers, in accordance with orders, executed my promotion to Vice Admiral."[10]

Three days of briefings from navy staff at Pearl Harbor were followed by a short holiday at the home of Al and Betsy Toulon on Kauai. Bud and Howard were struck by one particular omission during their briefings—they were getting a series of overviews involving blue-water carrier operations in Tonkin Gulf and Rolling Thunder air strikes against targets in North Vietnam, but there was little to no mention whatsoever of an in-country war for the navy.

The entourage next flew to Clark Air Force Base in the Philippines, where Mouza and the two girls settled into their quarters and registered for school. Bud and Howard were next off for their survival training in the Philippine jungles.[11] In a letter to his family, Bud described the training: "Ate lunch of fresh water shrimp, fish, wild game hen and wild coffee. Started fire without matches. Dishes cut from bamboo."[12] Bud found it a "fascinating experience," entering the jungle with nothing but salt and a pouch of rice, where "the Negritoes" taught him jungle survival. He thought it all "great fun and a piece of cake. . . . I thought it was being the Eagle Scout all over again."[13] Kerr offered a slightly different account. "The program was to set you down in the middle of the jungle. We had a machete, and that was it. We had Maguire, a small Negroid Filipino. And the admiral and I proceeded to survive in the jungle with this young man's skills. Without him I don't know what we would have done. We found that you could make coffee from nuts that fell off trees. That you could reach in amidst a tangled bunch of vines and find one that had fresh water in it. We found how to fish, without poles or string, in streams."[14]

By September 21, it was time for the final leg of Bud's journey. During the flight to Saigon in an Air Force T-39, Bud was reviewing briefing books and thinking about his upcoming meeting with General Creighton Abrams, commander of U.S. Military Assistance Command, Vietnam (MACV), who less than five months earlier had relieved General William C. Westmoreland. In the wake of Tet and Johnson's March 31, 1968, withdrawal from the presidential race, Bud shared Paul Nitze's view that the war was now lost politically, making it impossible to prevail militarily. It was only a matter of time before the political leadership would require a plan for disengaging the United States without damaging its interests in the region and upsetting the strategic balance.

Kerr thought his boss looked especially pensive and inquired what he was thinking. The presidential election was just two months away. If Hubert Humphrey won, Bud thought he might have as little as six months to one year to arrange the navy's exit. If Nixon was elected, he anticipated having as many as eighteen months. The primary mission of his command was likely to involve turning over the naval in-country assets to the Vietnamese, yet not a single briefing had mentioned the Vietnamese navy or the in-country war.[15] Vietnamization had yet to become the watchword of U.S. policy, yet Bud was already conceptualizing a plan for shifting responsibility to the Vietnamese. This was the "first time that I ever heard him express what he saw as his primary mission in Vietnam—to pave the way to get out," recalled Kerr.[16]

Arriving at Saigon's Tan Son Nhut Air Base on the morning of September 22, the youngest vice admiral in the history of the navy turned to his flag lieutenant. "Well, Howard, this is day one. Let's get on with it."[17] Bud Zumwalt was going into an area in which the two flag officers preceding him had not been able to achieve anything significant, either for themselves personally or for the navy.[18] Descending the airplane stairs, Bud eyed Admiral Kenneth Veth on the tarmac in his dress whites, a stark contrast with Bud's khakis. Exchanging common courtesies and small talk, the two men had little else to say. The awkwardness was undoubtedly fueled by Bud's third star.

The two men went from the tarmac to see General Creighton Abrams. Bud had met Abrams several years earlier when the then major general had served as the army representative of the Department of Defense task force for Cuba contingency planning. At the time, Bud was the assistant secretary of defense's ISA representative to the group. Neither could say they knew the other in any real way. Abrams kept the two admirals waiting for about fifty minutes, which seemed like hours as Bud struggled to make conversation. When Abrams entered the room, he walked directly to shake Bud's hand, totally ignoring Veth.

"I was present at the first meeting," recalled Kerr. "General Abrams walked into the waiting area, invited Zumwalt inside, leaving Veth and me outside."[19]

Bud benefited enormously from the fact that Abrams had just received a letter from his close friend, incoming secretary of the navy Paul Ignatius, advising that "one of the Navy's brightest stars would soon be arriving."[20] That was good enough for Abe to give the new arrival a chance, although he was inwardly steamed about not being consulted on the appointment. Abrams believed the navy had been useless in its strategic and tactical contributions to the in-country war. Admiral Veth had been dropped from the Saturday weekly intelligence estimates updates (WIEU) because he had nothing to contribute. "He was not putting any chips on the table," recalled intelligence chief Earl "Rex" Rectanus. "If you're not putting any chips on the table, then you can't play."[21] Abe did not know what needed to be done, but he knew the Brown Water Navy was missing the war. "I don't know what you are going to do, but I know that what you're doing now stinks," Abrams said to Bud.

Abe seemed completely comfortable letting Bud figure it out, or not. Abe's message was loud and clear. Bud could attend the WIEU meetings, but as soon as he stopped producing, he'd lose his place at the table. Bud understood that he needed not only to devise a plan for transitioning the navy out of Vietnam, but also to develop a strategy for getting the navy into the brown-water war. The army and MACV dominated the landscape in Vietnam because they saw it as their war, a war in which their soldiers were dying. Until he convinced General Abrams that he was there "to kill VC, to make some offensive moves," he was not going to have cooperation from MACV.[22]

The change-of-command ceremony was scheduled for September 30. Paul Ignatius was arriving in Saigon a few days later, and Bud was expected to brief the new secretary on his plans for the Brown Water Navy. These first days in Saigon had been sobering. "We came away from those days with a clear understanding that the staff in Saigon was cut off from the forces in the field," recalled Kerr. "It was just a sleepy, large, moribund staff which had fallen into a static pattern of reading message traffic."[23] Tennis matches and dinner dates were more important than fighting the war. In general, the staff came to work with no sense of urgency, planning their days around social events and five p.m. martinis. "Admiral Veth was a loser," said Bud in another moment of complete candor. "He was out there having a gay social time and was not invested in the war itself and his concept of the war was to keep his troops from getting shot at."[24]

Bud had much in-country homework to do, beginning with a tour of South Vietnam's major naval operational and advisory groups. He needed to meet as many component commanders and army commanders in the Mekong as possible. He also thought it best to get away from Admiral Veth until the change of command. In this case, the combat zone offered sanctuary because Bud knew Veth would not be there. After just two days in Saigon, a whirlwind tour of South Vietnam began, including twenty-seven individual flights around the country, from Da Nang near the DMZ to An Thoi on Phu Quoc Island off the coast of Cambodia.

Bud learned quite a bit during a meeting with his component task force commanders, Captains Robert Salzer, Arthur Price, and Roy Hoffmann. Each of these extremely capable field commanders had been yearning for an interdiction campaign that would hurt the enemy.[25] Salzer commanded Task Force 117, the joint army-navy Mobile Riverine Force involved in search-and-destroy missions. Art Price commanded Task Force 116, the River Patrol Force, whose mainstays were the patrol boats, river (PBRs), involved in severing enemy supply lines as part of Operation Game Warden. Roy Hoffmann commanded Task Force 115, the Coastal Surveillance Force, primarily engaged in Operation Market Time, a combined U.S. Navy and South Vietnamese navy effort by PCF Swift Boats to stop the flow of supplies from North Vietnam into South Vietnam.[26] Bud needed each of these men, promising promotion if they stayed with him for the battle ahead.[27]

Basically the naval strategy in-country from 1964 until 1968 had involved two relatively predictable and static patrol functions: Market Time (TF 115) patrol of Vietnamese coastal waters by Swift Boats and

Game Warden (TF 116) patrol by PBRs of major waterways, like the Mekong and Bassac rivers on a daily basis, searching for smuggled communist weapons in watercraft as small as sampans and as large as coastal trawlers.[28]

With its many crisscrossing rivers and canals, the Mekong offered a particularly appropriate area for naval riverine and interdiction operations. But what type of plan could be developed quickly and get the buy-in of so many disparate task force components? It was obvious to everyone on the ground that there were navy resources not being used. With the advent of General Abrams as MACV commander in 1968, the overall strategy changed to clear and hold rather than search and destroy.[29] Within that context, Bud sought to develop tactics that were more proactive in a strategy that took the fight to the enemy, interfered with their substantial logistics support, and rolled them back, holding the cleared territory. He would accomplish this by creating barriers and bases near the Cambodian borders and redeploying some Market Time assets to riverine interdiction.

The change-of-command ceremony aboard the USS *Garrett County*, moored at Pier Bravo in the Vietnamese Naval Shipyard in Saigon, occurred on Monday morning, September 30, 1968. After brief farewell remarks, Rear Admiral Veth turned to the commanding officer of the *Garrett County*, ordering, "Captain, haul down my flag." Turning to Admiral Zumwalt, Veth said, "Sir, I am ready to be relieved." Bud Zumwalt read his new orders, faced his predecessor, and replied, "I relieve you, sir." Vice Admiral Elmo R. Zumwalt, Jr., had now assumed command of the navy's operational forces and advisory teams in Vietnam, a command that included the coastal surveillance and gunfire support forces of Operation Market Time, the river patrol forces of Operation Game Warden, and River Assault Flotilla One, consisting of the navy elements of the Mobile Riverine Force and operations of the Naval Support Activities, Da Nang and Saigon. "On September 30 our lives changed forever," recalled Rex Rectanus. "The days of established routine were gone forever and the ZWIs (Zumwalt Wild Ideas) were here to stay."[30]

In the first of many letters he would write to Mouza about Bud's daily activities, Flag Lieutenant W. Lewis Glenn explained, "All went smoothly at the change of Command yesterday. Admiral McCain and General Abrams both had good things to say about the Admiral's past record and his new command."[31] Lew joked about his new boss being able to get back to quarters in time to watch television coverage of the ceremony. "Even though the television coverage started with Admiral McCain's picture and Admiral Zumwalt's voice, it all evened out in the end." Glenn comforted Mouza by saying that his boss hoped to get to the Philippines on the twenty-second in order to celebrate the couple's twenty-third wedding anniversary and that the stewards had already been briefed on the admiral's diet.[32]

In his first message to in-country naval forces, Bud relayed that he had just relieved Admiral Veth as their commander. "During the past week I had the opportunity to visit with many of you and observe personally the professional performance in your assignments under difficult conditions. The results of your efforts are plentiful and evident . . . the job that remains, like the job that has been performed, is enormous."[33] Few reading this first message could have possibly conceived the magnitude of the changes on the horizon.

Bud understood that Abrams's offer to sit at the table depended on having a product to sell. There was much work to be done, beginning with a radical makeover of the staff. Drawing on his years of BuPers experience and taking advantage of the twelve-hour time difference between Saigon and Washington, Bud was on the phone each evening at nine p.m. in Saigon, nine a.m. in Washington, demanding that detailers send him an A team, "people that I knew specifically had the talent, and then calling them and saying, 'you've just been volunteered, don't veto it.' "[34]

It was not necessarily a case of Veth's holdover staff not wanting to do a better job. The truth of the matter was that the low caliber of senior staff officers at the commander and captain level was a direct result of detailers not seeing brown-water service as career enhancing. "It was clear that they'd been sending a lot of the dregs out there," said Bud. "People that were clearly not going to go anywhere and clearly not highly motivated and ambitious."[35] Bud needed a staff that could respond at a higher level. With two notable exceptions, the holdovers were simply incapable of doing so.

The two exceptions were W. Lewis "Lew" Glenn and Earl F. "Rex" Rectanus. Lew had been serving as Veth's flag lieutenant and immediately appreciated what Bud was asking of him. "There were guys that either could respond to his fast pace on things, or you couldn't," recalled Glenn. "I think, in general, you were either fast on the uptake and could move with him, or you just weren't capable of moving at the motion and at the level that he wanted you to move."[36] Captain Rex Rectanus, assistant chief of staff for intelligence for the U.S. Naval Forces of Vietnam, thought that "as long as you were loyal and you tried your hardest to do what he wanted, he didn't care if you were a holdover or if you were somebody new."[37]

Bud's initial ministaff in Vietnam consisted of Kerr, Rectanus, Glenn, Captain Dick Nicholson, whom Bud had known as a lieutenant commander in the office of the secretary of the navy when Bud was Paul Nitze's executive assistant, and Captain Charles F. "Chick" Rauch, an engineer and nuclear submariner who had worked in Systems Analysis. They were soon to be joined by Captain Emmett Tidd, whose command of the destroyer tender USS *Everglades* was cut short by urgent orders to report for duty in Vietnam as aide to Vice Admiral Zumwalt.[38] BuPers had originally charted Rauch as chief of staff and Tidd as senior naval advisor to the Vietnamese. Once on the ground, however, Rauch and Tidd requested that Bud allow them to swap assignments because each felt like square pegs in round holes.[39] As chief of staff, Tidd worked tirelessly to impose order and clockwork precision to the staff operations.[40] His "green stripe" attached to a document meant Zumwalt wanted immediate action. Rauch made extraordinary contributions working with the South Vietnamese navy and with Commodore Tran Van Chon.[41]

The senior staff officers roomed together in the admiral's large villa, so their social life was a working one, bringing in for working dinners people from the field who had good ideas. Flag Lieutenant Bob Powers, who would replace both Kerr and Glenn, found that no one ever demanded so much in terms of time, energy, and substance but made you enjoy it so much.[42] Their day began at 5:30 a.m., with Bud leading a two-mile run around the compound. Bud had started jogging at the age of thirty-seven when assigned as executive assistant to the assistant secretary of the navy. He had been diagnosed with a spastic colon. Doctors attributed the condition to long work hours, high tension, and little exercise. Bud started running each morning, running not jogging, two miles in under twelve minutes and in later years, fourteen minutes. Indeed, at the age of seventy-eight, he experienced unusual shortness of breath after a 5K run in which he finished first in his age division and was diagnosed with his pleural mesothelioma. "I found that this brief physical exercise, taken at a time of the day when I could always count on getting it in, 6 o'clock in the morning, and almost never skipped, provided the necessary compensation for the long hours and high-tension jobs in which I continued to serve and was never again troubled with this spastic phenomenon."[43]

Running each morning did lead to one particular dilemma when Bud was serving as CNO. He had been invited to Houston for a dinner in honor of former governor John Connally. He was staying at the home of a friend, Rear Admiral Charles Howell, in Houston. As was his routine, he went out for an early-morning run the next morning, running mentally as well through a number of things he wanted to get done that day. Bud got so involved in the thought process he lost his way. Unfamiliar with the area, he saw a home being constructed where a worker had just pulled in. He told the man he was lost and asked if he knew where the Howells lived. The man didn't, but invited Bud into the construction trailer to look up the number in the phone book. As the man handed him the phone book, Bud realized he didn't have his reading glasses with him. He asked the man if he could look it up. This got a bit of a stare, but he did so. The Howells were unlisted. The man asked if Bud knew anyone else in Houston. He said only Governor Connally. That number was unlisted as well. So Bud called the operator and made a collect call to his aide Dave Woodbury at the Pentagon. "Dave, no smart remarks, just tell me the address of where I spent the night last night." Dave responded, "Admiral, I can tell you where you were supposed to have spent the night last night," and gave his boss the address. The construction worker took a look at the address and told Bud it was about a mile away. He offered to drive him there. During the ride, the man kept glancing over and finally asked, "What do you do for a living?" Sheepishly, Bud replied, "I run the U.S. Navy." The man, somewhat in disbelief, said, "Huh, you can't read, you don't know where you are, and you run the U.S. Navy?"[44]

The morning run in Saigon was followed by breakfast and morning briefings, which covered the previous night's activities, firefights, wounded, and casualties. If it had been really bad, Bud and an aide would be on their way to the airport by 6:30 for a flight into the delta in order to visit an area that had been hit the hardest the night before. Each morning Kerr or Bob Powers would read the casualty reports to determine where the VC had hit or where the firefights had been. After visiting the field, they would return to headquarters for additional briefings and paperwork. Afterward, Bud almost always worked the phones back to the States. "I guess the thing that you have to say about him is that when you can work with somebody seven days a week, 16, 18 hours, and you love that guy, he's got to be something special, because a situation like that where you never get away, could really grate on you if you weren't working for a very special human being," said Lew Glenn.[45]

Secretary Ignatius's arrival in Vietnam was scheduled for October 5, 1968. Bud tasked Rex Rectanus with preparing the briefing, knowing that they were already on the same wavelength. At his first meeting with Bud, Rex pitched his analysis of the enemy's logistics system, showing that Cambodia was a major supply depot. The enemy was thriving in the smaller waterways and canals of the delta. Rex offered a bold interdiction plan. "It did not take long for ADM Zumwalt to realize that the present passive/defensive strategy had no strategic validity," said Rex.[46] Bud and Rex sought to change the navy's in-country role to a more aggressive one, and they did this by reorganizing the naval forces to be more dangerous to enemy operations around the clock.

On October 14, 1968, just two weeks following the change of command, Lieutenant Junior Grade Michael Bernique violated the rules of engagement (ROE) by driving his Swift Boat at high speed on the serpentine Rach Giang Thanh River, a bit south of the Vinh Te Canal, which connected the Gulf of Thailand to the Mekong River along the Cambodian border.[47] The river winds back and forth, usually completely within Vietnam's territory, but in places the left bank is actually in Cambodian territory. Bernique had been in Ha Tien, a forward location that was being used as a base for supporting Swift Boat operations close to the Cambodian border. A friendly informant had passed on information that the VC had set up a tax collection station a few miles up the Rach Giang Thanh. The ROE strictly forbade Swift Boats from operating so close to Cambodia, but Bernique decided to investigate. When he discovered the collection site, Bernique opened fire; three VC were killed. After regrouping, the VC returned, and a fierce firefight ensued, resulting in two more enemy deaths and the recovery of supplies, weapons, ammunition, and documents left behind by the fleeing Vietcong.[48]

Because of the neutrality of Cambodia and the sensitivity of U.S. forces waging war on or near that neutral country, the navy had remained a healthy distance from the Cambodian border—three to five miles. Bernique was called to Saigon to face disciplinary actions, including a possible court-martial. Bud sent instructions that he wanted to meet personally with Bernique. During the interrogation, one of Bud's aides said to Bernique that Prince Sihanouk of Cambodia had accused him of shooting innocent children and civilians. "Well you tell Sihanouk he's a lying son-of-a-bitch," replied Bernique. Instead of a court-martial, Bud awarded his sailor the Silver Star. "When I got through talking to him, I decided first that he'd brought very valuable information, as had Horatio Nelson when he decided to cross the T in violation of orders." Bernique was "the kind of captain we need more of."[49]

Rex loved having an aggressive commander, but in this case disagreed with condoning and decorating Bernique, who had gone into the river without adequate intelligence information. "Zumwalt was damn aggressive and also terribly bright," mused Rex. Bernique was just damn aggressive.[50] Bernique's trip fundamentally altered naval strategy at the border. Bud later sent him with a flotilla of Swift Boats to transit the entire Vinh Te Canal, from Ha Tien to Chau Duc on the Bassac River.

Bud asked Captain Bob Salzer to meet with him in Saigon, ostensibly to seek his field commander's opinion on putting an operation into the canal about half way up the Cambodian border. Looking at a map of the area, Zumwalt saw an opportunity for waterborne interdiction of supplies in the Parrot's Beak coming out of Cambodia, with the Vam Co Dong and Vam Co Tay rivers running along Parrot's Beak. Pointing to the V along the Parrot's Beak, Bud said, "We can interdict it here, and then there's a canal running along the other border of Cambodia, where we can put those Florida boats, the outboards with the airplane motors on the back of them and use them in the areas where it's too shallow for the PBRs."[51] When he worked as Paul Nitze's executive assistant from 1962 to 1965, Bud had been involved in assembling the navy task forces. He and Nitze persuaded Treasury Secretary Fowler to give them Coast Guard WBPs that had been under the control of Treasury.[52] "Well, it's a breathtaking concept but I think it will work," said Salzer. Zumwalt was convinced that he could go all the way up the Vinh Te Canal for a strategic interdiction campaign that had never been done before, to be called Operation Search Turn.[53]

Two weeks following Bernique's brazen operation, Zumwalt called a meeting at the NAVFORV (Naval Forces Vietnam) headquarters conference room. The time had arrived to put his chips onto the table and get the navy into the war. The first signs that something was brewing arrived in a message from COMNAFORV (Commander Naval Forces Vietnam) headquarters directing that either the field commander or operations officer report to Saigon as soon as possible. Captain Art Price of TF-116 was on temporary duty in Coronado, California, and his replacement, Commander Wayne Beech, ordered Lieutenant Commander Tom Glickman to Saigon. Glickman caught the first Thai Air Force C-124 to Saigon and immediately after landing went directly to Bud's residential compound. "As I entered, the Admiral was exercising by rapidly going up and down stairs to the second floor," recalled Glickman, who was nursing a scotch and water while waiting. "Essentially he wanted to get the Task Forces working together on common goals to crack the hard nuts. He knew that we had been doing well in our respective general areas of operations, but he also surmised we had our own ideas about other areas we could and should expand into but had not been able to do so because of resource problems. His idea, integrated operation, could solve many of the resource problems."

Their conversation lasted for about an hour. Bud instructed Glickman to return to Binh Thuy, consult with his task group commanders, and come up with a shopping list for future operations. "He specifically directed me not to mention anything of our conversation to anyone on the NAVFORV staff."[54] Returning to the delta, Glickman went directly to headquarters in order to brief Wayne Beech. They were joined by Duane "Dewey" Feuerhelm, the intelligence officer; Bob Peterson, who ran the river squadron at Binh Thuy; Jack Elliot from Vinh Long; and the squadron commanders from My Tho and Nha Be. They started drawing up their list.

The clarion call from COMNAVFORV came soon thereafter with notification of a meeting at Naval Forces of Vietnam headquarters in Saigon on Saturday, October 26, 1968.[55] All three task forces were represented. From TF-116 were Wayne Beech, Dewey Feuerhelm, and Tom Glickman. TF-117 was represented by Captain Robert Salzer and his immediate staff; TF-115 was represented by Captain Roy Hoffmann and immediate staff. Rex was also present. "We sat at the conference table looking at each other," recalled Glickman. "Although we were all part of the same Navy, we really had never operated together cooperatively."

Bud entered the room and said he wanted to have the three incountry operating forces work together toward common goals to interdict VC supply lines and the VC/NVA themselves. He told them that the operation would be called SEALORDS, which stood for SouthEast Asia Land Ocean River Delta Strategy. Noting that "brown-water warfare was unsupported and likewise unencumbered by established doctrine," Bud asked them to use ingenuity and improvisation. "You have to make up riverine warfare as you go along."[56] He wanted a new game plan based on the principle that the enemy needed to be kept off balance and the best way to do this was to "keep changing the game plan. . . . You can get away with anything once or even twice, but you must change strategies frequently in order to keep the enemy from exploiting you."[57]

In October 1968, General Abrams was focusing on the dry-season campaign covering the entire delta. Bud understood that this was his opportunity to step into the game with multiple river raids and destruction of VC supply routes as a precursor to the major army campaign in the delta. He envisioned SEALORDS beginning with Swift Boat raids into VC sanctuaries, disrupting their logistical effort throughout the delta. "His words were like a breath of fresh air!" recalled Glickman, whose mind immediately went back to the Tet Offensive, when they had planned a joint TF-116/117 operation against Tan Dinh Island.[58] "I remember Bob Collins, my TF-117 counterpart, telling me, "Keep your god damn plastic boats out of the way of my boats. If not, we'll blow you and your silly black berets out of the water like we would the VC."[59] Bob Collins was now part of the working group, sitting at the same table as Glickman.

Before excusing himself so that the group could work freely in developing an operation plan, Bud urged his commanders to put aside their old competitiveness and animosities. Their brainstorming should be directed toward innovative ways of interdicting enemy infiltration routes whenever riverine forces found them. In his parting remark, Bud said he wanted a list of priorities and resources needed to accomplish the mission.

What especially attracted Rex to his new boss was this type of encouragement to challenge assumptions. "One of Bud's first activities upon arriving in Vietnam was to inaugurate a system of thought-stimulation," wrote Rex. Zumwalt Wild Ideas became the commander's invitation to one and all to be constantly innovative.[60] Bud called them ZWIs because he did not want staff to see them as directives. "We had a saying: 'Be careful what you propose to the Admiral because he might approve it and then you'd have to do it,'"[61] recalled Bob Powers. Chick Rauch offered another way of looking at it: "Bud was as good as anyone I know at nurturing creativity. I cannot remember him ever saying no to anyone who had a good idea."[62] As Bud would often say, "I have no prejudice against anything that makes sense."[63] In many respects,

Vietnam was a perfect environment for this type of thinking, because there was no real bureaucracy or political strata in OPNAV for air admirals, submarine admirals, and surface admirals to lobby in Congress against the idea. "You didn't have that in Vietnam."[64]

Some wild ideas were wilder than others. One was to airlift PBRs over to the rivers in Laos. "So we worked for about two weeks putting together a great deal of data on the subject," recalled Rex. "We finally convinced Zumwalt that to land a bunch of PBRs in the middle of the enemy, with no basing and that sort of thing, was questionable."[65] The process also encouraged staff dissent. "A great thing about Zumwalt, one might say, is that you could almost literally pound your fist on the desk and say 'Admiral, you can't do this. It's wrong. I don't believe what you're doing. You shouldn't do this.'" The staff also learned that they had better have their facts right, because "he could destroy your arguments in rebuttal, but you could literally tell him he doesn't know what the hell he's talking about, and he doesn't hold it against you," recalled Rex.[66] Bob Salzer offered a similar perspective: "The fact that I disagreed with Zumwalt on many occasions was amply documented. Once in a meeting I said, 'Could I differ with you?' He looked up and he said, 'I wouldn't know how to conduct myself if you didn't.'"[67]

Back in Naval Forces of Vietnam headquarters, it was time for another ZWI. After Bud left the room, Bob Salzer ran the meeting. The group went through a process whereby each task force identified its priorities and the resources needed for accomplishing its goals. At 2:00 p.m., Zumwalt returned for a briefing led by Salzer. Each task force was given the opportunity to explain and defend its recommendations. Bud asked a few questions, quickly making the point that this was exactly what he wanted. Before leaving the room, he said, "Write the operation plan!" Salzer tossed Glickman a legal pad and said, "Start writing."

Pacing around the U-shaped conference table, Salzer slowly dictated a basic operation plan in a point-by-point paragraph format: situation, mission, execution, administration, logistics, and finally, command and signal. Those six paragraphs are the basic plan and do not include annexes and other supplemental items. When Glickman developed writer's cramp, Rex relieved him as note taker. The operation plan came into being in about nine hours. The handwritten draft was sent directly to the communications center and prepared in message format. Later that day Admiral Zumwalt personally carried it to General Abrams for approval.[68] "Thus, we were there for the birth of COMNAVFORV OPLAN 111-69," wrote a proud Tom Glickman.

With their work completed, Glickman, Wayne Beech, and Dewey Feuerhelm returned to Binh Thuy in order to prepare a briefing for Captain Art Price, who was en route from San Diego. Glickman recalled, "We thought of opening with, 'Oh, by the way Boss, while you were out of town, Admiral Zumwalt had this idea, and . . .'" In fact, no other ZWI would have a greater effect on the navy's in-country forces than Operation SEALORDS. At last, the United States Navy's three operating forces in Vietnam combined their assets and capabilities and cohesively started operating together to attain common goals. The operational merger of the three in-country task forces offered a deltawide approach to control of the waterways—essentially a plan to move coastal Swift Boats and river patrol boats into the canal system of Vietnam in order to interdict shipments moving from the Cambodian border throughout the delta and to attack the enemy in previously safe sanctuaries.

The backbone of SEALORDS was the heavily armored riverine craft, the Swift Boats (PCFs) and the river patrol boats (PBRs). Four campaigns followed between November 2, 1968, and January 2, 1969—Operations Search Turn, Foul Deck, Giant Slingshot, and Barrier Reef all established interdiction barriers that successfully reduced the flood of goods from the north to barely a trickle. Giant Slingshot and other border interdiction campaigns, like Tran Hung Dao (named after the patron saint of the South Vietnamese navy), along with Barrier Reef, formed an interdiction barrier that stretched from Ha Tien on the Gulf of Thailand to Tay Ninh north of Saigon. General Abrams credited these operations with causing major disruption of the enemy's infiltration effort. East of Saigon, Operation Ready Deck used fiberglass PBRs to patrol areas of heavy VC activity. On the west coast of the delta, Search Turn and Breezy Cove provided arterial interdiction of enemy movements into the Three Sisters and Ca Mau base-camp areas. Concurrently, Operation Sea Float sought to pacify the Nam Can district, located 150 miles southwest of Saigon in An Giang Province, a haven for enemy arms shipments arriving by sea. The best defense in the area was the Cua Lon River, which bisects the peninsula. Into the middle of this swift, four-hundred-yard-wide river, the navy towed a complete floating support base built on pontoons, called *Seafloat*, and anchored it there. The base was sandbagged and heavily armored, with American and South Vietnamese ships protecting it. The presence of this base, with its Swift Boats and armored craft, so opened the Ca Mau peninsula that refugees soon resettled the area and profited from the lucrative woodcutting trade. This pacification process was seen by Bud as one of the most important keys to the success of a free South Vietnam. "As this effort began to clear out VC sanctuaries, we had refugees returning to the banks of the rivers. Instead of doling out bullets, the PCFs and PBRs gave the refugees water, medical assistance, and even delivered several babies aboard their craft."[69]

Commodore Bob Salzer was given the title of first sea lord of Task Force 194. "He did a magnificent job in pulling his assets together and quickly and efficiently moving the support ships and raiders into position," wrote Bud in a letter to his family.[70] The larger PCFs were taken from purely offshore coastal patrol and integrated with the navy's river divisions of PBRs and other larger craft. These craft were then sent on east-west patrols across the Mekong Delta from the Saigon River to the Gulf of Thailand on the canal system built by the French on the Vam Co Dong and Vam Co Tay rivers of the Parrot's Beak, and through the Song Ong Doc, Cua Long, Bo De, and other rivers of Ca Mau Province in the southern tip of Vietnam. Immediately upon commencing these patrols across III and IV Corps, the navy began making aggressive contact with the enemy units heading south or east to supply munitions and troops to the delta. Rectanus believed that had Zumwalt been in Vietnam to implement SEALORDS in 1966, "I'm convinced that the whole outcome of the war, certainly in the Delta, in III Corps, would have been entirely different."[71]

Lieutenant Bernique was soon given the honor to lead a mission, known as Operation Foul Deck, that became a key part of SEALORDS. From that point forward, the Rach Giang Thanh became known as

Bernique's Creek. Patrols were soon augmented throughout the length of the Rach Giang Thanh, from its northeastern head along the Vinh Te Canal all the way to the western bank of the Bassac River. Interdiction operations included not only Swift Boats, but also PBR units of the navy's Mobile Riverine Force and units of the South Vietnamese navy's River Assault Group 26.[72]

The plan for transforming the strategic use of naval capabilities in a land environment spawned other innovative programs. SEALORDS required a reconfiguration of the naval intelligence program supporting naval forces. The naval intelligence liaison officer (NILO) program had previously positioned a small number of naval officers as support intelligence officers for the tactical commanders who were performing the static patrols of Market Time and Game Warden. These officers reported directly to the tactical commanders, and their duties were at those tactical commanders' whims, so commanders missed the importance of those intelligence reports to the strategic "big picture" of enemy infiltration and operations in the delta.

Rectanus persuaded Bud to reconfigure the entire NILO program to flexibly position these naval intelligence officers as SEALORDS naval war, intelligence, and combat operations demanded.[73] In order to do this, the NILO program was restructured to place each of the NILOs on independent duty, reporting directly to Bud and Rectanus, supporting and answering direct strategic intelligence questions in addition to providing support to the naval assets and operational commanders. A reinvigorated naval intelligence program, which produced useful strategic and tactical intelligence, also fulfilled General Abrams's demand to the navy to bring chips to the intelligence table if it wanted to play.

Along the Cambodian border, where PCFs and PBRs set up nighttime ambushes to intercept enemy munitions and troop infiltration, the NILOs' reconnaissance observations and reports from their local intelligence networks provided information about potential enemy movements each night. The NILOs would brief the boat crews going out on night ambush as to where the likely crossing points and reported threats were.

On the staff side of the NILO program, Zumwalt encouraged initiative by emphasizing the personal side of intel. Whenever the navy took a casualty, Bud would ask, "Could this have been prevented?" Staff would then search to see whether they had overlooked a warning or report that could have made a difference. Bud's attitude meant that even those who wanted to cover their tracks in the intel process couldn't just sit on something. Information kept flowing with positive or negative or even neutral recommendations and comments. As a result, NILOs felt someone back at headquarters would pay attention to and read carefully any report they sent and would make sure the appropriate people would see it. They understood that their reports would not disappear into some black hole.

Besides obtaining reports from local agent networks, NILOs rode boat patrols to gain local intelligence. NILOs stationed at army air bases along the border flew visual reconnaissance flights—sometimes three or four times a day—to locate large enemy troop movements along the Cambodian border, the Rach Giang Thanh River, the Vinh Te Canal, and the Vam Co Tay and Vam Co Dong rivers, which formed the Parrot's Beak where the Cambodian border intruded close to Saigon. In the Parrot's Beak area, North Vietnamese army (NVA, or the Vietnam People's Army) main-force units in large numbers regularly crossed into Vietnam and could be seen in plain sight by air reconnaissance both day and night, then would come under fire from American and South Vietnamese boats and aircraft.

In the strategic intelligence area, NILO reports identified main-force NVA units and large enemy munitions storage and crossing points, allowing Bud and his staff to see a detailed picture of what the enemy was doing and to decide how to respond by redeployment of patrol forces. At the height of the Zumwalt tenure in Vietnam, there were no more than twenty to twenty-six NILO posts at one time. These NILOs were naval officers from very different backgrounds.[74]

All these officers, especially the line-officer NILOs, came from a naval tradition of independent command and direct operational action. The appointment as NILOs of naval officers in their twenties who came from this tradition of operational independence was in many ways a risky proposition. It was entirely conceivable to Bud and Rex that young men in their twenties could make potentially disastrous mistakes along the border of neutral Cambodia, which would have explosive consequences for them and the president they served. Nonetheless, they took this calculated risk and put these young men on independent duty as NILOs, and their bet paid off.[75]

Along the Cambodian border and the Song Ong Doc–*Seafloat* areas, the NILOs participated in direct combat action against crossing enemy units, in combat provocations of the enemy, which allowed heavyweight U.S. reprisals at and across the Cambodian border in hot pursuit, and in covert direct action with SEAL (sea, air, and land) teams, LDNNs (Lien Doc Nguoi Nhia, or Vietnamese navy SEAL teams), Kit Carson Scouts (Vietcong defectors used as scouts), Provincial Reconnaissance Units (PRUs, South Vietnamese–CIA direct-action teams), and the U.S. and South Vietnamese naval patrol units. In 1969 and 1970, these NILO combat activities actually cleared the Vietcong presence from the Ha Tien Peninsula at the western end of the Cambodia-Vietnam border. On the Vam Co Tay and Vam Co Dong rivers, there were numerous detections and decimations of large enemy crossing units, resulting in hundreds of enemy casualties and a rollback of VC units. On the Song Ong Doc River, naval activities allowed the establishment of the first allied land base there during the war. NILOs also participated in Bright Light POW rescue missions throughout the delta. NILOs ran intelligence-agent networks of the navy's Blackbeard Collection Plan Collection Team 5, and one reported the first sighting of North Vietnamese main-force units entering Vietnam's Mekong Delta at the west end of the Cambodian border.

The Zumwalt NILOs and Collection Branch staff officers went even further. A NILO was sent on special covert missions to the port of Sihanoukville, Cambodia, to undertake a naval port survey and surveillance to determine how to destroy the port if the new Lon Nol government changed allegiances and restarted enemy munitions supply through the port, to discover a Soviet eavesdropping communications-intercept site in Cambodia, and to negotiate a secret U.S. weapons supply agreement with the Cambodian navy.[76] NILOs were sent by Bud on temporary additional duty to act as intelligence officers for Market Time units participating in the Cambodian incursion along the Gulf of Thailand and as support for the naval invasion of Cambodia up the Mekong River to Neak Luong, Cambodia. Additionally, NILOs supported MACVSOG (Military Assistance Command Vietnam Studies and Observations Group) and its insertion teams, the "Roadrunners of the Ho Chi Minh Trail." NILOs were called upon regularly in the field to brief admirals and generals from high commands on the intelligence picture in their AOs (areas of operation).

The NILO program was innovative and adaptive within an aggressive strategy that resulted in the rollback of enemy forces in their areas of operation, especially in 1970 along the Cambodian border and in the Nam Can district.[77] The effectiveness of these programs undoubtedly contributed to the decision by the Vietcong to assassinate "the admiral in command of the U.S. Navy in Vietnam."[78] The attack that took place on May 11, 1969, had been planned for weeks by Trai Hai Phong and his covert team of assassins operating with near impunity in Saigon. Assisted by an informant working as a cook within the compound, who had been carefully monitoring Bud's daily routine, a plan was devised to throw a satchel charge over the thick seven-foot-high masonry wall of the compound during the admiral's daily noontime volleyball game with his staff. The court was just inside the wall, adjacent to one of Saigon's busy streets. The satchel hit center court, and the explosion rocked the compound, but the target had been called away for a meeting at MACV headquarters and the game had been canceled. The satchel charge exploded on an empty court.

One event secured Bud's seat at General Abrams's table. Early on the morning of Saturday, November 2, 1968, a number of senior officers and their aides assembled in the main conference room of the MACV headquarters. The air force and the navy were scheduled to brief General Abrams on how they were going to turn things over to the Vietnamese—to present Abrams with preliminary briefings that included their assessments of the time and resources needed to train and equip the Republic of Vietnam Armed Forces (RVNAF) to take an expanded role.

Howard Kerr and Bud took their seats along the wall behind the long, horseshoe-shaped table. The only people in the room were flag officers, the MACV chief of staff, and a couple of aides. Bud noticed that Abrams looked haggard and exhausted, probably because he had just returned from a series of meetings in Washington with President Johnson and the Joint Chiefs of Staff in the Pentagon. The president had warned Abrams that he intended to announce a bombing halt in Vietnam, hoping this carrot might induce the Vietnamese parties to start negotiations in Paris. Johnson still expected MACV to keep continuous pressure on the enemy and to get the ARVN (Army of the Republic of Vietnam) to do the same.

The air force briefer went first. "These two Clark Kent looking colonels got up to give the presentation," recalled Kerr. "They were right out of Hollywood casting. Their uniforms were perfectly tailored." The colonels presented slides with the Seventh Air Force crest, thunderbolts, and other impressive symbols to highlight their statistics for turning over aircraft and supporting facilities, logistics, and training programs.

During the air force briefing, Kerr became uneasy, thinking it was really impressive with all those bells and whistles. "Oh shit—wait until the general sees ours," mused Kerr, who had worked late into the previous night preparing handwritten charts for the briefing. "What we had was a flip chart, and either the admiral or myself or both had written what he was going to use as a guide in longhand on these flip charts. As the person who was going to do the flipping and the pointing, I was struck by the fact that they were going to think that we had just fallen off the turnip truck."[79]

Everything came to an abrupt halt when the air force planners stated that they envisioned an eight-year process, with complete turnover accomplished by 1976. General Abrams turned to General George Brown of the air force. "Look, when I ask you to present something to me, I want it presented to me. I don't want that kind of trash." Lew Glenn recalled that Abrams used a few other choice words.[80] Abrams placed his cigar in his mouth, slowly lifted his right fist, and brought it down onto the table with such force that the ashtray in front of him went up in the air, turned over, and landed with a crash. The briefer immediately fell silent. "Bullshit! Bullshit! Bullshit! That's all I ever get out of the fucking air force is bullshit," yelled Abrams. "Everybody was stunned," recalled Kerr. "I was mesmerized."[81]

All eyes were on General Abrams. The officers in the room considered ducking for cover. "Don't you people understand what's happening? Don't you have any sense for the pressure-cooker environment the president is in back in the United States? He has no consensus of support for this war. What support he has is dwindling. It's clear that the policy is to get us out of this war and turn it over to the Vietnamese. That policy change will be implemented by the incoming administration."[82]

Abrams paused for a moment before getting directly to the point. "I've got a letter in my pocket from the President of the United States whose final instruction to me was, 'I want every man in Vietnam who carries a peter mobilized.'" Abrams warned that anyone who did not get onto the same bus with him was going to be dropped from the team or simply ignored. "You are sitting there telling me that it's going to be 1976 before you can get these planes turned over to the Vietnamese? No way. Can't happen. The country won't give us that time. The incoming administration won't give us that time."

Abrams got up and left the room.[83]

At that point, Major General Corcoran, chief of staff at MACV, told Bud he would be better off not presenting his plan while Abe was in such a foul mood. "You want your briefing officer to go on after this? You can have another day or two if you want to," advised Corcoran. Kerr thought his boss looked "like the guy at the poker table who had just drawn the inside straight." He gave a slight smile and replied, "No, General. I'd like to go ahead and make my presentation." Corcoran said, "All right, Bud, if that's what you want to do, I'll go and tell Abe. Let's give him about 45 minutes to come down from the emotional level that he's at right now." Turning to his flag lieutenant, Bud said, "Howard, get the briefing charts!"[84]

Bud had grasped the fluidity of the situation. The commander in chief had told General Abrams that the war was going to be turned over to the Vietnamese. References to 1976 were simply no longer politically viable. Later, after the presidential election, Bud had reached the same conclusion, that the United States was looking for the exit. Public support was melting away. The only question was whether they had one year or three years. With Nixon's election, Bud was banking on three. After seeing Abrams's reaction to the air force plan, Bud instructed Kerr to revise their charts by removing each and every caveat. References to "maybe" and "perhaps" were replaced by the statement "If you give us this support, here's what the Navy can do."

Zumwalt and Kerr returned to face their MACV commander. Zumwalt personally conducted the briefing, with Kerr flipping the charts. Bud was the briefing officer, believing since his days in the Naval Academy in the persuasiveness of his own advocacy. Bud began by describing his plan for an accelerated turnover of equipment and training for the South Vietnamese navy that he called the Accelerated Turnover Plan (ACTOV). It involved a crash course in the English language for Vietnamese sailors, joint operations on the rivers of Vietnam, the rapid expansion of the Vietnamese navy through recruiting, and the turnover of all in-country naval assets to the Vietnamese. About halfway through the presentation, the army senior advisor said, "This is a bunch of bullshit. There is no way YOU can accomplish this. You don't have the resources."

Abrams glared at the major general and said, "Shut up, Goddamnit! He may be digging himself a big hole, but you're already sitting over there in a cesspool. . . . Look, this guy has something, he's the only one that's got a plan and the rest of you get hot." As he flipped through the charts, Kerr was thinking "that the admiral didn't just guess right: he analyzed correctly. He had seen what needed to be done. He had gotten out ahead. . . . The timing was almost perfect."

When Zumwalt was done, Abrams sat for a long moment before saying, "Admiral, I want to know the following. I want to know what you as Commander Naval Forces Vietnam can do on your own with the authority that you have and the assets that you have. And I need to know what support you need within the Navy chain of command to make this happen. I need to know what part of this plan requires the support of the Vietnamese that you currently don't have. And I need to know what I can do as Commander, Military Assistance Command Vietnam with the assets and authority that I have." Abrams asked Zumwalt what was the soonest he could have this information. "Monday, General."

Abrams turned to General Corcoran and said, "I'm not going to hear any more today." He started to leave, but then turned around and walked over to where Admiral Zumwalt stood. Putting his arm around Bud's shoulders, they walked together to Abrams's office. "This showed everybody that I was his boy for the day," said Bud.[85] It was a signal to everyone else in the room. "It turned out to be the thing that changed the whole course of our destiny out there," recalled Zumwalt. "From that point on, General Abrams was a Navy man."[86]

In the car on the way back to navy headquarters, Zumwalt and Kerr focused on developing their plan by Monday. Bud realized that he was in an awkward position, because he needed to inform CNO Thomas Moorer that he had just pledged to give away navy equipment, something Moorer hated to do. In a back-channel message, Bud informed the CNO that "General Abrams had come back from the White House with this stark guidance, and therefore this plan had been submitted." Bud recalled, "Admiral Moorer was furious." He had his assistant call Bud on the secure phone line "and question me whether or not I was telling the truth."[87]

Moorer grudgingly approved Zumwalt's turnover plan. Meanwhile, the staff worked all weekend, and by four o'clock on Monday afternoon, MACV had the ACTOV plan. Bud's relationship with General Abrams blossomed over the next few months and became as close a personal relationship as two flag officers can have. In many ways, the two men were kindred spirits and real war fighters. "General Abrams is a great military captain," wrote Bud in a letter to Nitze. "I'm already willing to place him alongside such historical military geniuses as MacArthur, Lee, Rommel, etc. He understands the multi-faceted nature of this war."[88] Abrams was also sold on Bud, so much so that when Nixon's new secretary of defense, Melvin Laird, made his first visit to Vietnam in the spring of 1969, Abe recommended that he consider Zumwalt for

CNO whenever the next opening occurred. Laird was impressed with the briefing he received from Zumwalt for an accelerated turnover program and communicated this to the new secretary of the navy, John Chafee, asking that "when you are making nominations for the next CNO, give me five names, make sure that Zumwalt is one of them."[89]

By late November, Flag Lieutenant Glenn sent a summary of Bud's activities to the Zumwalt family, ending with a P.S.: "We are having as much fun as it is possible to have away from one's family. *Little supervision—lots of room for initiative."*[90]

On November 29, 1968, Bud celebrated his forty-eighth birthday with his family at Clark Air Force Base. Before departing Vietnam, Bud made five stops in the delta to visit his men and pass out twenty-five Bronze Stars and thirty Purple Hearts. "It was one of the most moving experiences I have had in years," wrote Bud to his family. "Not one single man had a complaint about his bad luck, his pain, etc. Each one was concerned about family or buddies or 'how did we make out in this mission.' I've seen American youth at its finest. I'm mighty proud to be able to serve with them."[91]

Until Bud arrived, the navy got little respect in Vietnam. Larry Oswald was a navy supply officer in Vietnam. In the wake of the Tet Offensive, Oswald was having trouble getting spare parts for guns, radios, and engines. In the Mekong, these parts came from the army supply depot at Long Binh, commanded by an army two-star general. Admiral Veth's requests for additional parts were consistently ignored. Not inclined to make waves, Veth had literally left his sailors high and dry. One of the first things Bud did after arriving was to call an urgent meeting at Long Binh to discuss these critical equipment shortages. As Bud's helicopter arrived at Long Binh, the army general decided to have some fun by refusing permission to land. It was the army's way of sending a message to the navy's new commander, that no matter how many stars he had, the navy was still not a player. Bud sent his own message back, telling his pilot to hover about a foot off the ground adjacent to the general's office. Once again, permission to land was refused. Bud had the pilot tip the helicopter slightly to the left, with the left-side skid resting hard on the ground and the right-side skid about a foot off the ground. The pilot, on Admiral Zumwalt's instruction, flew outward spirals around the office, digging up all the grass around the compound. "Admiral Zumwalt got out of the Huey, went inside and advised the General that there was '*a new sheriff in town.*'"[92]

The basic style of Bud's leadership was intense loyalty downward. Bud knew that his aggressive offensive strategy would lead to higher casualties. He understood better than anyone that they were dying because his programs placed sailors at greater risk. "Chances are if it were not for SEALORDS, they wouldn't have been dead," said Rex.[93] Marine aide Mike Spiro recalled the night Bud was so concerned about a young ensign being alone on a remote listening-post assignment that they both went out to check on him, just to make sure he was OK. "Flag officers should not be doing things like this, but that was something you could not stop the admiral from doing," said Spiro.[94]

The army soon developed new respect for sailors going into harm's way rather than remaining in well-protected areas. The enemy also felt it, because the water was no longer safe for them. That is why each day there were touchdowns into remote and dangerous war zones. Once on the ground, Bud spent time with sailors in rap sessions, listening to them. SEALORDS was his strategy, but Bud could not direct tactics from Saigon. After firefights he wanted to go to the scene and speak with sailors, find out what had gone right and wrong and disseminate the lessons learned. He encouraged his sailors to be more like Bernique, not necessarily in violating rules of engagement but rather in the invention of tactics. "I was learning a lot and bringing back a lot of information to Saigon about things we needed to do better," recalled Bud.

Bud also learned from these rap sessions about what were known as Mickey Mouse regulations. His sailors understood patriotism; indeed, they were fighting at the behest of their country. His sailors were brave young men who adapted remarkably well to the various types of warfare in the delta and on the coast of Vietnam. Their spirit pervaded the entire naval organization. Few ever asked, "Why am I the unlucky guy to get wounded?" Instead, they wanted to know, "Why is it that I, out here in this unpopular war as a double volunteer to join the navy and to fight this war, am not permitted to have hair and beards that look like my contemporaries'?" Or "Why is it that we have to put up with this Mickey Mouse chicken shit?" None of these things interfered with fighting qualities, good order, or discipline.

Bud would find out what sailors needed, from the third-class boat-swain's mate to the ensign. If they wanted a bag of charcoal or a case of beer, Bud would instruct his aide, "Let's make sure that we don't return to that base without providing the things they wanted."[95] Since 1914, when the son of Josephus Daniels, secretary of the navy, became drunk on a navy vessel, navy regulations outlawed the consumption of alcoholic beverages aboard ships. In the delta there were no clubs along the river in enemy territory, so Bud began the practice of tying barges alongside ships to serve as floating bases for smaller craft and to supply beer for the sailors. In a letter to the family on January 28, 1969, Lew Glenn reported that Bud had just gone to an area near the Parrot's Beak border of Cambodia to congratulate men on the results of Giant Sling-shot. "We brought along three cases of beer to repay the sailors at Thuy Dong for their kindness on previous trip."[96]

Sailors never forgot what Admiral Zumwalt was doing for them. As supply officer, Larry Oswald was responsible for making sure that sailors patrolling the dangerous Vinh Te Canal had functioning flak jackets while on patrol. The body armor was covered by an olive drab nylon fabric that did not stand up to the heat and monsoon rains, quickly rendering the jackets all but useless. Oswald was responsible for order-ing new jackets and sent the request to the commander of naval sup-port activity in Saigon, who accused the crews of gross negligence in maintaining their equipment. The base commander sent a message to headquarters reiterating that he would not authorize new flak jackets; instead, he ordered crews to remove their jackets during the final hour of patrol and tie them to the backs of their boats in order to launder them. "I had never felt so alone and helpless as I was that day," recalled Oswald. When Bud read the message, he immediately reassigned the commander and made sure new flak jackets arrived. "Every time I see servicemen in flak jackets, I think about 'Z' and thank God that this one-of-a-kind man was our Boss," said Oswald.[97]

Jim Morgan, a senior patrol officer and operation officer in River Division 593, recalled that "during our three and one-half months on Giant Slingshot we got to know VADM Zumwalt and I gained great respect for him and his leadership that lasts until today. All of the river divisions on the two rivers were having hard times. Casualties were certainly higher than we were accustomed to, and there were a lot more firefights than in our previous areas. Every patrol had every crew member on a high state of alert. Men's nerves were stretched very taut. Into this atmosphere would come VADM Zumwalt periodically, probably once a month. He would bring members of his staff. I remember his supply and intelligence officers being along. He would call for an all-hands meeting, enlisted as well as officers, put his chin out and tell all assembled that he was the one who put them on these rivers, and if they wanted to take a swing at anyone, he was the one. He explained the strategy, thereby including the troops in the big picture, and he asked what the men were thinking and what they needed that they weren't getting. If one hasn't been in the service, one probably cannot appreciate how rare it is for a three-star officer to talk directly with front line personnel, to offer to take it on the chin from them, and to ask directly what they need to fight better and live a little better. In my opinion, it's extraordinary."[98]

Going into the field brought risks for Bud. Flag Lieutenant Bob Powers recalled the day he and the admiral visited a base at Ha Tien, at the west end of the Vinh Te Canal on the Cambodian border. Thinking the admiral was asleep, Powers negotiated a seat on an evening Swift Boat patrol. Out of nowhere appeared Bud, who said he was coming along. Powers strenuously objected, because the danger was just too great. "Who's the Admiral around here, Bob?" asked Zumwalt. A marine sergeant removed all signs of rank from the admiral, including his ID card and the three stars from his collar. "He ripped the name ZUMWALT off the Admiral's greens," wrote Powers. That evening they spent the night patrolling the Vinh Te Canal, watching and hearing the firefights. The entire evening, Powers kept thinking that if anything happened to Zumwalt, he would never be able to explain it in headquarters.[99]

Another evening, Bud was caught in an attack on the Swift Boats and forced to spend the evening in a muddy bunker with marines eating C rations. This was the only time Abrams ever showed anger toward him. When he returned from the ambush, Abe told him, "Admiral, we got the report that you've been out trying to be a goddamn hero. I understand you've been out in the goddamn boats all night. You are ordered not to do that anymore." Bud replied, "Well, General, I've got to be able to do that. If I'm going to be commanding these guys, I've got to know what they're going through. I can't know what needs fixing, and I can't know what their problems are morale-wise unless I'm out there."[100] Abrams admired Bud's bravery but in this case said, "Read this." It was a report that Don Starry, the army colonel Abe was fondest of, had just been shot down in his helicopter over Cambodia, a place he was not supposed to be. Abe had tears in his eyes and appeared heartbroken at the thought of losing his friend.[101]

Driving Bud's approach was a belief that a commander's responsibility was a commitment to the welfare of those who served under him. Bud believed in the old adage "Don't send men on missions you would not do yourself." Spending a night on a small naval patrol boat in the face of the enemy engendered immeasurable confidence and respect in the young sailors patrolling there, and Bud's helicopter became a familiar sight in visits to remote base camps to deliver food and mail. He came not only for information, but to be with the wounded and dying, often holding their hands and whispering words of comfort. Jimmy Bryant of the River Patrol Force under Art Price was in the Twenty-ninth Army Evacuation Hospital in Can Tho. When he visited him, Bud said, "Don't try to get up, Chief. I hear you had a pretty rough time of it." Bud put his hand on Bryant's left arm. "You will pull out of this, Chief. You've done a good job for me and the Navy. When you get back home and request your next assignment, use my name to help you get it. If that doesn't work, contact me personally."[102]

One death haunted Bud most of all but also demonstrated his sense of responsibility and leadership. John C. Brewton, leader of a navy SEAL team, was seriously wounded in a firefight on November 24, 1969, and evacuated to a field hospital in Saigon. He had almost no chance to survive, but he refused to die. Bud visited him eight or ten times, and because he could not be evacuated, his fiancée and father came out to be with him. Bud gave the Brewton family the guest quarters at his villa. Brewton died on January 11, 1970, and Zumwalt personally awarded him the Silver Star and posthumously promoted him to the rank of lieutenant. When Bud became CNO, he saw to it that the destroyer escort DE-1086, commissioned by Mouza in 1972, was named the USS *Brewton* in honor of the brave SEAL.[103] On August 21, 1971, Abrams wrote to Bud, "I remember so well his hard struggle to live and the dedicated and devoted efforts of all who struggled with him night and day."[104] Bud informed Abe about the recent launching ceremony of the USS *Brewton* in New Orleans. "It was a personal honor for Mouza and me to participate in christening this fine new ship and I know it will be a credit to the name of young Lieutenant Brewton."

As a child, Elmo Zumwalt III survived polio and surgery for a congenital heart defect. Doctors presented Bud and Mouza with the high-risk surgical option of repairing the small hole in one of the chambers of Elmo's heart, along with the option of doing nothing, meaning that Elmo would face a lifetime of fatigue and limited activity. Bud's first son had long dreamed of a naval career and feared that even with the surgery, he would be unable to serve. Bud assured him that "he could become a naval officer if he went through with the heart surgery and that I would be there to see him graduate."[105] Elmo replied, "Let's go for it!"

It was a proud day for the entire family in 1968 when Elmo was commissioned an ensign from the ROTC program at the University of North Carolina. His first orders sent him to the USS *Ricketts*, a guided-missile destroyer in the Atlantic, commanded by his father's good friend Captain Thomas Mullane. After several months on the *Ricketts*, Elmo knew he was in the wrong place. He wanted to command a riverboat in Vietnam. He yearned for the challenge of testing himself in war and the responsibility that came with it. He could not consider himself to be a worthy son if he did not do everything possible to see wartime duty. By now he had no interest in making the navy a career, but he did want to serve his country. "I am basically not made for 'the system.' I could write an essay on why I believe I cannot stay in but I would rather tell these to you in person," Elmo wrote to his father.

The Bureau of Naval Personnel received Elmo's request for reassignment to command a Swift Boat in Vietnam. This would mean that father and son would be serving in a war zone, as Bud and his father had done during World War II. In this case, however, Bud was commander of in-country naval forces, posing additional hazards for both men. Indeed, one of Captain Mullane's first questions to Elmo was whether he thought going to Vietnam would make his father's job more difficult? "I said there wasn't anything that man couldn't handle. He agreed."[106] Elmo's major concern about going to Vietnam was the hardship it would place on his mother. "I know there is no reason I can give her that will alleviate her fear of what might happen. I have told her that I am not going there to prove anything. I am going because of certain principles I believe in. A man has got to uphold what he believes in."

Elmo anticipated that his father might try to quash the request.

I think your first action is going to try and stop my orders. What would happen though if every father was able to stop his son from going to Vietnam? Look at all the poor bastards in the army who get thrown into the front line and have no choice. I am not going over there with the idea that it is going to be a John Wayne adventure. The truth is I am afraid of what could happen. I realize that I am getting myself involved in something I really cannot truly picture. But how many of the boys my age in Vietnam have the luxury of making sure their reasons are perfectly formulated and accurate? One of the many major reasons I feel I cannot stay in the Navy is because since there is an Admiral Zumwalt I will not come up the routine naval path. I will not be able to do what I want because I am the son of Adm. Zumwalt. What I want to do now is go to the swift boat in Vietnam. My orders are sending me there. Are you? I know they're going to check with you and please keep your cotton-picking hands off my assignment—I want to come.

On March 20, 1968, Bud wrote directly to Captain Thomas Mullane, enclosing a copy of Elmo's letter. "I am grateful to you for consulting on my son, Elmo III. I am attaching a letter from Elmo which will, I think, make you feel proud—as it did his mother and me. It also shows a certain suspicion that his old man might interfere. I cannot say that I am happy to see him coming because I know more than anyone, the risks he will run. But I would fail this boy as a father if I did not let him follow his own destiny. And I could not continue to send other fine young men to their deaths or permanent suffering if I had to live with this knowledge that I had used my position to save my own son from taking similar risks. I will look forward to having him in country. God bless you for giving me a chance to face up to this in my split roles as parent and COMNAVFORV."[107]

Elmo was soon detached from the *Ricketts* for Swift Boat and SERE (survival, evasion, resistance, and escape) training, undergoing grueling water torture and other tests in preparation for his assignment. By September 1969, Bud informed the clan that "Elmo arrived in Vietnam in August and has been in command of a Swift Boat engaged in coastal and riverine operations. From all reports, he is doing well and I am proud of the aggressive and positive manner in which he commands his boat."

Elmo was initially sent to I Corps but requested a transfer to IV Corps, where Swift Boats were more frequently engaged in stopping infiltration from Cambodia. In a letter to Paul Nitze, Bud observed that "Elmo III is probably our most aggressive Swift skipper. He is undoubtedly trying to live down the name and continues to take more chances than are sensible. He also has lots of advice for me each time I see him and I find that it really helps to have a trusted gent at the lower level."[108] In letters home to Mouza, Elmo explained how seriously he took the responsibility of serving as a Swift Boat officer in charge. "You make the decisions that may cost you and your crew their life," wrote Elmo. "I have been patrolling for the last month in the operation Sea Tiger on the Cua Dai River. It has been a long hot hard, dirty, dangerous, frightening, lonely, and life shaking experience. My crew is outstanding. They are capable and aggressive—I probably have the most aggressive crew in Da Nang."[109]

That aggressiveness paid off one evening when his boat was on a river that ran along the South Vietnam–Cambodian border. Elmo had told his father that the canal was being used by the Vietcong to smuggle equipment. Bud told Elmo he was mistaken, because his intelligence indicated otherwise. The next night, Elmo went up the canal into Cambodia, in violation of the rules of engagement, to prove his old man wrong. His Swift Boat sank a convoy of twenty sampans and killed several VC in the ensuing firefight. As he and the crew pulled AK-47s, mortars, rockets, and ammunition off the sunken sampans, Elmo kept saying, "I'll take these back and show my damned old man that they aren't using this canal."

In a letter to his mother, Elmo described the catch: "My boat caught a company of NVA trying to cross the river one night. We captured 11,000 rounds of machine gun ammunition, 20 B-40 rockets, 15 recoilless rifle rounds, 8 AK-47 rifles, 5 sampans, 25 lbs of explosives, 10 hand grenades, documents, food and uniforms. We also killed several of the NVA. This was the first time a boat had caught a sampan on this section of the river for over 7 months. It gave our boat quite a reputation."

Like Bernique and others, Elmo had violated the rules of engagement. "I was not the least bit angry with him," wrote Bud. "If the truth be known, I was proud of him, but I didn't show my pride as much as I wanted to because my job was to enforce the rules of engagement."[110] Fleet command awarded Elmo the Bronze Star for bravery. Bud thought that this was one of the few times being his son hurt Elmo. "In my judgment, his actions warranted the Silver Star, but I think the brass felt a Silver Star would be perceived as favoritism." Elmo and his crew received the Vietnamese Cross of Gallantry "for the enemy weapons and kills we got up by Cambodia."[111] In a letter to his father, Bud wrote, "All indications are that he's still giving Charlie hell!"[112]

One of Bob Powers's responsibilities as flag lieutenant was to listen in and take notes on Bud's telephone calls. Powers vividly remembers Elmo's calls from the delta, almost always offering advice on how to fight the war more aggressively. The son of an admiral himself, Powers understood the formidable challenge Elmo faced in being accepted by his Swift Boat crew and living up to his father's image and name. Volunteering for Vietnam gave Elmo the chance to define himself as a man and son. Behind his back, the crew often called him "the little admiral" and "brass brat III."[113] The only favoritism Bud ever showed Elmo was to read casualty reports from the bottom up, because the list was always alphabetical. He was commander to all, but father to one.

This lack of favoritism especially impressed the Vietnamese. Kiem Do, deputy chief of staff for operations in the South Vietnamese navy, remembered Elmo as "that young and daring PCF captain who rode his boat on the infamous Cua Lon River as he would do with his bike around the block." Do was "contemptuous of our high ranking Generals who used their power and connections to shield their children from combat positions. Here was a three star American admiral, whose role was only to help us and he sent his own son in that deadly area of Dam Doi."[114]

The endorsement of aggressive tactics was not without problems, most notably when some task force commanders took it upon themselves to expand the rules of engagement for SEALORDS in a way that jeopardized both men and boats. Zumwalt wanted his commanders to take the fight to the enemy in previously unpenetrated enclaves, but some commanders endorsed intense aggressiveness, directing boats to destroy everything, regardless of whether hostile fire had been received, including homes, livestock, and watercraft.

Several Swift Boat drivers complained about these tactics. At first, Bud did nothing, but soon the number of complaints flooded his command. Zumwalt hauled all the Swift Boat officers to Saigon for meetings and sent a message to all operational commands saying that no one was to again change the rules of engagement on their own in the field. Only the commander of naval forces had this authority. The naval message from Bud to his operational commands, dated January 17, 1969, noted that the "tempo of naval operations has quickened and we have become more aggressive and expansive in our endeavors to defeat the enemy." Commanders could not lose sight that certain types of aggressive tactics risked losing the support of local Vietnamese as well as the

U.S. populace. Recent actions threatened to undermine this support. "Some examples include firing into known friendly areas, taking targets under fire without assurance that they are the enemy, improper conduct of searches or inspections, and excessive boat speeds near indigenous craft. These examples are counterproductive to our efforts to achieve support and confidence. The danger of firing into our friends is particularly important and all hands must exert the upmost caution in taking targets under fire . . . we must insure the innocent are not harmed."[116]

After the meeting, Bud said, "I think Roy Hoffmann paid more attention . . . and things worked much better." Zumwalt respected Hoffmann because "he was in the forefront of those who were willing to see the Swifts exposed in the inland waters and waterways—something that was considered heresy by many of the more junior people and by some of his predecessors. He was convinced that they could survive. And I became convinced, after talking with him, that they could survive, and we did it."[117]

Meanwhile, as these tactics changed and the navy moved from major rivers and coastal patrols into the narrow canals and smaller rivers, sailors who had not been shot at very often suddenly found themselves in the thick of the action, along with soldiers.[118] The navy began taking a significant number of casualties in Vietnam, which meant that the average young naval person had a high probability of getting killed or wounded in a year's tour. Snipers preyed on sailors from their hiding spots along the riverbanks, perhaps ten to fifteen feet from their targets. The heavy density of jungle growth and the narrowness of rivers and canals gave the enemy protection.

Something had to be done about the foliage in the narrow canals. The army had been using Agent Orange for several years to destroy the jungles north and west of Saigon. Agent Orange offered a chance of improving observation, destroying the enemy's food supply, and clearing vegetation around fire bases, landing zones, and lines of communication. Defoliation also offered the promise of moving snipers back a thousand yards. Bud asked all the right questions. "The data we were given was that it had been tested and that there was not a health hazard," said Bud.[119] The Pentagon provided assurances that the ingredients in Agent Orange, 2,4,5-T and 2,4-D, were "nontoxic and not dangerous to man or animal life."[120] The chemical companies producing Agent Orange also provided the Pentagon with their own studies showing no harmful effects to humans or animals. Based on these assurances, Bud gave the order to strip the jungle terrain bare.

War is about surviving chaos and Bud did what he was trained to do in order to save the lives of his men. Agent Orange made it more difficult for the enemy to kill his sailors. Over 19 million gallons of Agent Orange and related defoliants were sprayed over approximately five million acres in South Vietnam, almost one eighth of the South's total area. Operation Sea Float was an extension of the original SEALORDS concept. Like the Vietnamese, those on Swift Boats were especially vulnerable. Sailors washed, walked, and waded in these waters. They ate fish from it. "Trees were stripped of leaves, thick jungle growth reduced to twigs, the ground barren of grass," wrote Elmo about seeing the result of Agent Orange spraying for the first time.[121] Ron Kirkwood recalled the first time he saw a defoliated area: "There wasn't a tree or plant standing except for only a few tree stumps jutting out of the terrain. It was as if someone had dropped a bomb and killed everything. Life was devoid of the area [*sic*]. There was nothing but red dirt and mud. It appeared to look as if a flame thrower had cleared an area about three miles in all directions."[122]

It took months of Agent Orange spraying and navy SEALs in the Rung Sat Special Zone (RSSZ) to get the VC forces eradicated. Known as the Forest of Assassins, the Rung Sat, a 405-square-mile maze of rivers, canals, waterways, and mangrove swamps on the upper Saigon River, was central to Bud's strategic plan for getting the "Navy's piece of the action" in an area that surrounded the main shipping channel into Saigon. The Rung Sat was extremely marshy, with vegetation covering most of the area, enabling small groups of VC to harass and slip quickly into concealment. The enemy controlled this area, and its attacks on merchant ships operating in the channel had increased rapidly since the Tet Offensive. Sapper groups of no more than five men armed with long-range rocket-propelled grenade launchers operated with virtual impunity.

The objective was to guarantee safe passage for mercantile vessels on the vital Long Tau shipping channel to Saigon. Bud specifically recruited Jerry Wages to serve as his senior naval advisor and commander of the Rung Sat Special Zone task group. The two men had met years earlier in the Baltic when Bud was on the *Dewey*. When Bud was in Hawaii with Howard Kerr on their way to Vietnam, he personally called Wages to join the team. Jerry agreed that he would do so, but only if Bud gave him a field command, not a staff position in Saigon.[123]

Commander Wages soon received orders to serve as the senior advisor for the Rung Sat Special Zone and concurrently as commander of the RSSZ River Task Force Group, TG-116.9.[124] General Abrams took special interest in the Rung Sat Special Zone and used it as a model for how to keep the pressure on enemy forces. Wages kept extreme pressure on the VC in order to prevent attacks on shipping coming up the Long Tau River to Saigon. In a letter home, Bud noted that "PBRs, helos, and SEALs have been successful in driving most of the VC back from the main shipping channel."[125] And the Long Tau shipping channel was frequently swept by small minesweepers (minesweeping boats, or MSBs).

During Christmas vacation in 1969, Elmo's fiancée, Kathy Counselman, flew to the Philippines to join the Zumwalts for their holiday break. Elmo and Bud flew from Saigon, joining Mouza, Ann, Mouzetta, and Jim. During this holiday, Bud offered his son a position as his flag lieutenant, that is, a chance to leave his Swift Boat crew and join Bud in Saigon for the daily helo rides into the fields. Bud most likely made the offer out of consideration for Mouza, who knew all too well what kinds of risks her son was taking on the rivers. Elmo indicated that he needed a few days to think about the implications, but no family member sitting in the room thought he would buy into Bud's plan. Before the trip was over, Elmo sat the entire family down to say that, while he realized what a terrific opportunity it would be to work daily with his father and that it would certainly be safer, he could not leave his crew behind to their fate without him. "Elmo always had to prove himself," recalled his brother Jim. "We all knew he was going to go back to his crew. Our father knew it too." Elmo returned to his boat for six more months of intense action and, of course, daily exposure to Agent Orange.

"I have continued to find myself consumed by the challenge here," Bud wrote to his mentor, Paul Nitze, on February 10, 1970. "The opportunity to make decisions and see the result come to pass is greater here than ever before because we are moving in double time. The first year was spent in rearranging operations (to get the navy into the war along the Cambodian border) and laying the plan. This year we are doing the implementing and facing the rough places."[126] Nowhere were these opportunities clearer than in shifting responsibility to the Vietnamese.[127] The Accelerated Turnover to the Vietnamese (ACTOV) plan involved expansion and improvement of the South Vietnamese navy as well as turning over craft and logistics supplies. Navy personnel would remain as advisors for some time into the future, but the goal was "to leave here as rapidly as possible, but to ensure that we leave the Vietnamese navy as an effective fighting force."[128]

Captain Paul Arbo was the initial senior naval advisor to Admiral Tran Van Chon, but it soon became apparent that Arbo and Bud were not on the same page. Arbo wanted the Vietnamese better trained before the turnover occurred, ignoring the political window of opportunity that was closing. Arbo's error was to keep challenging Bud on the timetable for turning over naval assets. He would be reassigned and replaced by Chick Rauch, who became the action officer on Vietnamization and naval assets, the senior naval advisor to the Vietnamese navy, and deputy chief of staff for Vietnamization. Rauch was the one who conceptualized the business of ACTOV. He was also the best available person, aside from Bud, to work with the remarkable Tran Van Chon, head of the South Vietnamese navy.

Chon had been born at the seaside town of Vung Tau, where as a child he dreamed of going to sea. He attended the French Maritime Navigation School in Saigon, starting as an officer on a merchant marine ship at the age of twenty-one. By thirty he was in the first group of students to attend the Vietnam Naval Officer School, and at thirty-seven, he was appointed commander of the South Vietnamese navy. He had been the valedictorian midshipman of the first session at the Vietnamese Naval Academy and the first Vietnamese to enter the U.S. Naval War College. He acquired fluent English and returned to Vietnam, where he worked through a succession of jobs at the Defense Ministry, the general staff of the armed forces, and the Regional and Popular forces command headquarters. At the age of forty-six, he returned to the navy in the position of CNO on November 1, 1966. He was revered by his officers and men.[129] "We really became more like brothers," said Bud. "He was, I think, a magnificent individual and a magnificent leader."[130]

In their first week together, Bud "sucked him dry" because of the wisdom and understanding he possessed. "He was a very devout Buddhist—a man of very high ethical and moral principles. There was not the slightest doubt about his incorruptibility." It was Chon who put Bud on to the importance of Operation Sea Float for establishing a presence in the resource-rich area at the tip of the Ca Mau Peninsula. "He was a strategist and experienced tactician," said Bud.

Chon had extensive cross-cultural training, from associating with senior advisors for years and from a year at the Naval War College. Chon insisted that when in front of his Vietnamese sailors, all conversations with Bud be in English. "I, at the very outset, always believed and knew that I had to present myself in the presence of other Vietnamese as seeking Chon's advice and giving him advice in a subordinate-to-senior role, even though he had one star and I had three, initially. And I was meticulous about that, and, although he never acknowledged that he knew I was doing it, he always permitted me to do it in the presence of his juniors. It was a very helpful thing to him."

Chon had immediately grasped the significance of what Bud was trying to do. The United States was leaving Vietnam, and his job was to prepare the Vietnamese navy for the responsibility of carrying on the war without the Americans. The Vietnamese were the ones in it for the long haul. "Early in the game it became obvious that a rapid increase in manpower would be necessary and that training centers would need to be enlarged and upgraded in their methods," Bud wrote to his family.[131] The Vietnamese Naval Training Center at Cam Ranh Bay was completely remodeled and began turning out large numbers of sailors qualified to serve in the Brown Water Navy. The Vietnamese Naval Academy at Nha Trang soon had an output five times greater than originally, producing a cadre of motivated and qualified Vietnamese officers and men. "On the whole, I'm very optimistic about the increasing capability of the Vietnamese to take over the responsibility for the operational conduct of the war. With additional training, their logistics and maintenance capabilities should produce a highly effective, self-sufficient combat force," wrote Bud.[132]

It was a momentous challenge that also involved expanding the size of the Vietnamese navy from seventeen thousand to over forty thousand. Vietnamese sailors would need to be trained on boat operations, under way maintenance, shore-level maintenance, public works, store keeping, supply, and accounting. The plan involved a turnover of about thirty ships and a thousand small craft during the three-year window. It would require construction of more than forty new facilities, such as operating bases, repair bases, and supply facilities—all accomplished while continuing to fight a war and removing thirty-seven thousand U.S. officers and sailors, who had been carrying the brunt of the fight in brown water and by naval support of the marines. The concept of "sequential turnover" required putting a Vietnamese sailor alongside an American sailor and training him while continuing to prosecute the war.

What most impressed Bud was Chon's "we can do it" attitude from day one. At first it was extremely hard. The Vietnamese needed to be taught Pidgin English while learning such terms as port, starboard, fire, cease fire, beach, and get under way. A boat school was set up to teach enough basic English to serve aboard a PBR or PCF. Mockups of engines, guns, and other basic items were brought to the school to make certain the new skills were utilized. A personal response program had been created for dealing with cultural clashes and problems. Run by a navy chaplain with a degree in sociology, the program was designed to educate all sailors concerning cultural differences and possible areas of misunderstanding. "In summary, we have many fine officers and men engaged in a most challenging job. All indications are that the enemy is being hurt, and the Vietnamese Navy is well along the road to becoming a very effective force."[133]

Vietnamese sailors went directly from school to combat and then returned to teach at the school. "This gives us the opportunity to get additional instructors quickly, and it enabled the Vietnamese sailors to work with and observe U.S. sailors in the direct fire environment," wrote Bud in a letter to his sister, Saralee. With only a five-member crew on most patrol boats and one South Vietnamese sailor per boat, the Vietnamese needed to pull 20 percent of the load. "This effort has been very rewarding and has given us good insight into the fine quality and high level of competence the Vietnamese Navy develops when given good leadership and training."

In other letters home, Bud marveled at the initial success of the transition and the performance of the Vietnamese. Admiral Chon's visits to Sea Float were nostalgic "because he was so visibly, professionally excited by what was going on there, and so fully felt that his judgment was being fulfilled that the people were coming in and settling along the bank there, and that the resources were beginning to flow up to Saigon and improving the GNP of the country." In Operation Sea Float, Vietnamese sailors were serving side by side with their American comrades. Several of the operations had combined command systems, with a U.S. Navy commander and a Vietnamese navy deputy. Eventually, all operations would have this arrangement, preparing for the day when boats would be manned totally by Vietnamese and Vietnamese officers would take the top spot in each command. Admiral Chon was telling his young officers, "Learn well, because soon it will be yours." The intent was to create a Vietnamese navy capable of continuing to do alone what they had been doing together with the Americans.[134]

Bud and Chon agreed that the standard of living of the Vietnamese navy man and his family needed to be improved. ACTOVLOG was a plan to provide the logistic support infrastructure needed by the Vietnamese navy. This involved constructing new bases and training programs for maintenance and supply personnel. It was the birth of a new Vietnamese navy.

Operation Helping Hand was a program to build adequate houses for Vietnamese navy men and their dependents. Another project was animal husbandry, setting up pig and chicken farms at bases. The intent was to increase the amount of protein in the diet and greatly improve economic conditions. Vietnamese were trained in the proper feeding of the livestock. New breeds were introduced with the hope of providing farmers with a future as well as an immediate relief to their food problems. Bud delighted in the news that two pregnant sows at Cam Ranh Bay were named Admiral and Commodore at a formal pig-farm commissioning ceremony. "The number of programs for upgrading the Vietnamese navy that Commodore Chon and I have been successful in implementing has been staggering," wrote Bud in a letter home. Arriving in An Thoi, the admiral was greeted with a sign, Z-HOUSES, PIGS, FISHES, CHICKENS—A BETTER LIFE FOR THE VIETNAMESE NAVY.[135]

After one year of ACTOV, the Vietnamese had taken control of approximately one half of the armored boats, one third of the PBRs, and one quarter of the PCFs. Chon and Zumwalt exulted in the largest single turnover to date—more than seventy-five armored boats. "On the whole, I'm very optimistic about the increasing capability of the Vietnamese to take over the responsibility for the operational conduct of the war," wrote Bud. "With additional training, their logistics and maintenance capabilities should produce a highly effective, self-sufficient combat force."

Not everyone was happy, though. Rex Rectanus was demoralized by ACTOV, because he had to turn over his intel to the Vietnamese navy. "That's a demoralizing feeling, totally demoralizing to see everything that you've worked for abandoned. We all knew intellectually we had to do it, and not only that, the commander was saying, 'Not only do you have to do it, you're going to do it and do it well.' "[136]

Bud's primary concern and focus by early 1970 was making sure that ACTOV met the goals of Nixon's Vietnamization program. "We will remain as advisors for some time into the future, with the exact time being dependent on the political situation at home and the progress of the Vietnamese Navy. Our goal is to leave here as rapidly as possible, but to ensure that we leave the Vietnamese Navy as an effective fighting force. In relation to the latter, Commodore Chon is proving to be a very capable CNO and the professionalism of his navy is reflecting his excellent leadership. I regard him as our *Ace-in-the-Hole* and feel confident that his efforts will enable us to redeploy just as rapidly as it is humanly possible."[137]

Bud envisioned the Brown Water Navy diminishing in importance as the navy continued to turn over craft and logistic matériel. Whatever its shortcomings may have been, ACTOV was a program designed to meet the requirements of political authority. Certainly the program would have benefited from the luxury of more time for training, but political reality dictated otherwise. The transformation from adolescence to maturity was not without its share of growing pains. Repair and maintenance facilities were not up to par; training the Vietnamese in shipyards took longer than in the States. Expertise in leadership and in technical abilities can be gained only through experience, and experience requires time, but with the rapid pace of expansion, time was not an ally. The Vietnamese, by necessity, continued to rely on American advisors for support in such technical areas as logistics and training. "Hopefully we will soon provide for our own needs," Chon wrote to Bud.[138]

Bud had completed eighteen months in command of naval forces in Vietnam when Secretary of the Navy John Chafee requested that he fly to Washington to discuss the war and options for his next assignment. Bud wanted to remain in Vietnam so he could help the South Vietnamese navy complete its transition. He had changed the scope of riverine operations, but much work remained to be done with respect to the turnover of operational boats and other craft. Bud hoped to be selected as the first nonaviator in more than a decade to command the Seventh Fleet, a plum assignment that would also keep him heavily involved in the fighting in Southeast Asia. There was much speculation that Vice Admiral Ike Kidd was going from command of the First Fleet in the Eastern Pacific to the Sixth Fleet operating in the Mediterranean. Bud feared getting Ike's old job, a less challenging assignment.[139]

Chafee began their meeting by saying he was interviewing a number of flag officers about future assignments. The secretary's first question led Bud to think he had the Seventh Fleet. Chafee wanted to know Bud's attitude regarding the tradition of naval aviators commanding overseas fleets. It was a softball question. Bud spoke about excessive parochialism and said that he believed there was no excuse or rational way to explain it. Bud proposed a rotational basis as a best way of achieving optimum training, development, and sense of community.

Chafee next asked Bud to explain or describe the problems in the navy and how they should be dealt with. Bud saw four interrelated problems, starting with the transition to an all-volunteer navy. Current practices in personnel administration would not work once the draft ceased to provide bodies. Therefore an entirely new approach to personnel administration would be needed in order to improve reenlistment rates, which were at an all-time low. He also made a strong pitch for changing the navy's image to attract more qualified recruits.

The second major problem involved the modernization of ships and weapons systems. Bud told Chafee he thought the navy had become obsolete since World War II. The dramatic and impressive growth of the Soviet navy posed a serious threat to the U.S. capability to use the seas in war and therefore represented a real threat to the deterrent impact of the U.S. Navy. Modernization was going to require serious belt-tightening as the navy gave up a significant number of current ships and aircraft in order to invest in new weapons systems.

A third problem was the modernization of strategic forces necessitated by the rapid growth of the Soviet Union. The Polaris and Poseidon submarine missile systems were approaching obsolescence—by 1980 the oldest boat would reach the end of its twenty-year life span. Bud thought the second and third problems were interrelated in that the more that was spent for the navy's contribution to the nation's strategic second-strike capability, the less there would be for modernizing the nonstrategic part of the navy.

The fourth major problem involved retaining sufficient power vis-à-vis the Soviets. The phaseout of operations from South Vietnam, the expenditures necessary to achieve an all-volunteer navy while simultaneously increasing investments in modernizing the nonstrategic and strategic navy posed a serious leadership challenge. The naval capabilities of the United States were deteriorating in comparison to those of the Soviet Union, and expenditure patterns needed to be reshaped.

Chafee listened intently while Bud answered the questions but kept returning to the field of personnel—people. Speaking passionately about what he had learned from his sailors in Vietnam, Bud provided specific ways of improving morale and reenlistment rates. Bud thought the interview went very well, "although when I left the interview I didn't know any more than [when] I went in about what my future job might be."

Bud was next ushered into the office of Chafee's boss, Secretary of Defense Melvin Laird. Bud had last seen Laird a year earlier in Vietnam, during the secretary's initial visit in the spring of 1969. Laird came away impressed by the vice admiral, who briefed him on the navy's strategy in the delta and coastal war as well as on the ACTOV program for the South Vietnamese navy during the transition.[140] The fifteen-minute briefing had also addressed the changing riverine operations.[141] Laird wanted Bud to meet Henry Kissinger, the national security advisor. Laird called Kissinger to ask if he knew Zumwalt, who "is in charge of Naval Ops in South Vietnam." When Kissinger said no, Laird filled him in. "He is the person I lean towards for CNO. He is down the rank a bit—a Vice Admiral." This did not bother Kissinger, who said, "I don't know the name, but I don't think it's a bad idea to go down a few ranks." Laird said that President Nixon thought the same. Since Zumwalt was in town, Laird asked Kissinger to meet with him so that "you can size him up a little bit . . . this is a great guy. He's the best of all the Vice Admirals we have. There is no 4-star Admiral I can recommend." Kissinger replied, "I would like to."[142]

Bud was soon in the White House office of Henry Kissinger. After the brief meeting, Kissinger told Laird, "I liked him. I talked to the President about him and he said if you liked him and I liked him we should proceed."[143]

CHAPTER 9

THE WATCH BEGINS

The U.S. Navy has been extraordinarily fortunate to have had a leader with your vision, energy and courage during these rather trying times. The Navy is not an easy institution to move, even when it's for its own good, but you have literally succeeded in bringing it into tune with the times and giving it an up-to-date sense of purpose.

—ADMIRAL WILLIAM J. CROWE, JR.[1]

While eating breakfast in Saigon on April 12, 1970, Bud received a call from Captain Stansfield Turner, executive assistant to Secretary of the Navy John Chafee. Turner told Bud that the secretary was on the other line waiting to speak with him. Bud knew the call involved discussing his next assignment. Chafee instructed him to catch the next commercial flight to Washington, dressed in civilian clothes. So great was his desire to remain in Vietnam in order to complete the ACTOV transition that Bud visited Abe before departing in order to request that he be allowed to tell Chafee that General Abrams wanted his commander to stay on for another year. "Sure, Bud, you can tell them that, but first listen to what they have to say," said Abrams. "You never know when those civilian fellows are in trouble and might decide you're the only man who can do a certain job for them."[2]

Flying commercial air in civilian clothes and accompanied by his aide, Flag Lieutenant Mel Stephens, a wartime riverine commander who had earned a Purple Heart and a Silver Star, Bud arrived at Dulles. He was taken to the Georgetown home of Undersecretary of the Navy John Warner while Stephens checked into the Marriott. Bud was under orders to remain incommunicado. Chafee soon arrived at Warner's home, informing Bud, "You are the one Mel Laird and I have nominated to the President to relieve Tom Moorer as Chief of Naval Operations when the President appoints Tom to relieve General Wheeler as Chairman of the Joint Chiefs."[3]

Chafee and Laird both wanted a nonaviator as CNO and someone of a younger generation. Laird and Chafee consciously selected someone they hoped would be an agent of change to lead the navy through the later part of the century. Both men shared Bud's values on the most central issues of reform. "Whether or not they were right can be argued, but it should not be argued that I had pulled the wool over anybody's eyes. I was the alternative chosen and the programs I had said I would initiate, I proceeded to initiate."[4]

Upon hearing the news from Chafee, "My heart leaped and my mind raced. . . . There was the thrill of achievement that there would be a capstone, reward and challenge to a professional career. There is a traumatic sense of how lucky I had been so often to be in the right place at the right time, to be observed by my superiors who were able to do something about their perception. There were immediately thoughts of opportunities to carry out programs that had been stored away in the memory tanks throughout the years. The feeling of excitement at the opportunity to pull together a team of people with whom to associate in great work. There was at the same time a feeling of regret that I would have to leave brave comrades to complete our work in South Vietnam, and that the team that had been assembled out there would have to be left for the moment behind. There was an instant reflection on how proud my family would be. . . . I have to confess there was a feeling of pride that the news would justify the confidence of people I revered had put in me over the years and there was also a feeling of concern as to whether or not I would be equal to the challenge."[5]

Admiral Tom Moorer, now slotted to chair the Joint Chiefs of Staff, opposed Bud's selection. When learning that Chafee was leaning toward Zumwalt, a shocked Moorer instructed his aide, a future admiral, Captain Harry Train, to "get Jerry King up here." Jerome King was Moorer's protégé, one of the earliest Navy ROTC graduates to achieve the rank of three-star admiral.[6] Moorer took King by the arm into Chafee's office. "You want somebody young. He's somebody young."[7] In a desperate attempt to prevent the nomination, Moorer wrote to Chafee on April 10,

1970, under the heading SENSITIVE—HOLD CLOSE. Moorer was "gravely concerned and have been thinking about this matter almost continuously for the past three months. Our sole objective should be to do what is best for our Navy." Moorer recommended another protégé and confidant, Admiral Chick Clarey, as his first choice. Clarey had been Moorer's deputy when he commanded the Pacific Fleet and then served as Moorer's vice chief of naval operations. "I think it would be detrimental to the Navy as well as to Vice Admiral Zumwalt to bring him into this position as this time. He requires more experience—he would be forced to retire at a very early age—and, in my view, he simply is not ready for this assignment."[8] By "not ready," Moorer meant that Bud lacked command experience at sea and should be given command of either a numbered fleet or the U.S. Pacific Fleet. Moorer believed that senior officers would be especially troubled by the appointment.[9] Chafee ignored this advice from the incoming chairman of the JCS. Chick Clarey was assigned as commander in chief of the Pacific Fleet.[10] When he learned of Chafee's final decision, Moorer kicked the wastebasket across the room.

Bud Zumwalt was a black-shoe navy man, a nonaviator, the first since Arleigh Burke to serve as CNO. He had also been deep-selected; seven admirals and twenty-six vice admirals had been passed over. He hoped that the Old Guard would give him a chance but was prepared for rough sailing. Melvin Laird recalled how impressed he was with Zumwalt's briefing during his first trip to South Vietnam. Bud's overview of navy strategy in the delta and his vision for accelerated turnover were precursors for Laird's concept of Vietnamization. More important, Laird was committed to leading the Defense Department into the equal-opportunity era, most especially in rooting out racist practices and other injustices. At the time, there were only three black captains with sea commands in the entire navy. One of Laird's top priorities was to have selection boards allow him to appoint the first black admiral and Laird sensed he had the right person to make it happen.[11]

Laird's private life reflected his public commitment. Laird had joined the Kenwood Golf and Country Club as a young congressman because it was within walking distance of his home and it had a swimming pool for his children. In 1968 waiters at the club refused to serve his guest, the mayor of Washington, D.C., Walter Washington, a black man. Laird demanded that they serve him. Afterward, the president of the club told Laird that this had been a onetime exception and that no blacks would be served again. Laird decided to fight the policy, joining with Senator Frank Church to lead a petition drive. By the time the case reached the courts in 1970, two prominent Republican members of the administration had already resigned from the club in protest of discriminatory practices, Laird and Secretary of State William Rogers.[12]

Secretary of the Navy John Chafee had backed Nelson Rockefeller against Richard Nixon for the 1968 Republican presidential nomination, and John Warner, who was then serving on the Nixon transition team, had been lobbying for Chafee's position. Warner's father-in-law, billionaire Republican contributor Paul Mellon, was advocating on behalf of his son-in-law, but Laird had already made up his mind on Chafee. Laird needed Chafee as much as they now both needed Bud. "Nixon hates him!" Warner protested. "He's a Rockefeller man. Don't you understand that? He's a Rockefeller liberal! You can't have that guy in there! I'm the guy that worked with Nixon for eight years!"[13]

Warner was given the position of undersecretary of the navy, with the assurance that he would succeed Chafee. When Chafee departed in 1972 in order to seek public office, Bud wrote him a personal letter saying, "The 22 months we worked together were some of the most enjoyable of my life. I always felt that we were in harmony, and convinced we did good for the Navy. You are a wonderful person to work with. It was fun as well as exhilarating to be with you."[14] All this would change when John Warner became secretary of the navy.

Bud was instructed to remain at Warner's home, where that evening Warner told him the details of Moorer's opposition to the nomination. The next day, Bud was scheduled to meet privately with President Nixon at the Pentagon prior to the announcement of his nomination. After that, Chafee wanted Bud to stay in the traditional quarters for the CNO, known as Quarters A, a stunning 1893 Victorian sitting atop a grassy and wooded hill on the Naval Observatory grounds off Massachusetts Avenue. The home was, of course, currently occupied by

Admiral Moorer. Chafee told Bud he was to stay at the home as a visible display of Admiral Moorer's support of the selection.

On April 13, 1970, Laird placed a call to President Nixon. "I flew him in and have him hidden away here," said Laird. Nixon agreed to see Zumwalt "for two or three minutes."[15] The next day, Bud was driven to the basement of the Pentagon and spirited up to the defense secretary's office, where the president told Bud that his age would be an asset for the job. Bud asked whether Nixon was aware of his ideas on personnel administration. Nixon said he was familiar with the ideas and looked forward to reading even more of them. The two men discussed their childhoods in California. Nixon recalled picking lemons in the town of Lindsay, not far from Tulare. It was all small talk until the president "observed that he was quite concerned about the growing Soviet maritime balance—Soviet maritime capability." Laird had already told Nixon that Bud possessed a healthy appreciation for the situation. Bud mentioned that he hoped his navy budget would be increased in order to meet this danger. As the meeting ended, Bud noted that as naval advisor to the president, he anticipated communicating directly with his commander in chief. It would be three months before they met again.

Returning to Warner's home, Bud requested permission to call his family. The president's announcement was scheduled for the next day, so absolute discretion was required, particularly because of the secrecy surrounding Admiral Moorer's appointment as chairman of the JCS. "Therefore, we were selecting a new CNO in secret, but even interviewing other people besides Zumwalt," recalled Stan Turner, who helped pick the list of candidates. "Thus, the public campaigns and the pressures that come from retired admirals and others were nonexistent at the time."[16]

Mouza was still in the Philippines, unaware that Bud was even in Washington. Meanwhile, in Vietnam, Bud's aide Mike Spiro thought his boss was in the Philippines with Mouza and their daughters. The phone in Warner's personal residence was not secure, forcing Bud to speak in code. After informing Mouza that he was in Washington, Bud mentioned that he would be "taking her to the house that Aunt Tina had predicted we would live in."[17] Mouza shrieked with delight,

understanding the reference from two decades earlier when Christina
Wright, the aunt of their neighbors and dear friends, Jim and Caroline
Caldwell, whom everyone called Aunt Tina, had taken them to meet
the Marshalls in Pinehurst. A few years after that meeting, CNO Robert
Carney's wife had invited Tina and the Zumwalts to dinner at their
home. As the evening drew to a close, Tina said, "I'm so glad that you
kids were able to come because I was very anxious for you to see your
future home." Mrs. Carney added, "Now isn't that a sweet thought."
Mouza understood the code, and yelled out, "Bozhe moy!"—My God!
Then, agreeing with Moorer, she said, "Do you realize that that means
we have to retire in just four years?"[18]

The next day, Bud moved into the guest room at Quarters A. Admiral Moorer was out of town, so that evening Bud was entertained gra-
ciously by Moorer's wife. When Moorer returned the next day, the man
who had exiled Bud to Vietnam spoke candidly about his opposition to
the appointment, saying it was not personal, just business and in the
best interests of Bud's career. He elaborated on each one of his reservations, focusing especially on the fact that Bud's career in the navy would
be over at age fifty-four.

For the next four years, Bud Zumwalt and Tom Moorer were able to
forge a solid although rocky relationship. They were in general agreement on navy issues and budgets, but Moorer was less supportive on
social and people programs. They were at their strongest when taking
on common adversaries, like the Soviets, Hyman Rickover, Henry
Kissinger, and even the president of the United States, and they stood
shoulder to shoulder during the crisis over the admirals' spy ring.

On April 14, 1970, President Richard Nixon announced his intention to
nominate Admiral Thomas Moorer to succeed Earle Wheeler as chairman of the Joint Chiefs and Vice Admiral Elmo Zumwalt to replace
Moorer as chief of naval operations. The *Washington Post* described
Zumwalt as the navy's "modernizing intellectual."[19] The *Wall Street Journal* noted that "Admiral Zumwalt's elevation will bring an even more
dramatic change in style to Naval leadership . . . he is known as a man of
aggressive intellect, a zealot for hard work and long hours, and one who

is extremely determined about getting his own way." The writer mentioned that Zumwalt had a reputation as "something of an upstart in the tradition-oriented Navy, and not all senior officers are his admirers."[20]

Letters of congratulations poured in from old friends and shipmates. In a Western Union message, Andy Kerr joked, "This will look good on your record." From California, Governor Ronald Reagan offered congratulations: "Your native state is proud of your achievement, and gratified that one of its own will head the world's largest and finest navy."[21] Former Annapolis roommate George Whistler was so excited that he wrote directly to Admiral Moorer. "Bud Zumwalt was my roommate at the Naval Academy, and I can vouch for his qualifications as your successor. Not only will he be an eloquent and logical spokesman for the Navy in these austere budget days, but, despite his relative inexperience, he will be one of the most brilliant and imaginative naval strategists in the history of our wonderful Navy."[22] In a true act of diplomacy, Moorer replied to "Dear George" by saying, "There is no question with regard to Bud's qualifications—that is why he was picked to head up the greatest Navy in the world."[23]

One of the first congratulatory cables had come from General Abrams. "The place was buzzing with the news of your nomination. The atmosphere was like a small town where news has come that one of 'their boys' has made good. I join with your many friends in acclaiming this selection. More than this I am aware of the awesome burden of leadership that will soon be yours. For this I pray that God will bless you with the health, the patience and the wisdom you will need to fulfill this responsibility."[24] In reply Bud wanted Abe to know that "your thoughtful message was the one I needed most. For 19 months, I have been inspired by history's greatest combat leader and have sought no greater reward than to be counted on your team. I will leave South Vietnam with regret that I cannot be close to you in your great work. You will always have my support and gratitude."

From the University of North Carolina, Bud's son Jim wrote, "What took so long to do it? You never told me that you were ranked 275th in conduct. . . . I don't want to sound too corny or anything, Dad, but I can't express in words how proud I felt when the news came out. I just hope that I will someday be able to become as successful in the career I end up following to make you proud of me." Bud replied, "There are fewer [sic] things in life closer to a father than to know he has his son's admiration for his accomplishments. Words are not necessary. I can only add, 'Keerist, what a sensation.' . . . As to my standing 275th in conduct, chalk it up to youthful exuberance. After all, one can't be perfect in every category. 'All work and no play.' "[25]

Bud was scheduled to be frocked with four stars on May 15, following the COMNAVFORV change-of-command ceremony in Saigon. The CNO change of command was scheduled for Annapolis on July 1. Rumors were spreading like wildfire that one of Bud's mandates was to tackle an OPNAV organization that fostered parochialism and internal fighting. Bob Powers wrote that he had already heard "senior gentlemen exuding away," but there was greater enthusiasm from the "zoomie community."[26] Powers urged his former boss in Vietnam to move quickly on the shake-up, getting rid of the old admirals and getting younger people into these positions. This would mean using his "big broom to sweep away the cobwebs" by altering the daisy chain, leading to the selection of a number of people who were outside the traditional patterns. "He promulgated guidelines that said in essence, 'I want some iconoclasts,'" recalled Admiral William J. Crowe, Jr., who was one of those promoted under Bud's shake-up.[27]

A former ministaffer in Vietnam, Captain Dick Nicholson, wrote, "The famous announcement on 14 April, has triggered a new and stimulating fever in the young officer and enlisted rank. . . . A refreshing spirit de corps [sic] will come about now."[28] Dick warned that "the Grand Old Gentlemen are worried, the Big Men will come forward, and the young are in the aisles waiting for the mediocrity to be banished." He added that "the list of complaints is long and almost all known to you. However the most repeated comment throughout is that we harbor mediocrity in the senior petty officers and officer ranks. Many instances have existed where our people have lost faith in their senior officers because of this mediocrity and sometimes unsat [unsatisfactory] performance. The morale is poor in the surface navy." Navy leaders had given lip service to the concept of people being their primary resource, but they had no idea how to nurture that resource, because "people oriented" was simply not viewed as the same type of asset as "hardware oriented." Like so many former members of the Vietnam team, Dick offered to join the cause because "I believe in your thinking and in you as a man, therefore I am available anytime I can serve in any job large or small." In reply Bud let his loyal aide know that "I have a couple of jobs in mind for you and will be in touch after I take over in July."[29]

Rear Admiral Earl "Buddy" Yates, Commander, Fleet Air, Whidbey Naval Air Station, wrote that "some people are concerned that there will be a mass exodus of the people you were promoted over. Some people feel that your selection as CNO should have been delayed until you had a major fleet command. Some people feel you are too young. But I have heard no one who has suggested that you do not have the capability to handle it with ease. So, in spite of some minor professional jealousy created in some areas, I feel certain that except perhaps in isolated cases, you will have a tremendous amount of personal loyalty and support, probably more than anyone else in the Navy, regardless of rank. Consequently, I feel that you can move into the job with considerable confidence that your policies and actions will be genuinely supported, that your directives will be positively and aggressively followed, and that your power base will be sturdier than any naval officer since Radford. In short, I feel that you will be the most effective CNO we've ever had, and not just because of your ability—but a combination of that ability and the unusually strong support you will get both from the officers of all ranks and the civilian echelon. I have always felt that in spite of your great talent and your outward display of confidence, you have worried too much deep down inside that you would fall on your face. Perhaps this is healthy to some extent and keeps one alert, but in the job you have coming up, complete confidence and an inner peace with yourself is essential."[30]

Poignant letters also arrived from Bud's counterparts in the Vietnamese navy. Lieutenant General Le Nguyen Khang, commandant of the Republic of Vietnam Marine Corps, wrote, "When you leave my country you may go with the knowledge that you have made a great contribution to the welfare and freedom of the people of Vietnam."[31]

Admiral Tran Van Chon, whom Bud considered a brother and from whom he had learned to love the Vietnamese people and culture, wrote, "I would like to take this opportunity to assure you that we are grateful to you and your Navy for the sacrifices you have made for us and your assistance in our fight against the Communist aggressors."[32] Chon spoke of their friendship and the bonds between the two navies. "Your work in our behalf has been truly heroic. You frequently took considerable personal risks to advance our cause. Truly, events have proven that you have always had the larger vision."

The Senate Armed Services Committee agreed to schedule expedited confirmation hearings on April 16 so that Bud could return to his command in Vietnam. Chairman John Stennis, who personally agreed to waive the seven-day waiting rule between nomination and hearings, began the hearings by reading into the record the statute bearing on the role of the naval chief: "The Chief of Naval Operations is the principal naval advisor to the President, and to the Secretary of the Navy on the conduct of the war, and the principal naval advisor and naval executive to the Secretary on the conduct of the activities of the Department of the Navy."[33]

A large contingent of supporters was in the hearing room, but noticeably absent from those speaking or present was Tom Moorer. Secretary Chafee told the committee that "other" commitments prevented Moorer from being present, but Moorer was still seething about Bud's nomination. In an act of pettiness, Moorer was unwilling to allow Bud to select Bob Salzer as his replacement in Vietnam. "I don't know why. He was also unhappy enough with my own selection that I think he was being a little bit childish about it, and that may have been the main reason why."[34] Instead, Bud was told to offer three or four names, from which Moorer would make his selection. One of the names on Bud's list was Jerry King, whom Moorer selected.[35] Moorer was completely within his prerogative because he was still CNO. This was a case of Moorer's loyalty downward in promoting an officer with whom he had a strong personal relationship, something Bud did many times throughout his career.

King was the person Moorer had pushed on Chafee for Bud's job as CNO. Bud's revenge would be to move King as soon as he could, making him deputy chief of naval operations for surface warfare. Bud then moved Salzer into the Vietnam command. Bud felt that King "did a very poor job, in my judgment, primarily because he was the kind who simply never understood the Vietnamese culture. He was, by nature, a driver rather than a leader. And he turned off Admiral Chon almost completely. It was a very standoff relationship with, you know, directions being sent, and was totally counterproductive."[36] Naval historian Paul Stillwell offered a more nuanced perspective. "Vietnamization became frustrating to King because it wasn't the same desire to victory that had existed before. He presided over the diminishment of American capability there and was not always confident of the South Vietnamese ability or willingness to take over the equipment and the roles" of the U.S. Navy operating in rivers and canals.[37]

The Senate confirmation hearings amounted to a stroke fest. Bud had arrived without any prepared statement, but Chairman Stennis wanted something on the record. He told Bud not to be concerned about substance but to just say something because "we want you to feel like you are a member of the family here, but we've got to pass on you, and make a definite recommendation, so just make such a statement as you see fit." Speaking without notes, Bud took a few minutes to thank the committee for waiving its rules so that he could return to command. He noted that in succeeding Tom Moorer, he would now be working with a chairman of the JCS who knew more about "my" navy than he did. "I look forward to the challenge. I promise to do my best."[38]

Senators Scoop Jackson and John Tower followed Bud's brief comments with words of praise. Jackson recalled meeting Captain Zumwalt several years earlier during the complicated and sensitive Paul Nitze confirmation hearings. "I had nothing but admiration and respect for his very broad knowledge of national security matters. We are dealing with certain aspects of the situation that far transcended the narrow confines of naval responsibility. I have nothing but respect for his ability as a good generalist, to deal with the problems that he will have to face, particularly as a member of the Joint Chiefs of Staff." Senator Tower expressed special delight that a boatswain's mate in the U.S. Naval Reserve was passing judgment on a new CNO. "I am personally delighted to see a black-shoe sailor take over as the Chief of Naval Operations. I think it is time us surface sailors have something to say about what goes on." Tower added that he recalled "the many fine things that General Abrams had to say about Admiral Zumwalt and his work in Southeast Asia."

Bud was asked one important question during the hearings by Senator Peter Dominick of Colorado involving the Nixon position on a "one and a half war strategy." Bud's reply foreshadowed his own strategic concept for a surface navy that needed to be able to defeat the Soviet navy in all oceans. "Our problem is that we can never be local in the Navy as long as we face the Soviet Union as a primary threat. . . . One of the things I want to do is to take a look at the present structure of the Navy and see whether I would recommend to Mr. Chafee and to the President any changes."

Senator George Murphy of California next captured the sentiments of the committee: "Since I have had the privilege of serving on this committee, I know of no nomination that has been received with as much enthusiasm as yours. I think this speaks very highly." Murphy was scheduled to be in Tulare that Saturday and promised to announce the nomination from Mineral King, a glacial valley and backpacker's paradise whose nearby peaks are the highest points in Tulare County.

With that, the hearings concluded, taking just twenty-five minutes. Bud was on his way back to Saigon the same evening. On May 7, 1970, the full committee voted unanimously in favor of the nomination. Bud Zumwalt was confirmed by the Senate on May 15, 1970.

The change-of-command ceremony at which Admiral Elmo R. Zumwalt, Jr., was relieved as commander of naval forces in Vietnam by Vice Admiral Jerome H. King, Jr., occurred on May 15 aboard the USS *Page County* moored on the Saigon River near the Vietnamese navy headquarters in Saigon.[39] Mouza, Ann, and Mouzetta flew in from the Philippines to join Elmo for the ceremony. Guest speaker General Creighton Abrams awarded Bud the Distinguished Service Medal and bestowed the Navy Unit Commendation on behalf of the secretary of the navy for "exceptionally meritorious service." In a touching tribute to his friend,

Admiral Tran Van Chon spoke about all that Bud had done for the Vietnamese navy.

When Bud's opportunity to speak came, he looked back at the past twenty months and could see the map of Vietnam's waterways changing from blue to green, from U.S. to Vietnamese navy, all along and throughout the rivers of the South. He noted that the South Vietnamese navy had just demonstrated its progress by joining the U.S. Navy in opening the first thirty miles of the Mekong, then without U.S. support "made a dramatic movement to Phnom Penh and then overnight to Kampong Cham with a three-inch [*sic*] gunship and armored boats." This represented a tremendous feat of professionalism and navigation during which the South Vietnamese sailors removed nine thousand refugees and escorted merchant ships that had been denied passage back down to the harbor.

Turning to unfinished business, Bud reminded his relief, Admiral King, that the job was only 35 or 40 percent done. "There remain in this year 29 bases to be completed to replace U.S. Navy ships, there remains 7,500 repair technicians to be trained in these bases, there remains the job of upgrading the training of these beginners to the point where they relieve our senior petty officers and junior officers and take over their own middle management." In closing, Bud looked to the future as CNO by sharing what he had learned from his sailors. He felt "re-qualified in youth," meaning that from his sailors he had learned "their aspirations, the pressures under which they operate, the inducements to be discounted, the courage with which they participate nevertheless to the fullest in the support of their country—and I pledge myself to represent them in my leadership of the U.S. Navy."[40]

Vietnam and Bud Zumwalt would be forever linked. As he was preparing to depart Vietnam, Bud wrote Commander Pham Manh Khue in the Second Coastal Zone, Nha Trang. "As the time draws near for me to depart Vietnam, I feel a great sense of nostalgia for . . . the pursuit of our common goal of victory in Vietnam. I assure you that my support for the Vietnamese navy will never cease."[41] As CNO Bud would do everything possible to support ACTOV and the training of the Vietnamese navy. Admiral Chon's son, Tran Van Truc, would enter the U.S. Naval

Academy as a plebe in the summer of 1970, and Bud promised to keep
a special eye on Chon's son. The Zumwalt home became Truc's home
during holidays. Bud wrote regularly to the superintendent of the academy, inquiring about Truc's progress, and passed this information on to
Chon.[42] Four years later, both Bud and Chon attended Truc's graduation
from the academy.

The other link to Vietnam was through Bud's son Elmo, scheduled
to be released from the navy in early August 1970. Bud tried convincing
Elmo that he should remain in the navy as a career, just as Bud's father
had convinced him to do, but Elmo had other plans. Elmo was ready to
leave Vietnam. "The sense of adventure and self-testing that had first
drawn me to Vietnam had long since gone. Now I just wanted to get out
alive and marry Kathy."[43] Bud wrote to Dr. James Caldwell at the University of North Carolina, thanking him for all the help he had given
Elmo and his family. "And who am I to stand in his way?"[44] In an August
4, 1970, note to Lew Glenn, Bud reported that "Elmo returned healthy
and fit from his tour in Vietnam during the latter part of June."[45] No one
knew of the ticking bomb that Elmo carried inside his body.

In order to consult with his commanders in the Sixth and Seventh fleets,
Bud departed Vietnam via a circuitous route. He went to Japan to meet
with a senior military leader, Admiral Shigeru Itaya, who twenty-five
years earlier had been a commander in the Battle of Surigao Strait when
Bud was a young destroyer lieutenant. This visit was followed by meetings with Chiang Kai-shek in Taipei and a briefing from Prime Minister
Lee Kuan Yew in Singapore. From there he flew to Naples for consultations with the commander of NATO's southern forces, Admiral Horacio
Rivero, Jr. In Brussels he consulted with General Andrew Goodpaster,
supreme allied commander, Europe.

Bud planned on going to Vienna to check in with his mentor Paul
Nitze, who at the time was serving as the American delegate for the secretary of defense at the Strategic Arms Limitations Talks (SALT). Secretary Chafee nixed this idea because Admiral Moorer might once again
become concerned about the Zumwalt-Nitze axis. Chafee did not want
his new CNO to alienate the new chairman of the JCS. Instead, the JCS

representative, Lieutenant General Roy Allison, met Bud in Brussels for a limited briefing. Bud knew that from Nitze he would have received a wider vision of the dynamics of world and superpower politics, as well as insights into relevant domestic considerations.

From Brussels Bud went to London and then for a much anticipated and needed family vacation in Bermuda. He had a month to get ready for his new position, but suddenly fate threatened to upend everything. Bud became very ill while in Bermuda, where he was enjoying a few days of well-deserved rest and fun with his family. He was so sick that he needed to be medevacked from Bermuda to Andrews Air Force Base. Waiting on the tarmac was his personal physician, Bill Narva. "You look like shit," said Narva. "I feel like shit," replied Bud.[46]

Narva saw immediately that Bud was dehydrated and in need of fluids. He personally made the decision that Bud must go directly to Bethesda Naval Hospital. In order not to raise press attention, Bud would go in Narva's car, not by ambulance. Marine aide Mike Spiro took Ann and Mouzetta home while Mouza stayed with her husband. Tests showed that while in Vietnam he must have contracted giardia. He then had an allergic reaction to the medication, so bad that doctors thought he was having a heart attack. "I had the sinking sensation that after all these years of striving for a goal and having been appointed to the top spot now, for physical reasons I might never get there."[47] The first question Bud asked Narva was, "Will I be able to take the oath in ten days?" Narva assured Bud they would both be there.[48]

One of the most colorful and tradition-bathed of naval ceremonies occurs at Tecumseh Court at the Naval Academy for changes of command. At 11:00 a.m. on July 1, 1970, the ceremony at which Bud Zumwalt relieved Admiral Thomas H. Moorer as chief of naval operations began. On the way, the Zumwalt limo broke down on Route 50, three miles from Annapolis. Dressed in full regalia, Bud and Mouza hitched a ride with a young sailor, who could not believe he was escorting his new CNO to the change of command.

In the audience were Bud's father, Mouza, and all four children, who over the years had shared the pleasures and borne the hardships and separations with patience and good humor. The day belonged to them as well. The ceremonial party—consisting of Secretary of Defense Melvin Laird, Secretary of the Navy John Chafee, Admirals Moorer and Zumwalt, the retiring chief of chaplains, the superintendent of the Naval Academy, the judge advocate general, and the new chief of chaplains—marched onto the stage in front of Bancroft Hall. "Here in the open area with the famous statue of Tecumseh directly in front of our stage with Bancroft Hall behind us and the Naval Academy Chapel off to our far left and the beautiful Severn River on our right, it seemed truly appropriate for this important and symbolic moment to be taking place," Bud recalled afterward.[49]

After the national anthem, Secretaries Chafee and Laird offered brief remarks, commenting on the superb performance given by Admiral Moorer during his tenure as CNO and expected in the higher station to which he was being called. Both secretaries welcomed Bud Zumwalt to his new command of the navy. Admiral Moorer gave his farewell speech, reading orders detaching him as chief of naval operations and ordering him to the job of chairman of the Joint Chiefs. This was followed by the traditional nineteen-gun salute, based on the superstition that gun salutes should be an odd number preceded by four ruffles and flourishes.

With the honors completed, Bud Zumwalt's time arrived. The oath of office was administered by the navy's judge advocate general. Placing his hand on the same Bible that Admiral Farragut used aboard his flagship, the USS *Hartford*, during the Civil War, Bud took his oath of office. Admiral Moorer then turned to Captain Harry Train and said, "Captain Train, break my flag." The flag was then hauled down. Bud read his orders, turned to Captain Train, and said, "Captain Train, break my flag." After the flag with four stars was broken at the flagpole, Bud turned to Admiral Moore, saluted him, and said, "Admiral Moorer, I relieve you." Moorer returned the salute, saying, "Very well."

Turning to Secretary Chafee, Bud saluted. "Sir, I report for duty as Chief of Naval Operations." Chafee replied, "Very well." Turning to the secretary of defense, Bud said, "Sir, I report for duty as a member of the

Joint Chiefs of Staff." Laird replied, "Very well." This too was followed by the nineteen-gun salute, preceded by four ruffles and flourishes.

Bud Zumwalt was now officially chief of naval operations. His three predecessors had been aviators; the person he was replacing, who now was his boss, had not supported the nomination.[50] In his relieving remarks, the new CNO focused on looming challenges, which included a Soviet navy increasing in size, versatility, and quality and confronting the United States in areas of the oceans once considered Free World Lakes. National priorities were changing, and the navy needed to compete for personnel and resources with new and innovative programs. Meanwhile, the navy had to simultaneously continue the war in Vietnam while planning and transitioning toward the navy of the future. In closing, Bud said that the one special virtue in his selection was that only nine years earlier he had been a commander and for the past twenty months he had been closely associated with the young officers and men of the Brown Water Navy. He was in tune "with the problems, hopes and aspirations of the young at a time when inadequate personnel retention is becoming our greatest problem." He promised to tackle and reverse these current adverse personnel trends by making the navy a place where young people wanted to be. "Let us assign my marks based on results during the next four years."[51]

With that, the Navy Band played "Anchors Aweigh."

During the transition period, Secretary Chafee had shared his thoughts on matters that would be affecting both of them.[52] "The most important thing that Chafee did to reshape the Navy was, of course, to appoint Zumwalt. It had to be a revolutionary thing in his time," recalled Stansfield Turner. "I think it was Chafee's frustration with the personnel management procedures of the Navy and the flag detailing procedures that tempted him to take a radical step toward a Zumwalt."[53]

Chafee urged Bud to think carefully about "where the Navy is going," because "as soon as we can decide just what kind of Navy we want, then our fulfilling that requirement is made much easier." Chafee knew this question was still unsettled. "We just haven't made up our mind to date as to what type of a Navy we want. Is it to be a navy with a broad thin base that can be filled out in mobilization (many ships lightly manned) or is it to be fewer ships adequately manned and equipped? This isn't an easy question but one we have got to wrestle with soon."[54]

Chafee framed the next issue as one of "greater quality control on our personnel." Considering the marines, he said, "Their philosophy that it's a privilege to qualify for the reduced Marine Corps has much merit." Chafee wanted to focus on "retention of just the best as we go into these big draw-downs of personnel. I think we can put greater challenge to the men to measure up to the Navy requirements rather than our sounding the alarm so much on our problems of keeping people." The secretary raised the challenge of giving people greater responsibility. Focusing specifically on new criteria to be used when selecting captains for promotion to rear admiral, Chafee noted the navy had always selected men of character, distinguished by loyalty, physical and intellectual courage, tenacity, respect for the opinions of others, kindness to subordinates, candor in rendering opinions, and capacity to inspire and lead. "We seek in addition to the above characteristics, men of high professional competence who have the ability to think logically, express themselves clearly both orally and in writing, and finally who evidence a capacity for growth in future years."

The new navy also needed to be prepared to fight, and therefore "we need men who will be superb leaders in wartime." Youth should not be a deterrent, and the standard should not be whether they are potential CNOs. "I would hope that you would select a few iconoclasts—original, provocative thinkers who would be unlikely CNO material but would stimulate the Navy to constantly re-examine its premises and whose selection would encourage those in the lower ranks to do likewise with the realization that they are not just tolerated but in fact welcomed. Finally, full consideration must be given for service in Vietnam. This is the crucible where men are daily being tested and those who perform well are worthy of every consideration for higher Navy responsibility."[55]

Chafee had given his new CNO a broad menu. "I am really looking forward to the years ahead with you and I know we can meet the challenges that are certainly there." On July 1, 1970, his first day on the job, Bud wrote Chafee that "in the years before us the problems in both the personnel and material areas will be great. I am sure that together we can help make the Navy ever better for the people and for our country."[56]

In order to meet these challenges, Bud created a small study group to look at the present status and future possibilities of the navy. Project Sixty, named for the sixty-day deadline imposed by Bud so that he could report to Secretary Chafee, was revolutionary in every respect. Its main theme was modernization of both equipment and people. The project reflected Bud's vision for the long-term future of the navy by rebalancing the surface, subsurface, and air components in order to counter the growing threat of the Soviet navy.[57] No previous CNO had ever tried to undertake this type of reform agenda, involving strategic modernization within the context of maintaining a high-quality all-volunteer force with sufficient capability during the modernization process for the navy to continue to perform its mission.[58]

Project Sixty also reflected Bud's philosophical approach to institutional change. On his own initiative, he circulated to all flag officers a 1950 article written by Elting E. Morison titled "A Case Study of Innovation."[59] The essence of the article was that an entrepreneur "imbued with an overriding sense of social necessity" was needed to lead revolutionary change in the navy. Otherwise, the bureaucrats and naysayers would win. Bud saw himself as that type of agent and envisioned his flag officers as joining the brigade. Bud also knew that if he went through normal channels, Project Sixty would need to be renamed Project 365 because it would take at least a year to get a report through the OPNAV bureaucracy's "chop and approval" process. Moreover, whatever emerged would be diluted of anything worthwhile. The freedom for strategic creativity that he had in Vietnam when designing SEALORDS did not exist in Washington, where each branch had its own lobby and interest in Congress. Bud therefore directed Project Sixty to the secretary of defense rather than OPNAV. He would make his case first to Secretary Chafee for the buy-in and then to Secretary Laird for the commitment.

Bud wanted his longtime associate and friend Worth Bagley, a destroyer man and not an aviator or submariner, to lead the study. But Bagley needed a few weeks to disengage from the Seventh Fleet. Bud could not wait, so he asked Captain Stansfield Turner to serve as interim manager until Bagley came aboard. Turner was a destroyer officer, a captain already selected for rear admiral then serving as executive assistant to Secretary Chafee. Turner became one of Bud's strongest advocates for revolutionary change. "Just before becoming CNO, Bud called me in and told me what he wanted me to do, what he was going to call Project 60. It was overwhelming, as I sat there and listened to him with my four stripes on and two stars in my back pocket," recalled Turner. "I said to myself, 'What this man wants me to do is to tell him what the shape of the Navy should be, its rationale, its definition, its purpose, and its composition for years to come.'"[60]

Turner had been keeping a binder of suggested changes and ideas for Chafee, with recommendations like developing antimissile defenses, lessening strategic submarine vulnerability, increasing the length of command tours for commanding officers, improving wartime fleet exercises, and eliminating the intense competition among warfare communities. "I would sit home at nights and work up ideas for the Secretary when I was working for him. Then, when I saw Zumwalt coming on the horizon, it was just a marvelous opportunity to get my ideas to the front man," recalled Turner, who gave the binder to Bud.

As Turner later recalled, "There was absolutely nothing that one would possibly want to say about a navy that wasn't in the charter that he gave me to do in sixty days."[61] Bud urged his associates to "work against the mindset" prevalent in the navy and to ask questions "that fell between the cracks" of the surface, air, and submarine officers' fiefdoms.

Turner described the forty or so days he worked on the project as the "most frantic" period of his career. He was asked to develop both general and specific proposals. The general proposals involved ones of philosophy and direction for the future that would fit with the CNO's views but were also practical enough to find consensus in the ranks.

Turner would often send Bud a two-page decision paper, the first page laying out the problem, the next page presenting a few courses of action. Bud would then call small group meetings that included only himself, Turner, the flag officers affected, and Emmett Tidd, the newly appointed coordinator of decisions. Tidd would be responsible for monitoring the implementation process, which Bud knew was the burial ground for many good projects. Over the years, Tidd had developed a system for tracking staff work and expediting program development. Using multicolored tasking directives, each color signifying a different level, staff officers were given a specific and strictly allotted time to act on a directive. Everyone in OPNAV staff was expected to march to Tidd's drum because he had the CNO's complete backing. Tidd became Zumwalt's SOB. Implementation meant convincing the action officer to get it done. A green strip on the edge of a piece of paper indicated a CNO/VCNO decision, based on Arleigh Burke's advice to Bud that "if you really want a decision carried out, send for the action officer after you have made it and convince him it's a good idea."

Project Sixty identified and prioritized four major missions for the navy of the future: assured strategic retaliatory potential, sea control, projection of power ashore, and overseas presence in peacetime. "The significance of this mission prioritization should not be missed."[62] Power projection—carrier and strike aviation—had previously dominated the navy's thinking. By moving sea control ahead of power projection, Zumwalt was signaling "his intent to shift the Navy to the direct Soviet maritime threat and to U.S. dominance at sea."[63] From Project Sixty came the concept of sea-control ships. A high-low concept of weapons-system procurement was developed, whereby the navy would purchase small numbers of highly effective ships and aircraft—such as another nuclear carrier, nuclear submarines, and high-performance jet aircraft—while at the same time developing a new family of low-cost ships—such as the patrol frigate and sea-control ship, giving the navy new offensive weapons platforms and systems to meet global commitments. A Resources Analysis Group was created in the office of the CNO to provide overall review and analysis of major weapons systems, get better control of costs, and improve credibility with Congress. Secretary Laird thought the system worked so well that he imposed it on all services.

There was always one formidable exception when it came to building consensus for the programs and priorities outlined in Project Sixty. Vice Admiral Rickover came to meeting after meeting with point-by-point arguments favoring nuclear propulsion for most vessels. The surface navy was the greatest sufferer under what Bud described as "the Rickover malady."[64] Project Sixty planners had endorsed inexpensive gas-turbine propulsion systems, believing that nuclear power for low-capability ships was too costly. Rickover's main concern was that new programs for combatant ships might not fall into the Rickover criteria for mandating nuclear power. And the sea-control ship was one of those. "I was rather familiar with all of the fine grain of the points that Admiral Rickover had made because he and his staff had thrown them at me in spades during that period in 1967 when as the navy's director of Systems Analysis, I was responsible for the major fleet escort force level study supplement on endurance."[65]

Bud offered to support two nuclear programs in return for Rickover's support on Project Sixty. "In my position, I could and did try to minimize contacts with Admiral Rickover," recalled Bagley. "I dealt with him necessarily at times, but it was always difficult and usually unproductive to have a discussion with him. You often had to listen to a unilateral discourse rather than reason on an issue. His mind was always made up."[66] Admiral Moorer had another way of dealing with Rickover. "Moorer never called Rickover and whenever Rick called him he'd say, 'Yes, Rick, Yes, Rick, Yes, Rick,' and then hang up and never do anything about it." Bud's style "was to take him on, take Rickover on, if I disagreed. It caused a hell of a lot more friction."[67]

The climax of Stan Turner's six weeks on Project Sixty came on August 26, 1970. In the CNO's conference room, Turner led the briefing of general conclusions and specific recommendations of Project Sixty to a group of navy flag officers, including Rickover. "Bud had me put on a one-man, unsupported, unhelped slide show presentation to all of the admirals in Washington, including Rickover—not all, but all who could fit into the conference room at OpNav, all the barons and czars of the

Navy. Here I was, still a captain, wearing captain's insignia. I got up and presented what we had come to in Project 60."

Turner was departing the next day for his daughter's wedding and to a flotilla command in the Mediterranean. He received few questions or comments from the assembled group of admirals because "they knew Zumwalt was behind this." Few had any desire to challenge the new CNO. "I walked out of that room into a car," recalled Turner, "and drove to the airport and literally changed my shoulder boards myself in the automobile. I got on the airplane as a rear admiral and flew to my daughter's wedding." A few months later, Bud wrote to Turner, saying that "much of what we are doing each day here is following the sound precepts you set forth in the Project 60 Concept and supporting program. Your work is of the highest importance in setting a new direction and I am tremendously grateful for the outstanding job you did. It reflects faithfully your energy and intellectual capacities which I look forward to relying on again before too long."[68]

Between August 26 and September 10, Worth Bagley took charge of preparing Bud for his briefing of Secretary of Defense Laird and Deputy Secretary of Defense David Packard. In a memo dated September 16, 1970, to all flag officers and marine general officers, Zumwalt reported that Project Sixty had been completed, that Secretary Chafee and he had made the presentation on September 10 to Laird and Packard. "I consider that the substance of this presentation sets forth the direction in which we want the Navy to move in the next few years. The decisions that we make, and implement, at the command levels of the Navy should be consistent with these concepts."

The September 10 briefing covered philosophy, missions, capabilities, and problems, with the backdrop of the substantial Soviet naval and nuclear threat. Bud emphasized the declining state of American naval power in the first official navy paper to explicitly articulate these four types of capabilities—assured second strike potential, sea control, projection of power ashore, and overseas presence in peacetime. Bud endorsed increasing navy force levels to a point "commensurate with two-ocean needs." In reviewing the wartime role of the navy, the CNO emphasized sea control over the projection of power ashore and proposed specific initiatives, like development of a type of aircraft-capable ship smaller than most carriers, the sea-control ship (SCS); development of patrol frigates; development of a deep-ocean mine with integrated sonar, able to fire an encapsulated torpedo (CAPTOR) when it detects a foreign submarine; and development of an antiship missile system that came to be called the Harpoon.[69]

The Project Sixty paper offered a dynamic statement of the direction that the navy would move in order to ensure that a balance of power at sea was maintained and that a navy of adequate size existed for the future. "Implicit in achieving this goal was reversing the trend in the cost of weapons systems and the implementation of the high-low balanced-force concept. Project Sixty provided a rationale and codified priorities that made more efficient the manner in which the navy adapted to unavoidable reductions."[70] The navy declined from an active fleet of 769 ships in 1970 to 512 by the end of Bud's tenure. Bud liked to joke that "Admiral Moorer and I sunk more U.S. ships than any enemy admiral in history."[71]

Bud would be disappointed with his degree of success because "he had hoped to put his stamp on the navy and shift its direction and especially change the navy's orientation from carrier battle groups and the power projection mission towards increased emphasis on sea control during a period of declining force levels." The Project Sixty recommendations were referred to the CNO Executive Panel (CEP) program-analysis group "as the primary guideline for their deliberations in advising me on actions we should take and on the suitability of current programs." At the first meeting of CEP on October 24, 1970, the CNO said that Project Sixty "expresses some changes of direction, but not as many as I would have liked."[72] Most of the changes were the product of hard-fought bureaucratic battles and political compromises with the Pentagon. Two major innovations—the air-capable or sea-control ship and the surface-effects ship were never authorized or funded.[73]

Détente had become a household word by 1970, but the possibility persisted of major big-power confrontations involving U.S. military forces. American dependence on overseas sources of raw materials increased while the historically land-oriented Soviet Union deployed an oceango-ing navy second in fighting power only to that of the United States. As Bud Zumwalt assumed command, major decisions loomed as to the size and composition of American naval forces. Long-overdue fleet modern-ization delayed by the extended war in Southeast Asia was complicated by the increased cost and complexity of replacement ships and aircraft.

After a decade of burgeoning military spending and entanglement in foreign conflict, the nation welcomed the vision of lower defense bud-gets balanced by a reduction in American involvement overseas. For the new CNO, the challenge was fulfilling obligations with fewer re-sources. The cuts forced the navy to look carefully at its goals and how to accomplish things with fewer resources. Divesting itself of obsolete ships and aircraft, discarding procedures of less value, and enhancing the quality of its people through greater attention and sensitivity were the goals.

A window of opportunity existed for the navy. The Nixon Doctrine assigned the navy greater responsibilities and prominence in further-ing American foreign policy.[74] As the Nixon Doctrine evolved, the navy was at the forefront of American foreign policy designed to repel Soviet geopolitical adventures. The Nixon Doctrine presented an opportunity to restore naval sea and air power to its place in national strategy and foreign policy. With the drawdown of national forces and the reduction of land bases in other countries, a new navy could be the best and most logical vehicle for the protection of our national interests, the display of national power, and the implementation of foreign policy.[75]

The president had articulated a national defense concept that main-tained the nuclear deterrent and called for a conventional arms capa-bility to deal with one major and one minor war simultaneously. This step down from the traditional concept—two major and one minor—was clearly a recognition of restraints on resources. These restraints required a strategy of quick-reaction forces, deployed from a minimum number of centrally located positions with assurance that lines of com-munication could be made secure.

The Soviet Union was placing a high priority on development of strategic offensive and defensive forces, especially the deployment of antiballistic missiles (ABMs), new intercontinental ballistic missiles (ICBMs) with multiple independently targeted reentry vehicles (MIRVs), and construction of nuclear-powered ballistic missile submarines. The remarkable thing was the quantity of resources the Soviets managed to put into sea-based forces. Both the United States and the Soviet Union were enhancing their strategic strength by investing heavily in ballistic missile submarines along with an increasing dedication of resources to conventional naval forces since they learned their bitter lesson in the Cuban Missile Crisis. This included not only their growing nuclear-powered submarine fleet, but also their impressive buildup of multimission surface forces. This sea power was being deployed in all oceans and seas, supported by naval logistical forces that afforded these ships long endurance at sea relatively free from reliance on land bases.[76]

By turning to the sea, the Soviet Union had leaped beyond its land frontiers. For the first time since the missile crisis of 1962 the Soviet Union had accepted the risk of direct confrontation with the armed forces of the United States. This was a reversal of policy, illustrated by multination naval exercises in which the Soviets had deployed over 180 ships and submarines under centralized control and coordination. Bud did not see the Soviet Union as a power preoccupied with the idea of employing naval forces in defensive roles. The change in strategy meant that "for the first time in a quarter of a century U.S. capability to control the seas is challenged. . . . If the Soviets ever felt that they could defeat the United States Navy in a war limited strictly to the seas, they could place us in a very difficult position between economic strangulation on the one hand and resort to full scale nuclear war on the other."

All these factors weighed heavily on Bud as he prepared for his first meeting with the president on August 18, 1970. During a JCS meeting on July 29, the "true measure of the need for me to pay great personal attention to the strategic situation [was] driven home to me." The chiefs were briefed by General Bruce Holloway, commander in chief of the Strategic Air Command and of the Joint Strategic Target Planning Staff, on the results of his latest evaluation in the field of strategic war—outcomes of strategic exchange based on very detailed computer wargaming. The briefing covered threat and force levels, damage analysis, and the consequences of an exchange of nuclear weapons. "It painted a grim picture on the relative strengths of the two sides on the outcome and on the relative position of great disadvantage that we would have after an exchange," recalled Bud.[77]

For Holloway's calculation, the assumption was made that no weapons would be held in reserve; that is, there was no planned strategic reserve. Bud thought it dangerous to assume that those weapons that did not get off would then be held in reserve. "It seemed to me to be quite clear that after a massive exchange if one side had weapons left, knew where they were and knew it could control them, and the other side did not, that would be the side that won. And I set myself the task of getting this national assumption changed."[78]

Admiral Moorer was already forecasting that the meeting with President Nixon was likely to be more important than any in his previous four years on the JCS. In preparing for the meeting, Bud worked with Laird, Moorer, Charles DiBona, now president of the Center for Naval Analyses, and Rear Admiral Rembrandt Robinson, military liaison between the JCS chairman and Henry Kissinger in the White House.[79] Bud's analysis of the data showed that the United States had a 55 percent chance of winning a major conventional war at sea, and he forecast a 45 percent chance by July 1, 1971, and a considerably smaller chance by July 1972. Using probabilities on prospective outcomes in war with the additional forecast cuts, the chances of winning would be reduced 30 percent.

The declassified records from the August 18, 1970, briefing of the president reveal that Bud focused on two significant changes of the last few years in the power equation between the United States and the Soviet Union: the rapid growth to parity of the Soviet strategic nuclear forces and the rapid development of Soviet maritime power. Drawing attention to the importance of a sea-based deterrent, Bud argued that control of the seas with conventional naval forces made it likely that our ballistic-missile-system submarines would survive a conventional naval war. The huge expense of the Polaris/Poseidon program had reduced the size of the conventional navy appreciably. The next generation of ballistic-missile submarines could not be funded by the navy's general-purpose-forces budget because this would have disastrous effects on the ability to control the seas.[80]

The majority of reductions since 1968 had been in sea-control capabilities in order to save projection forces for South Vietnam. Sea control provided combined capabilities of naval air, surface, and subsurface systems to defeat the enemy at sea. Projection forces were those applied to overseas land areas—aircraft from carriers, marine forces via amphibious landings or airlifts, and the 96 percent of the logistics for all services projected via commandeered merchant marine vessels.[81]

In each measure, Soviet capabilities were improving relative to those of the United States. The Soviet merchant marine had already overtaken that of the United States in numbers and tonnage. The Soviets had significantly more ships ten years old or less and they were overtaking the United States in nuclear submarines.

The question of balancing sea-control forces and projection forces was central to planning for the navy. For the new CNO, the bottom line was that naval forces were already at the lowest possible level for restraining the Soviets from contesting control of the seas explicitly or through proxies. Moreover, the Soviet navy was emerging as a balanced and modern global navy, growing from its restricted operations in the form of a large submarine fleet. In honoring NATO commitments, the U.S. Navy needed to be prepared to defeat Soviet or Soviet-sponsored interference with sea lines of communication and to overcome determined resistance to projection of power ashore.[82]

After the briefing, Nixon called Bud aside to say that he understood that Bud was nervous, but the first goal was to nail down strategic superiority in the SALT negotiations, get out of the war in Vietnam, and then get to work on getting sufficient budget support to turn the maritime balance around. It made sense to Bud so long as strategic superiority was achieved in the negotiations.

The Limited Nuclear Test Ban Treaty of 1963 had been a first baby step, a contract between two sides who did not trust each other but were willing to see what might be possible. This resulted in the passage of a nuclear nonproliferation arrangement, a hotline, a ban on weapons of mass destruction in space, a limitation on the use of seabeds for strategic weapons, and further down the road, a separable first-stage disarmament proposal—later known as SALT.

Nitze was back in Washington during the SALT recess in June when he told Bud that "our government was offering too many changes in our position in too short of a time."[83] Nitze shared his grave concern about the optimism felt within the Nixon administration about prospects for the SALT negotiations. "It was his view that the only basis for optimism about concluding negotiations was under the assumption that the United States would continue to erode its positions until they became undesirable for the U.S. point of view and acceptable to the Soviets."[84]

On Bud's first day in office, he issued an administrative edict that "a competent study be undertaken on strategic arms limitations and their relation to naval forces. It is felt that inadequate analytic information and substantive analyses are available to support the current Navy rationale."[85] Bud had been prepared for the complexity of the strategic-arms issue but not for the way the arms-control bureaucracy would stampede normal staff processes. He had served a tour as the director of arms control in the office of the assistant secretary of defense for international security affairs and, of course, there was his service during the 1962 Cuban Missile Crisis. After coming so close to the brink, both sides wanted to reduce the chances of annihilating the world. Bud believed that the Soviets had a strategy to negotiate a test ban treaty while also initiating the largest strategic maritime construction program in history. They were trying to reduce the risks of nuclear war while working to shift the correlation of forces in a way that made it possible for Soviet foreign policy objectives to be supported by an increasing Soviet military capability relative to the West.

In a letter to Abe some sixty days into his job, Bud provided his former boss with an overview of his first two months as CNO: "As you know, I relieved Admiral Moorer in a ceremony at the Naval Academy on the first of July and began to face the major challenges unique to the era of the early 70s. The major problem for the Navy, of course, is that the Soviets have built a magnificent Navy along with arriving at parity in the strategic arms race while we have been engaged in Southeast Asia. As a significant measure of this change, last year their Navy had more ships in the Mediterranean than ours when only ten years ago their Mediterranean presence was negligible. At the same time, as you are well aware, the severe budget restraints have caused dramatic decreases in our Navy's ship population. At the same time as a result of poor image of the military and an unfavorable domestic atmosphere, we are facing one of the worst personnel retention problems that the Navy has faced in the last decades."[86]

Bud offered an optimistic assessment from initial briefings of the president. "The Chiefs had an opportunity to discuss briefly their views with the President in the middle of August and I was able to point out some of these problems to him. I was much impressed both by what he already knew in the strategic area and how fast he was able to take aboard the points I made."[87]

Bringing Abe up-to-date on related matters, Bud explained he was "investigating means of living within budget constraints and still molding a tight efficient Naval force that can maintain a pretense of control of the seas in the face of the Soviets' build-up; and we have kicked off many programs that we hope will eventually have an impact on the personnel retention problems." Junior officers had already been brought in to brainstorm the retention problem. "Some of their suggestions have been unique and all of them have been mature and worthy of our addressing in some manner. Their primary concerns seem to be the standard Navy problems of family separations, a desire to be challenged with meaningful assignments, and a chance to be heard. I feel so strongly that we need to turn this downward retention problem around (our first term reenlistment on carriers is down to 3%) that I have taken steps to implement as many of their suggestions as we can within general constraints on good discipline."

In transitioning to an all-volunteer force, it was apparent that the navy faced a particular set of problems, not only in competing with the other military services, but in maintaining public support for an adequate navy. It would be necessary to broaden the base from which people were recruited and to make the navy more of a microcosm of society at large. The navy would be linked more closely with the mores and lifestyles of the civilian population from which it was recruiting. This would be Bud Zumwalt's next battlefield.

CHAPTER 10

ZINGERS

Your Z-grams have had an electrifying effect upon those of us lacking in excesses of gold braid and upon the men with whom we work and live so closely with. Even more than the printed Z-grams and their messages, the realization that "The Man" cares for his men.

—SAILOR P. M. MCDERMOTT[1]

Each of the candidates John Chafee had interviewed for the position of CNO recognized the challenges posed by the Soviet navy, but only one was able to articulate the belief that existing policies and practices in personnel administration posed an equal or even greater danger to the navy's future. The disaffection of the nation's youth and the loss of credibility the military had suffered as a result of the war in Vietnam required a concerted investment by those in leadership positions. In transitioning to an all-volunteer force, it was apparent to Bud that the navy faced a particular set of retention problems, not only in competing with the other military services, but also in regard to maintaining public support for the concept of an adequately staffed navy.[2]

When it came to identifying inducements for making naval service more attractive and enjoyable, Bud endorsed words rarely used to describe military service; *fun* and *zest* became the watchwords for addressing regulations pertaining to personal behavior, beginning with dress and grooming but also including operational schedules, homeporting, and job rotation—all of the things that affected the long family separations that had defined "the trauma as it affected me, my wife and our children."[3] The navy had acquired some irritating and unproductive barnacles over the years that had unnecessary and unfavorable impacts on people and their families. It was those barnacles to which he took the chipping hammer.

Three weeks into his watch, a message went out to all commands that henceforth any policy or guidance emanating directly from the CNO would be identified by Zulu (military word code for the letter Z) series numbers—Z-1, Z-2, etc.[4] The messages were quickly dubbed Z-grams. Bud's decision to use Z-grams was his tactic to make himself the lightning rod for the controversy that inevitably accompanies dramatic reform. The basic purpose of Z-grams was to permit the general policies of the navy to meet the general needs of the majority of navy personnel. Family separation had long been understood as one of the main disadvantages of navy life. The program for overseas homeporting enabled the navy to meet deployment commitments without sharply increasing time away from home. The program benefited men aboard ships as well as reduced additional deployments for the rest of the navy.

Z-grams highlighted the interdependence of human-relations goals with the requirements for order and discipline. They were designed to correct racial and gender discrimination and to fulfill legitimate aspirations for justifiable changes. They were also intended to make clear that "I desired the kind of leaders whom men would want to follow rather than those leaders who had to drive their men. . . . Good commanders put fun and zest into the daily routine. Creative leadership can produce ways of making even the dullest routine palatable."[5]

Bud's Vietnam experience convinced him that patriotism, fighting effectiveness, and morale had little to do with the length of a sailor's hair or allowing beer in barracks. The need to constrain a few potentially troublesome individuals would no longer be allowed to drive navy policy. Z-grams did not change the basic conditions of military service or the authority of commanding officers in the chain of command. What had changed for those in the chain of command was the need to pay greater attention to a sailor's individual needs, aspirations, and capabilities. Therein lay the problem, since many in the chain of command were incapable of enlightened leadership.

The Z-grams came in a flurry; 69 were issued in the first six months of his term[6] and 113 over the first two years, but only 8 in the next twenty-four months, when the focus was on implementation. Of all the names given to Z-grams, Bud liked Zingers best,[7] which was what Lew

Glenn's crew aboard the USS *Tattnall* nicknamed them.[8] "Each morning everyone looks forward to the good news," wrote Glenn. Even Bud's harshest critics were caught up in the enthusiasm of the moment, as evidenced by a note from Ray Peet, commander of the First Fleet: "Your Z-grams are having a tremendous impact out here. All the youngsters and the COs of ships are enthusiastic about them. They are serving as a catalyst for communication between seniors and juniors. The JOs are really rejuvenated."[9] After reading the Z-gram on homeporting, Mrs. Eric N. Brueland of Camarillo, California, a new navy wife, wrote, "It brought tears to my eyes for I have never seen such sincerity for others in the military expressed. . . . Just knowing that you are aware of how we wives and families fight our own little war of waiting at home is most gratifying to me."

The most publicized and prominent Zingers were Z-57 of November 10, 1970, "Demeaning or Abrasive Regulations, Elimination of," and Z-66, "Equal Opportunity in the Navy," issued on December 17, 1970. Others were equally controversial, like Z-48, "Programs for People," October 23, 1970, which created a new office in the Bureau of Naval Personnel, Pers-P, the second P standing for people. Z-48 represented Bud's attempt to institutionalize the reforms. "It is one thing to promulgate new programs, but quite another to sustain and nourish their forward progress."[10]

On November 10, 1970, Bud issued Z-57, eliminating what he often referred to as Mickey Mouse or chicken regs. Indeed, Bud had intended to title the Z-gram "Mickey Mouse, Elimination of" until his vice chief, Admiral Ralph Cousins, suggested replacing it with "Demeaning and Abrasive Regulations." The day before, two messages were sent to all officers in command alerting them of the next day's Z-gram: "It is not in any way intended that this and other NAVOPs usurp your prerogative as an officer in command—I do not desire nor will I accept sloppiness and indifference but I believe we can maintain our high standards while allowing our people to conform to today's styles."[11] Z-57 stated specifically, "I am not suggesting that a more lenient attitude toward irresponsible behavior be adopted."

Z-57 liberalized regulations and practices in twelve areas: style of hair, beards, sideburns, civilian clothes, uniforms for trips between home and base and when visiting commissaries and snack bars, attire for enlisted men at officers' clubs, salutes, motorcycles, conditions for leave, and overnight liberty. Its impact spanned the oceans. Inside the Hanoi Hilton, longtime POW James Stockdale tapped on the wall in Morse code to ask a recent arrival if there was any news from home. "Got a new CNO, named Zumwalt. No more Mickey Mouse or chickenshit."[12] Stockdale could not know that Bud was wearing a POW bracelet imprinted with Stockdale's name. Once home, Zumwalt gave Stockdale the bracelet. "No single memento of my return or imprisonment will have as profoundness [*sic*] of meaning comparable to this emblem of your faith in me," wrote Stockdale.[13] On October 11, 1995, Stockdale gave Bud a copy of his book, *Thoughts of a Philosophical Fighter Pilot.* The handwritten inscription read, "For my Boss at a crucial time, Bud Zumwalt. It was he more than any other man who gave me a boost when I came out of prison, and the confidence to press ahead in the Post-Vietnam years."[14]

Bud saw no evidence that listening to popular music or wearing neatly trimmed beards, mustaches, and sideburns affected sailors' ability to operate ships and stations in a disciplined, seamanlike manner. This did not stop critics from mocking the loosening of standards and claiming that such appearances fostered permissiveness. From San Diego, a bastion of the growing anti-Zumwalt fervor, retired vice admiral Lorenzo S. Sabin, Jr., wrote Bud that a prominent La Jolla citizen was shocked "to see Navy personnel with grotesque facial hair and head hair, wearing sloppy uniforms in the streets, some with trousers tucked into clips roaring the streets on motorcycles."[15] Blanche Seaver was especially bothered when attending a change-of-command ceremony and seeing beards on half the band, making them look like "hippies from Haight-Ashbury."[16] In reply, Bud pointed out that several of the earliest CNOs would be out of compliance or nonregulation by contemporary standards.

The authorization for beer in the barracks was limited to those quarters housing officers and senior enlisted personnel, but that did not stop critics.[17] Ruth Collins of Cincinnati wrote, "Concerned citizens are aroused over the possibility of increasing drunkenness in the Armed Forces. We need alert men guarding our country. We remember Pearl Harbor!"[18] R. Hagmaier followed suit, "Surely you don't think beer makes a better sailor? I remember hearing that Washington's army was able to beat the British on a Christmas Eve because the British were 'in their cups.'"[19] After watching Bud's guest appearance on *The David Frost Show*, Mary Carr wrote, "I must say that I think you are going for publicity and popularity and don't seem to give a damn if you wreck the navy. Long hair, beards, beer on board ship means sloppiness, laziness, a hippie cult among the men—gradually 'pot' and eventually stoned and discredited United States navy. How could you betray your men so?"[20]

Retired commander Edward Loftin wrote directly to Secretary Chafee, saying he was "appalled to read the latest Z-grams permitting long hair, rock music, liquor in barracks, and several other 'mod' innovations by the new CNO. . . . I am not a traditionalist, but traditions have been built over a long period of time and bear examination quite well. . . . These Z-grams are getting more radical as they increase in number. I implore you to replace or subvert the new CNO before the state of discipline is reduced to shambles throughout the navy. History has proved that an undisciplined military force become a rabble which throws down its arms and flees in the face of the enemy."[21]

Bud was on safe ground, because he had the support of Secretary Chafee, who shared Bud's view on all of the changes being implemented. "One's ability to fight isn't affected by the length of his hair . . . actually it's going to improve our fighting ability in that being a more attractive Navy, moving with the times, we'll be able to keep more of our good people," said Chafee. "Some of the greatest fighting men in the Navy had hair considerably longer than even under these regulations."[22]

While some critics focused on appearances, others addressed a much more serious issue, involving erosion in the chain of command. The argument was that young enlisted men thought they could now go directly to the CNO with grievances, bypassing the petty officers. This manifested itself during the racial disturbances aboard three ships in 1972, but in some ways it was miraculous that in a navy of over six hundred ships, with all the changes going on, there were incidents on only three ships. Rear Admiral Joe Stryker, who had been in charge of Bud's summer plebe class at Annapolis, wrote, "You will never know how much I approve what you are doing to humanize the Navy and make it fun again. . . . The only adverse remarks I have heard on your plans have been from old, fat, retired Army officers. Most retired officers tend to become Colonel Blimps as far as anything you have proposed and if you get any static from them for anything you have proposed, I would pay no attention to them. More power to you and I am sure you are on the right track."[23]

Restoring zest, fun, and satisfaction in a navy career meant removing the continual pressure under which people lived. This effort led to examining the root causes of instability and then a search for solutions. To outsiders, some Z-grams may have seemed trivial, but to those inside they were revolutionary.[24] Z-04 gave thirty days' leave between assignments, alleviating some of the hardship in back-to-back deployments. It took one year and three separate Z-grams to do something as simple as allow a sailor to store civilian clothes aboard ship. Z-05 created a pilot program allowing first-class petty officers to keep civilian clothes on ship to be worn on liberty, something only officers and chief petty officers had been able to do in the past. This program was such a success that a few months later, Z-68 extended the privilege "to all petty officers on all ships."[25] Six months later, Z-92 let all nonrated men do the same.[26]

The enlisted men of Attack Squadron 145 were excited enough by these changes to all sign a letter of thanks to their new CNO: "We are happy the Navy's leaders are changing standards to correlate with the changing times and the responsibilities of a petty officer. We hope that this latest change will instill additional incentive in our non-rated men to advance and attain the responsibilities of a petty officer. We wish to forward our 'Well Done' to you."[27]

Another change involved uniforms worn by sailors below the rank of chief petty officer. The uniform being replaced had been introduced in the previous century, the design dictated by weather and the working conditions of the times. After many surveys and focus groups, a change was made, eliminating service dress khaki for officers and chiefs, service dress blue and white jumper uniforms for junior enlisted men, and service whites for chiefs. The traditional bell-bottom trousers and jumper uniforms were replaced by a new uniform. The change ignited a firestorm of controversy. One reassuring note arrived from an early mentor, retired admiral James Holloway. "You are doing so many things for our Navy—things that had to be done under present day operating and other pressures, and which would never have been done without your perception, and the power and prestige of the Chief of Naval Operations in directing and underwriting them."[28] Bud appreciated Holloway having his back on this issue of enlisted navy dress uniforms, stating, "I strongly believe that this change will go a long way to giving added dignity to our individual Navyman, and increase the attractiveness of our service as a place to make a rewarding career."[29]

Internal polls taken six months after the first Z-grams showed strong disapproval among 15 percent of the flag officers, captains, and chief petty officers; the discontent attenuated rapidly as one went down to the more junior ranks. Years later, in a letter to a friend, Bud observed, "I think we took the staid old institution for about as much as we could move it at the time. I like to say that I have a wonderful list of friends and a wonderful list of enemies as a result, and I am very proud of both lists."[30]

The young men and women in the navy loved these changes, especially the aviators. Admiral Jerry Miller recalled that when Bud first became CNO and started issuing Z-grams, "His popularity with young people was fantastic." The venerable aviator Admiral John Hyland was scheduled to receive the Tailhooker of the Year award, which goes annually to an individual with a distinguished career in sea-based aviation. In 1965 President Johnson went over the heads of seventy-two more senior rear admirals by promoting Hyland to vice admiral and commander of the Seventh Fleet. Soon thereafter, Johnson promoted Hyland to commander in chief of the Pacific Fleet, and he received his fourth star.[31] Hyland had been a hero to young aviators.

Bud found Hyland to be likable, but "he just didn't seem to me to have four-star competence. Neither then, nor when he came out to Vietnam, did he express any interest in what was going on there. He didn't seem interested in my requests for help on various things. He seemed to me to be a very competent naval aviator, four-stripe captain, who had, on the Peter Principle, exceeded his level of competence. He just didn't seem to have a grasp."[32]

Bud ended up wrangling with Hyland after just one day on the job, which senior aviators saw as the CNO's attempt to take apart aviation leadership in the navy. In a "Personal and Private" handwritten letter dated July 1, 1970, to "one of the finest bosses I ever had," the new CNO asked Hyland to retire so that new blood could come into the flag-officer ranks. Hyland had been fighting Zumwalt to the very end on the new daisy chain for early retirements. "A lot of people used to say, in discussing Zumwalt, any guy who had the guts to fire Johnny Hyland had the guts to do anything. No holds barred," recalled Miller. In a personal handwritten reply, Hyland wrote, "I don't want to retire. I can hardly imagine not participating any longer in what we do in our great Navy." Hyland resented being forced out. "When your extraordinary appointment was announced, it was clear that some of the 4-stars would have to go." Hyland was only fifty-eight years old and beseeched Bud to reconsider, telling him, "I've received a remarkable number of expressions of regret over my 'fate.' Obviously it is nice to receive such things, but they do testify to my service reputation, and I think it follows that I can still be a valuable member of your team. I very much want to be."[33] On October 15, Bud sent another personal handwritten note, describing the decision as "the most personally painful I have ever had to make."[34] Hyland was ordered to retire as of January 1, 1971.

While all this was going on under the radar, Bud had been invited to be the principal speaker at the annual Tailhook banquet in Las Vegas. At the last minute, Hyland refused to attend the award ceremony because Bud, not Hyland, was the featured speaker at an aviators' event. The Tailhook meeting and award ceremony went on without Hyland. "There must have been 25–30 Admirals there to honor Hyland, along with hundreds of sailors, but Bud was now the drawing card," recalled Admiral Miller. The sailors started chanting, "We want Z, we want Z." They were soon "standing on their chairs, swirling their napkins around and so forth. They wanted him. He had that much appeal, but you can imagine what the reaction was among the aviation admirals, some of the senior people."[35] When Zumwalt entered the room, he gave "that perfected Tarzan yell. Here he is, the brand-new CNO, and he's got these Z-grams out, and he tells this great story about Tarzan and he ends up with this tremendous Tarzan yell. And it just brought down the house. It was absolutely fantastic. He couldn't have established himself better with the young aviators."[36] Meanwhile, the older admirals seethed, plotting their revenge.

One of the most important of the early Z-grams was about the creation of retention study groups. When Bud became CNO, first-term reenlistment rates in the antiwar, antimilitary era were at an all-time low—9.5 percent overall and less than 4 percent on aircraft carriers. The navy was clearly experiencing a hemorrhage of talent. Z-01 convened a junior officer retention study group to focus on the reasons why so many officers and enlisted men were no longer making careers in the navy and to advance solutions. The idea of retention study groups grew from the Vietnam ACTOV Personal Response Program that sought to identify the causes of poor morale and low retention. In a letter to Chon, Bud discussed the "exhilarating and challenging" program for the retention of officers and enlisted personnel and wanted Chon to know how much he had learned from him, evidenced by the creation of the post of assistant chief of naval operations for benefits and services, which was patterned after Chon's deputy chief of staff for political warfare.[37]

The idea was for groups to meet in seclusion and brainstorm ideas for improving the quality of navy life and thereby improving retention and morale of naval personnel. "He was as good as anyone I know at nurturing creativity," recalled Chick Rauch. "I can't remember him ever saying no to anyone who had an idea."[38]

"The excellence of our people has long been our heritage—it is my source of strength. I intend to further enhance this reservoir of strength by assuming as my first task the improvement of many aspects of the naval career," Bud announced in his first all-navy message. "There is much that will require the support of my civilian superiors and Congress. These changes will take time. There are other improvements which we can make within the uniformed Navy. These can come more quickly." The retention study program would be co-chaired by two junior officers, Lieutenant William Antle and Lieutenant David Halperin. Halperin spent eighteen months in Vietnam, part of the time in Operation Sea Float. As an operations officer, he became close friends with Elmo. Halperin planned, executed, and supervised an extensive area-development program involving both the U.S. and Vietnamese navies. The net effect of these innovative programs was to increase economic activity and improve the general well-being of inhabitants. After four arduous months in one of the VC's most remote and controlled areas, Halperin was assigned to the Plans Division on Bud's staff, where he was given the job of designing a time-phased plan shifting responsibility for certain types of in-country air support to the Vietnamese air force. Bud thought Halperin exhibited "the rarest sort of practical judgment and the broadest analytical skills" and found him to have "an exceptionally keen and inquisitive mind" and to be very conscientious. In a letter to the Harvard School of Law, Bud rated Halperin "#1 in 10,000."[39]

Zumwalt did not know Lieutenant William Smoot Antle nearly as well. In 1970, Bud asked the chief of naval personnel to select three of the most outstanding young officers from each of the major officer communities to head the retention task forces. Antle reported in August 1970 as the submarine representative and came to the job after a distinguished career in the Naval Academy. He had graduated 28th out of 868 and qualified as an engineer officer on a nuclear-powered submarine. Antle exhibited "the rarest sort of poise and discretion" and was "as articulate and perceptive as any junior officer I have worked with in the past several years," wrote Zumwalt. Bud believed that if Antle had remained in the navy, "he would rapidly be promoted through the ranks to Admiral."[40]

The concept behind retention study groups was to assemble a group of about a dozen officers and enlisted men from each part of the navy. Wives were soon involved, doubling the size and number of recommendations. On July 20, 1970, the first group convened, comprised exclusively of junior officers from the aviation community—some who had already decided to get out, others who were undecided, and some who had committed themselves to naval careers. This group came up with sixty-three specific recommendations, including beer machines in barracks, a twenty-four-hour mess line on carriers, squash courts on all bases, permission to wear flight suits anywhere on base, a goal of six months at home before deployments, giving aviators an opportunity to learn seamanship by assigning them as navigators to cruisers or destroyers, establishing career-counseling programs, and requiring that fitness reports be reviewed and signed by the candidate before being forwarded to higher authority.[41] "Your message on retentions was well received by the young officers in the fleet and hopefully it will start the trend line up rather than down," wrote Lew Glenn. "We definitely need a positive program to help our retention problems."[42]

Retention study groups eventually included destroyer and mine-force officers, amphibious and auxiliary officers, POW/MIA dependents, WAVES (Women Accepted for Voluntary Emergency Service), civil engineers, minority women, ROTC midshipmen, and enlisted persons from the aviation, service-force, amphibious, submarine, destroyer, and mine-force categories. The format for the groups evolved through trial and error. Members met for a week trying to develop a consensus on ways of improving the situation; they were instructed not to make a gripe list. Rather, their focus was to develop discrete, implementable recommendations that addressed specific problems by improving conditions. The CNO did not need to hear long eloquent speeches about what was wrong. Each group was promised time with Bud personally, so that the system could not dilute their recommendations. The chairman of a retention study group was to moderate these one-week seminars and then brief Zumwalt, the secretary of the navy, and flag officers.

Bud instructed his staff that they were not to prescreen or drop wild ideas or in any way limit what he heard or what he was exposed to. "Each group was allowed to blow off steam at the beginning, but I then made them see that they only had a short time with Zumwalt and they would blow it if all they did was repeat what he knew—that things were bad," said Halperin. "They needed to come up with practical remedies."

Halperin put them through a dress rehearsal the night before meeting with the CNO, making sure that the recommendation was framed properly and that each member got a chance to participate.

At the end of the week, the group assembled in the SECNAV conference room, where Bud and senior flag officers and staff were briefed. "For myself, a meeting with the retention study group was likely to be the high point of my week," wrote Zumwalt.[43] Secretary Chafee and senior staff came to as many meetings as possible. Each study group presented between sixty and eighty discrete recommendations. "Green stripes"—approved ideas—were drafted as orders and issued to action officers. The action required took many forms. For some ideas, it was "Do it now," as in changing the name of the BT rating from boiler tender to boiler technician. For others, it was evaluating and considering alternatives and costs with a plan of action and milestones or perhaps just preparing a background memo for the CNO on the problem and the recommendations. Emmett Tidd followed up on the implementation—those implemented at the Washington level resulted in Z-grams, pilot programs, proposed legislation, or changes in instructions. The others were referred to the appropriate level.

Most Z-grams came from these retention group studies. A Young Turk program, soon called the Mod Squad (Destroyer Squadron 26), was placed under the command of Richard Nicholson, a protégé of Bud who served in a staff position in Vietnam and commanded the Operation Market Time coastal-surveillance task force. This demonstration program attached each officer to a billet one rank below the norm. It demonstrated to the entire navy the benefits of giving young officers more responsibility than they would normally have at their career stage.

Howard Kerr chaired the lieutenant commander retention study group. After receiving the report, Bud wrote his longtime aide and friend, "I concur with your premise that the high quality junior officers today are troubled because of some of those problems which they realize they too will encounter if they decide to remain in the navy. As you know, my position is that we must endeavor to identify and move our best officers to key positions as soon as they are ready in order to maintain the challenge for them."[44]

From aboard the USS *Constellation* on April 12, 1972, Jerry Carr wanted his CNO to know that he had always considered himself a short-timer on first enlistment but that after the Z-grams started coming out, "I decided to reenlist and am intending to do so again when the time comes. Thanks to you, the Navy can finally be a very challenging and rewarding career. I have nothing but the greatest admiration for you and the job you are doing."[45]

Just three months into Bud's first year, Dick Nicholson wrote that "the young people (officer and enlisted) are excited about our new Navy and their future in it. My comment can only be 'They ain't seen noth'un yet.' . . . It has been exciting here in left field watching the stones in Washington being lifted and the worms running for dark corners. For the first time in my 23 years, someone is actually giving more than lip service to our personnel problems. I'm sure happy I was able to stay around to see it."[46] First-term reenlistment rose from below 10 percent during fiscal year 1970 to 32.9 percent during fiscal year 1974. David Halperin learned many lessons under Bud's mentoring. "By your own example I have come to understand my own limitations," wrote Halperin in a personal letter to his mentor. "Each time that I have had to acknowledge that the system, the bureaucrats might somehow win in the end, I have intuitively known that you would be able to do whatever I could not do, and that however I might knuckle under compromise—in your finer sense of judgment and principle you would hold the line. John Kennedy was fond of saying, 'Some men see things as they are and say, why? A few men dream things that never were and ask, Why not?' I think that I will always think of you in that way, Admiral: a visionary in an age without heroes."[47]

Racial relations and sensitivities within the navy were not at the top of Bud Zumwalt's list of major issues to be addressed at the start of his watch, but they inevitably surfaced during the transition to an all-volunteer environment. Not until he participated in retention-group meetings did he absorb the effects of institutional racism on individuals. He was visibly shaken by the stories he heard. Afterward, he was ready to fulfill a lifelong ambition "to throw overboard once and for all the

Navy's silent but real and persistent discrimination against minorities."[48]

The navy had an ignominious history on race, starting in 1798, when Secretary of the Navy Benjamin Stoddert banned "negroes and mulattos" from service, a ban that lasted until 1812. Substantial numbers served in the navy during the Civil War. In fact, despite bans on enlistment of black sailors, they continued to serve on ships of the fleet, numbering about 1,500 to 2,000 men throughout the first decade of the twentieth century. President Truman's executive order in 1948 established a policy of "equality of treatment and opportunity for all persons in the armed services without regard to race, color, religion, or national origin." The navy essentially ignored the order; enlisted blacks remained in the stewards branch, serving whites. Truman's Committee on Treatment and Opportunity in the Armed Services, headed by George Fahy of Georgia, sought to determine why the navy was doing such a poor job with respect to recruitment of black sailors. One vice admiral testified that this was because blacks were "not a seafaring people."[49]

The navy devised many ways to circumvent Truman's executive order by emphasizing qualitative recruitment and merit promotion. Bud always recalled his days detailing surface warfare lieutenants in the Bureau of Naval Personnel. At the start of the job, he was given a briefing on how to use the system to deny black sailors the opportunity to become officers, that is, how to get black sailors to wash out of the navy: As soon as a black was commissioned, assign him to the Recruiting Service. It was undeniable that the navy needed black officers in the field recruiting in black areas. After completing this normal tour, extend him for another year, thus giving the sailor a lengthy shore tour as opposed to the normal sea tour as first duty; thus black officers would already be falling behind. Then, assign the black officer to an amphibious or auxiliary ship, which in those days was considered less professionally challenging than assignments to combat ships. The result would be that by the end of this tour, the black officers, having had less rewarding assignments and less professionally acceptable assignments than their contemporaries, would be passed over by selection boards. "It was suggested that those few who escaped this screen by promotion could then be similarly hazarded as lieutenant commanders."[50]

The navy practiced tokenism right up until Bud took over. Bud believed that racism was "endemic" in the entire structure of the navy—from the smallest boats to highest headquarters. "In my judgment there's absolutely no doubt that Admiral Anderson, Admiral McDonald, and Admiral Moorer all sought to maintain a lily white Navy. Ike Kidd used to tell me that when he was aide to Admiral McDonald, when I was aide to Paul Nitze, that it just shivered him to listen to some of the things that were said in his office. I hasten to add that Admiral McDonald is a man whom I consider to have a basically decent Christian attitude toward people. But as late as when Mouza and I went down to visit Admiral McDonald when I was CNO and he was retired down at Ponte Vedra, he was telling me about a black fellow that was interested in going to the Naval Academy and said, "I think I'm going to help that fellow. He's a pretty good nigger."[51]

Less than 1 percent of the officer corps were African American, yet 12 percent of the population was black. In 1971 blacks made up 5.3 percent of the navy but only 0.7 percent of the officer ranks.[52] All of the other services had much better representation. Looking at those numbers, Bud realized that with an all-volunteer force, black recruitment would be a benefit for them as well as for the navy. Bud created two retention study groups, one for black officers and their wives and the other for black enlisted men and their wives. The unfiltered feedback left little doubt about what needed to be done. "I found myself absolutely astonished at the extent of my own ignorance as to the subtle ways in which the navy was discriminating. The lack of black beauty aids and soul foods in the commissaries and exchanges, the subtle forms of discrimination with regard to housing, the contention that housing would be available when telephone calls were made only to discover that housing was not available when blacks showed up in person. All were heartbreaking evidence to me that the navy, both internally and with regard to its external relationships with civilian communities, was far from a fully integrated organization."

The retention meetings were especially emotional because the stories came from the heart and personal experience. "Prior to these meetings, I was convinced that, compared with the civilian community, we had relatively few racial problems in the Navy," said Bud. At the sessions, people spoke about the total recruiting process, from the time someone came into the navy, and how because you were in the navy you were unable to do certain things, like going to a barber shop, because barbers did not know how to cut the hair of blacks. From things as small as that to the inability to get a good navy job because the good assignments were closed to blacks. There were no detailers or aides in the fast-track sections of BuPers to help blacks with assignment options.

Zumwalt's successor as CNO, and his former vice chief, Admiral James Holloway, thought that Bud "was so overwhelmed by the reports of discrimination, harassment, and downright brutality coming out of these sessions that as he told me later, it left an indelible scar on his consciousness."[53] One poignant moment during a retention feedback session remained seared in Bud's memory. One of his finest white flag officers, after hearing from a group of blacks, stood up to say that he had always been concerned for his stewards, that he had always done the right thing: "My boys have always been very happy," said the admiral. This led the wife of one of the black officers, Esther Fisher, to stand up and ask, "Admiral, let me ask you—how old is that 'boy'?" The admiral said, "Oh, I guess he's 25 or 26, why?" Mrs. Fisher said, "Do you see what I mean?" The white flag officer didn't, but Bud Zumwalt did.[54] "I had never seen him quite that way but he was visibly shaken," recalled Bill Norman. "And he said he had been in the Navy all these years and until this particular moment . . . when he heard people who were part of the Naval community, people who had served extremely well who all they wanted to do was be respected for their individual merit and worth as persons and no more, and they weren't even receiving that. And he said, 'I think I've seen it but I haven't seen it.' "[55]

But Bud Zumwalt needed help. Thirty-two-year-old Lieutenant Commander William Norman had tendered his resignation after a decade in the navy, feeling that "the unceasing strain of the conflict between being black and being Navy" was no longer worth bearing. One of the navy's most outstanding officers, he rarely received voluntary salutes, had been called a "goddam nigger" by a petty officer, and when teaching at the Naval Academy could not find anyone to rent him a room. He had tolerated it for as long as possible but decided that being navy and being black were incompatible. Norman had a friend at the minority recruiting desk who, upon learning that Norman had tendered his resignation, went to Bud to recommend that he meet with Bill Norman, even though Norman was not enamored with the idea of meeting another white CNO. "As far as I was concerned he was just another one of those people making all those promises and didn't intend to do anything," said Norman.[56]

Realizing he had nothing to lose, however, Norman agreed to the meeting, although he became irritated upon learning that the CNO had budgeted just fourteen minutes for their meeting. Norman prepared a list of twenty action items and presented them to the CNO.[57] He had decided to test the new CNO's commitment by asking him to end the navy's practice of sequestering Filipinos in the steward rating, dating back to 1919. Navy admirals loved their Filipino stewards, could not envision a navy without them, and had rigged a racially segregated rating system. If Zumwalt was serious about reform, Norman told him to start there. Within a short time, Bud came out in support for hundreds of Filipino stewards, earning praise from Taylor Branch, who wrote in support of "your plans for dealing with the problem of frozen steward ratings."[58]

Bud recalled the meeting with Norman thus: "He told me, in effect, 'J'accuse!' "[59] Knowing that he had only about ten minutes of Bud's time, Norman had prepared "[the] hardest, tightest set of notes I have ever seen and he proceeded to tick off in staccato terms a whole set of ways in which the Navy was not doing well by its black personnel." Bud hired Norman on the spot, telling him that he would accept the challenge but he needed Norman at his side.[60] It was through Norman's eyes that Bud began to see what it was like to be a minority member in the navy. The study groups revealed that blacks believed the navy was a segregationist service that cared little about the well-being of black or other minorities. "They were poignant meetings and I used it as an educational tool as well as everything else," recalled Norman. "The thing that made it so good was that he understood it." Blacks could not avoid perceiving the unequal promotion opportunities and the indifference to issues like housing at new assignments. Bill Thompson was present at the briefings and noticed the "body language of several senior officers and flag officers evincing embarrassment or discomfort; but there were a few who were defiant and essentially shrugged, 'So what?'"[61]

Norman and Zumwalt developed a close personal friendship and mutual respect. Said Norman, "As I grew to know him better, I began to see many of his human qualities, his sense of humor, and his extraordinary sensitivity and candor. He also has a relentless tenacity once he's committed to a cause, and that was evident as he followed up his directives to make certain they were implemented. He insisted on fast facts and real numbers from people."[62] They had breakfast alone every Tuesday morning at Bud's home and then drove to the office together.[63] Mike Spiro, the marine aide on Zumwalt's staff, was given clear instructions to make sure "that his Minority Affairs Officer, Bill Norman, had easy access to the CNO, traveled with us whenever and wherever feasible and ensured there was always time on his [Bud's] schedule to meet with minority groups and to be briefed on minority affairs. The same policy extended to Mrs. Zumwalt who always met with minority groups while traveling with her husband."[64]

Norman took the lead in writing Z-66, "Equal Opportunity in the Navy." He first assembled a group of people he could trust. "I made certain that no one was going to stop it," recalled Norman.[65] Before releasing the Z-gram, Norman was obligated to run it by Bill Thompson, Jack Davey, and Robbie Robertson, all of whom felt it had to be toned down. Norman became as intransigent as he had ever been in his life. After much wrangling, they agreed to rephrase a couple of things and came up with the final line of the Z-gram: "There is no black navy, no white navy—just one navy—the United States Navy."

On December 17, 1970, the CNO issued Z-66, directing every base, station, and aircraft squadron commander and ship commanding officer to appoint a special assistant for minority affairs with direct access to the chain of command. The new special assistant was to have direct access to the commanding officer and was to be consulted on all matters involving minority personnel. Zumwalt asked Norman to visit every major naval base in order to consult with commanding officers and minority personnel and families. "By learning in depth what our problems are, I believe we will be in a better position to work toward guaranteeing equal opportunity and treatment for all of our navy."

Thus with one stroke of the pen, Bud Zumwalt put the navy on notice that he would not tolerate existing discrimination in housing, promotions, and opportunities. Zumwalt ended Z-66 with a personal pledge: "It is evident that we need to maximize our efforts to improve the lot of our minority Navymen. I am convinced that there is no place in the navy for insensitivity. We are determined to do better. Meanwhile, we are counting on your support to help seek out and eliminate those demeaning areas of discrimination that plague our minority shipmates. Ours must be a navy family that recognizes no artificial barriers of race, color or religion."

Z-66 served as the foundation for a revolution. Almost 2,800 new positions were created in the chain of command for the purpose of advising commanders on the issue of equal opportunity. Some critics argued that establishment of the minority affairs officers bypassed the chain of command and thereby created a lack of discipline and order. The minority-affairs assistants established by Z-66 had direct access to their commanding officers. The assistant was to be the consultant for minority affairs, but 99 percent of the commanding officers were white and had little experience handling such issues. The system worked when the CO used the minority-affairs officer as a consultant and then implemented a solution through the chain of command. It did not work in cases where the CO used the minority-affairs officer to "solve all problems relating to minorities." That type of unenlightened leadership did not supply the person with the necessary power and support to accomplish the task.

Black navymen memorized Z-66 because it served as a symbol of hope and empowerment from the first boss who ever really cared about their welfare. This is perhaps best illustrated in one handwritten letter from a self-identified "Black Petty Officer of the world's No. 1 Navy," writing only to say, "I Sir am extremely proud and near Tears as I thank you from the bottom of my heart for your wisdom, your clairvoyance and your great courage to stand tall and tell it like it's gotta Be."[66]

In 1971 Samuel Gravely, Jr., became the first African American to achieve the rank of rear admiral and in 1976 was the first to be advanced to the rank of vice admiral. In looking back on Zumwalt's role, Gravely explained, "The normal way things happen in the Pentagon is that the CNO comes up with an idea, checks it out with his deputies and they sit around and chew the fat [until] the CNO says, 'Well, I don't give a damn how you feel about it, but this is the way I am going to go.' And he does it and the guys try to back him to the fullest. [I believe] some of the things the CNO felt strongly about had never been tried on some of these deputies before they heard, 'here comes a program.' The automatic thing was to resent that."[67]

In January 1971, Bud established the ad hoc CNO Advisory Committee on Race Relations and Minority Affairs. He appointed Bill Norman as executive director. The committee's report, *Navy Race Relations and Minority Affairs Programs*, stated that the goal of the navy was to "create and maintain a Navy image of equal opportunity and treatment regardless of race, creed, religion, or national origin" and to "increase and intensify the Navy's efforts to attain and retain the highest quality officer and enlisted volunteers from the minority community, thus seeking to achieve increased representation of minority personnel in various categories and grades of service." As John Sherwood noted, "The document made not only equal opportunity but also affirmative action a Navy goal."[68]

Perhaps the defining moment in Zumwalt's transition to being a revolutionary in the racial area occurred in June 1971, when he was scheduled to speak before the National Newspaper Publishers Association in Atlanta. Bud intended to give a major speech on race in the navy. Norman flew down early and met with retired sailors and minority officers who would be in the room, many of whom had served in World War II as stewards. They still saw little evidence that the navy had changed, "because all they could see was the black faces replaced by the brown faces of the Filipino, and blacks still were in demeaning jobs, and they hardly had any officers compared to the other services."[69] They doubted that Zumwalt was really any different, but Norman pleaded with them to give this white guy from Tulare a chance.

When Bud arrived in Atlanta, he was more nervous than Norman ever recalled seeing him. Norman noticed this from the kinds of questions Bud was asking during the car ride from the airport. He was feeling a bit insecure and asked Norman to sit on the platform with him. Norman urged Bud not to read his speech, because he was so much better speaking from his heart. The man who issued Z-66 gave the speech of a lifetime, turning an audience of doubters into believers. He closed with a story about Hannibal preparing to cross the mountains. The passage was going to be treacherous with many obstacles, and Hannibal went to his men and said, "It is on this wind that we succeed or die." Zumwalt tied the quotation to the contemporary problems of inequality and discriminatory practices. But he was talking about a navy that was going to be *one navy*, and he was going to succeed. "He meant it and they knew he meant it," recalled Norman. "For the first time in my life I felt proud to be in the Navy. We had black officers with tears because it was the first time that they thought they could remember that they felt proud. That they didn't feel they had to apologize for what was happening to other blacks in the navy."[70]

In the eyes of the audience, "He became 'superman,'" said Norman. The ripple effects were obvious to Norman. Minority officers started willingly giving up their navy careers in order to recruit for Zumwalt's new navy, to work as minority-affairs assistants in race relations, to work together for the greater good. This spawned extraordinary community outreach programs aimed at bringing the face of the navy into minority communities.[71] This new navy came from Watts, the South Side of Chicago, from Fourteenth and U in Washington, D.C., and from Harlem. "What we started bringing to the navy was a piece of real American life with all its problems and everything else. These kinds of men and women were not going to tolerate the injustices and the prejudices as people had before," said Norman. That was because Bud Zumwalt had said there's not going to be any discrimination.

"We made some institutional changes that will never be reversed without the navy being torn asunder," said Norman.[72] Bud considered Bill Norman's contributions to the navy as being "beyond measure," but even more important was the fact that Norman's efforts had

"unquestionably, spared the Naval service from racially generated difficulties of unprecedented magnitude and, in fact, have helped transform
the forces of division into mortar with which to build. He has, almost
singlehandedly, laid the foundation upon which a new Navy family that
recognizes no artificial barriers of race, color, sex, or religion today is
being built."[73]

This tide of change was most evident in the way Z-66 was being
implemented on ships under the command of enlightened leaders. Lew
Glenn, executive officer on the USS *Vreeland*, wrote about how the *Vreeland* was working on implementing the policies behind the Z-grams.
The VRAS (*Vreeland* Racial Anthropology Session) met three times
weekly with about thirty crew members and had been extremely "successful in making the crew aware of their true feelings."[74] Bud responded
by noting that the *Vreeland*'s "progress in [the] human relations area is
most heartening." He was especially excited about the establishment of
the striker (enlistees who had qualified for a rating but had not yet advanced to petty officer third class) selection board, calling it enlighten-
ing because "I believe this is a most viable way to solve one of the more
serious aspects of discrimination, that of denial of equal opportunity
in assignments. Job satisfaction is such a basic human need that I feel if
we could bring complete fairness to our assignment practices that we
would have licked a major fraction of our racial equality challenges."[75]

As commander of the USS *Hawkins*, a destroyer in the Mod Squad,
Howard Kerr was a long way from the day he and Bud had touched
down in Saigon. The Mod Squad was intended to give hard-charging
officers a chance for early promotion. Bud envisioned the program
from his own life experience, envisioning a youth movement in the
ranks of admirals. Officers detailed to certain squadrons were picked
one rank lower than normal in order to increase opportunity for early
advancement. On September 13, 1971, Kerr issued an "all hands" memorandum, "*Hawkins* Commitment to Social Justice, Equal Opportunity
and Meaningful Human Relations." It began, "Equal opportunity and
social justice has not been a navy tradition." Sounding like a philoso-
pher, Kerr noted that despite swearing the same oath and working and
eating together, "meaningful human relations, the day to day social

intercourse, one man to another, one group to another, has somehow escaped us." He wrote that the hard facts of life were that through misunderstanding, prejudice, ignorance, and inaction, "we have created an environment and a structure of life aboard ship that denies social justice and equal opportunity to all our shipmates and saps the foundation of meaningful human relations." Kerr added, "We live in a time of turbulent social change" reacting to "a generation of injustice. . . ." Kerr said that although the navy was affected, it "was changing for the better because of 'great leadership.' We are casting off traditions that no longer serve us. . . . Now is the time to dedicate to a new mission of social justice! We should do this because it is right. We do this because the Navy and the quality of our individual service to our life's commitment is at stake. We have demonstrated the capacity to deny justice, we must now demonstrate the capacity to forge a new justice—one that recognizes no artificial barriers to race, color or religion. Opportunity must be equal . . . we cannot allow inequality and what it breeds to work at dismantling our Navy. Our mission is too vital."

Months later Kerr wrote Zumwalt that he had learned all this from conversations with minority groups aboard the *Hawkins*. "Although the words are mine, the feeling, the resentment and particularly the hope expressed, came from the men who expressed themselves with sensitivity and emotion. My experience with *Hawkins* causes me to firmly believe that the vast majority of our people of all races desperately want to see the artificial barriers brought down and true Equal Opportunity realized. We are sensitizing our antennae, SIR!"[76] Bud replied, "I am convinced that we must eliminate the artificial barriers that impair effective race relations and ensure equal opportunity and treatment for all our personnel."[77]

"Although the Navy was a racist institution, I found it easier to deal with racism than with sexism," recalled Zumwalt. "It takes longer for a white man to come to believe that a white woman is his equal than it does for him to come to believe that a black man is his equal."[78] Indeed, President Nixon was one of them, telling Zumwalt, "I guess I can put up with this race thing, but don't push so hard for women."[79]

There was also much more opposition within the navy to Zumwalt's programs for equal opportunity for women. Congress had passed a law forbidding women to serve on fighting ships or in fighting planes, so by law women did not serve on an equal basis with men. Women were also not permitted to attend the service academies. At the time, it was legal to assign women only to hospital and transport ships. Bud understood that the culture believed women should avoid aggressive activities, but again he took another view: "I had no problem supporting women in combat for two reasons: 1. I remember well my great grandmother's stories about fighting off the Indians along with her husband as they crossed the Plains; 2. The most vicious and cunning enemy I have ever had to fight was the Viet Cong women."[80]

In mid-1971, Bud convened a WAVE retention study group that revealed general dissatisfaction with the reality that women were still being assigned primarily as receptionists and coffee runners, rather than receiving assignments based on their competence and ability. In a December 1971 letter to flag officers, Bud noted that it was "demoralizing and disheartening" to a young WAVE who graduated at the top of her class and "is then assigned at her new command to such stimulating duties as running the ditto machine and keeping the office mess going." One of the attitudes at work might be "the professional jealousy of the male supervisor who cannot admit that the woman can do the job as professionally as her male counterpart, or the complete bewilderment of the division officer who has never had a professional woman working for him before and doesn't quite know what to do with her! In the former the misuse is deliberate, in the latter it is thoughtless—but in both it adds up to a real waste of talent."[81]

Anticipating passage of the Equal Rights Amendment, Bud was intent on being ahead of the curve. Writing to Pat Quaglieri, president of the National Organization for Women, Tampa chapter, Bud observed, "I am hopeful that with passage of the ERA and continued vocal support in the public sector, the remaining institutional restrictions in all career choices will soon be eliminated. Then all we have to eliminate is personal bias!" Z-116, "Equal Rights and Opportunities for Women in the Navy," August 7, 1972, was pure Zumwalt. The message stated that all men and women should be permitted to serve their country in any way they chose. This meant utilizing navy women in responsible positions, whether ashore or at sea. This would strengthen the navy in meeting its worldwide commitments and in the defense of our shores. Women were to have open and equal access to all jobs, schools, college officer-candidate programs, and service schools and were to be assigned to sea duty based on their qualifications. Z-116 sought to achieve gender equality by removing almost all restrictions on opportunities in ratings, the ultimate goal being that women would serve on ships at sea as officers, crew, and as combat pilots.[82]

The USS *Sanctuary*, a hospital ship that already had living quarters for women, was overhauled so that it could be the first ship to set sail with women officers and enlisted sailors helping to run it. Some navy wives saw it as a threat to their marriages, while the retired community conjured up images of unisex showers and floating orgies. Bud pushed hard to conduct the trials on board the *Sanctuary*, which showed that women could perform well in areas of seagoing ratings, leading to the gradual broadening of the areas into which women were assigned. He also initiated the pilot test program, which showed that women, notwithstanding the wails of male aviators, could perform well in flying roles. Years later, Captain Rosemary Mariner, a retired naval aviator, found a cartoon titled "Old Guard Bar and Grill." Sitting at the bar were two admirals crying in their beer above the caption, "If God had wanted women at the Naval Academy, he would have made them men!" Like many wonderful but human institutions, the navy would never have altered course toward racial equality nor staffed the fleet with the best-qualified male and female citizen-sailors under its own momentum, said Mariner. "Adm. Zumwalt's methods, however unpopular, were the only way to cast off old lines and allow the Navy to steam into the future. In my mind's eye, I see the always dignified admiral elbowing up to a heavenly Old Guard Bar and taking his seat as one of history's truly great Americans and naval leaders."[83]

Bud was especially proud of the day Alene Duerk was selected as the first female admiral. When a photo of Bud kissing the new rear admiral on the cheek appeared on the front pages of the country's newspapers, Mrs. John Malott of Garden Grove, California, wrote, "First it's Booze & Rock Music in the barracks to corrupt the boys—now it's a kiss for the LIB—Are you sure You don't represent the French Navy? No small wonder we can't win the war with the likes of you in Charge—May God save America in spite of you."[84] Bud liked to joke in response to this type of criticism, "You must understand one does not become CNO without having kissed a lot of admirals."

The tenth-anniversary issue of *Ms.* magazine, in the article "*Ms.* Heroes—Men Who've Taken Chances and Made a Difference," saluted men engaged in fundamental humanizing change, men who were both symbols and real people. Bud was one of those honored. "Admiral Elmo Zumwalt, former chief of US Naval Operations: for advocating the repeal of the Combat Exclusion law, thus allowing women to volunteer for combat positions on the same physical-capability basis of men."[85]

Two days before his term as CNO ended, Bud sent a general message to all naval personnel, noting that of the 121 Z-grams, 87 had either become or were in the process of becoming directives in the navy system; the remaining 32 were either informative in nature or statements of policy. Only 2 had been canceled. Then, in accord with tradition that allowed his successor leeway, he stated that "Z-NAVOPS 01 through 121 are hereby cancelled for record purposes."[86]

There was actually one more "symbolic" zinger issued: Z-gram 122 to Roberta Hazard in recognition of a magnificent naval career and extraordinary service to Bud's career. Hazard did not meet Bud until he became CNO. She had been in the audience in November 1970 at the Naval Academy to see the man behind the scuttlebutt who was "working hard to transform the Navy into a more caring, a more respectful of the average man (and woman) organization—one with fewer barriers and more encouragement to personal achievement and contribution."

She was so inspired by hearing Bud's vision for the navy that two days later she called to express interest in the protocol assistant job in the CNO's front office, a job she had twice declined as stereotyping and unchallenging. "Thus began what I still refer as the most instructive and the most informative of my navy years."[87] Hazard believed that both Bud and Mouza had been able to produce positive and enduring changes not only in the navy but in America. Both had demonstrated "their pervasive caring and concern for sailors, their commitment to necessary change and to a clearly envisioned role for our navy."[88] In 1984, Hazard made rear admiral.

CHAPTER 11

ROUGH SEAS

*The Thermidor of my four year term as CNO came shortly
after the half-way point in the fall of 1972.*

—BUD ZUMWALT [1]

In an early 1972 message to all flag officers, commanders, commanding officers, and officers in charge, "The Commander and Equal Opportunity," Bud noted that the navy had made great strides "from its World War II position as a service regarded by blacks and other minorities as highly discriminatory to one now looked at as truly receptive to Equal Opportunity issues." Nevertheless, he recognized that "the navy still has many potential sources of racial tension, and that only continued personal attention by officers in command can provide the key to diffusing [sic] these obstacles to true racial harmony."[2] With retention-group sessions seared in his mind, Bud implored his commanders to realize that "it is sometimes difficult for many of us to comprehend the sensitivity and degree of awareness our minority Navymen and women have regarding discriminatory practices." Bud urged members of his command to understand that "whether or not unfair or discriminatory practices actually exist within a command, if they are *perceived* to exist by either whites or blacks, discipline or morale will suffer."[3]

The navy might not be the best laboratory for social experimentation, but it was on the way to becoming a microcosm of society at large. The navy could no longer afford to draw a line between the values and practices of the community outside the gate and those within. In effect, Bud was challenging those in command to recognize that it was their duty to create an atmosphere conducive to true equality of opportunity so that institutionalized inequality and racism could be eradicated. It sounded clear enough, but the high seas would offer no respite in the struggle for racial equality.[4] It would all come to a head on two aircraft carriers in the Pacific, posing the greatest challenge in Bud Zumwalt's career and providing old-guard traditionalists as well as the president of the United States the opportunity to turn back the clock. Bud fought back with everything he had.[5]

The March 30, 1972, Easter Offensive across the demilitarized zone (DMZ) separating the North from the South was the largest conventional military attack launched by North Vietnam during the entire Vietnam War, designed to deal a crippling blow against ARVN forces. President Nixon believed the enemy had committed itself to a "make-or-break campaign." The attack occurred during the planning stages for Nixon's summit with Soviet leader Leonid Brezhnev, and in fact, Henry Kissinger was scheduled to visit Moscow secretly from April 20 to 25 to make arrangements for the summit.

With Vietnamization at full throttle, the president could not send troops back into the war. Instead, on April 1, he ordered the bombing of North Vietnam within twenty-five miles of the DMZ. By April 14, he ordered air strikes up to the twentieth parallel. The Paris peace talks were again suspended. It was a period of great duress for Nixon, who recorded in his diary on May 2, "I decided now it was essential to defeat North Vietnam's invasion." On May 8, Nixon convened a three-hour meeting of his National Security Council, outlining plans for mining Haiphong Harbor and renewing the bombing of Hanoi and Haiphong. He planned to announce his "go for broke" plan to destroy the enemy's war-making capacity in a televised speech that evening. "Those bastards are going to be bombed like they've never been bombed before," Nixon told his National Security Council. "I have the will in spades," he declared.[6]

Speaking to the nation, the president outlined the steps he was taking. "There is only one way to stop the killing. That is to keep the weapons of war out of the hands of the international outlaws of North Vietnam. I have ordered the following measures, which are all being implemented as I am speaking to you. All entrances to North Vietnamese ports will be mined to prevent access to these ports and North Vietnamese naval operations from these ports. United States forces have been directed to take appropriate measures within the internal and claimed territorial waters of North Vietnam to interdict the delivery of any supplies. Rail and all other communications will be cut off to the maximum extent possible. Air and naval strikes against military targets in North Vietnam will continue."

Bud was also involved in the planning for the mining of Haiphong Harbor. "That was one time during the four year period when Tom Moorer (the primary advocate for mining) had to rely totally on me," recalled Bud. "He had been instructed by Henry Kissinger, whom he feared, to ensure that there were no leaks."[7] Bud claimed that he and Rex Rectanus developed the operational plan for Moorer. "It was approved as we had submitted it, and was carried out as we had submitted it."[8]

The invasion and mining placed extra stress on the navy, leading to doubling the number of ships off Vietnamese shores during a period when the United States was reducing its overall forces and budgets. But through all these turbulent and uncertain times, the navy met its commitments. During the aerial phase of Pocket Money (as the operation was named), navy planes dropped mines into the port of Haiphong and six other harbors of North Vietnam. The mines were activated on May 11. Admiral Moorer wrote his field commanders, "We do not expect to lose this one, consequently, must bring as much air and naval forces to bear as possible in order to give the enemy a severe jolt . . . we have received increased authorities and must make full use of them at every opportunity. Our objectives are: to ensure that the North Vietnamese do not endanger remaining U.S. forces, to provide maximum assistance to the South Vietnamese in their efforts to destroy the invader and to prevent the North Vietnamese from interfering with Vietnamization plans."[8]

The tempo of these naval operations required extended deployments for those at sea. Up to six aircraft carriers were on the line at Yankee Station in the Tonkin Gulf, and these required additional support ships. Not since World War II had the navy been so intensely deployed.[9] Not only were deployments extended, but turnaround time between deployments had been significantly reduced. It was not unusual for sailors to work sixteen-to-eighteen-hour days when on line, reminding Admiral

Bill Thompson of the old cliché, "I joined the Navy to see the world and what did I see? I saw the sea."[10]

To meet these manpower needs, a new program, From Street to Fleet, rushed new minority recruits through basic training. Recruitment targets were met by allowing into the navy individuals with very low test scores, high school dropouts, and those with criminal records.[11] Once aboard ship, the new recruits were assigned the most basic and menial assignments, like food service and deck and bilge cleaning, with little hope of ever gaining better assignments. "The new crew included a much larger percentage of non-trainable young Black enlisted men, considered so because their low-test scores indicated that they were not capable of handling service school curricula," explained Admiral Thompson. "They were disgruntled because they had been assigned to the traditional deck and engineering divisions, a lot of grunt work and chipping paint."[12]

Minorities did score low, but not because they were less smart than white recruits. The Armed Forces Qualification Test (AFQT) was the test that determined a sailor's category ranking, and it was later determined by the Center for Naval Analyses to be culturally biased against blacks and other minorities.[13] Moreover, most minority recruits had never received the necessary technical or educational opportunities afforded others. The result was a powder keg. It was a period of societal unrest outside the navy, and the new recruits came under the supervision of an almost entirely white command structure within. Racial tensions simmered on the lower decks. Future secretary of the navy John F. Lehman recalled that during his service on the *Saratoga* in 1972, "No white officer would walk unescorted on the second deck where the enlisted men's mess was."[14]

The *Kitty Hawk* was on her sixth war deployment, supporting round-the-clock air strikes against the Vietnamese Easter Offensive. The *Kitty Hawk* had been involved in some of the most intense activity of the war, with 164 days of consecutive air combat operations, and more sorties were flown from the *Kitty Hawk* than any other carrier in Vietnam.[15] As the Easter Offensive dragged into the summer, the *Kitty Hawk* was extended beyond its scheduled deployment. The ship had been in the western Pacific for almost eight months, 239 days of separation from the crew's families.

The *Kitty Hawk*'s commander, Captain Marland Townsend, had assumed command on June 5, 1972, and his executive officer (XO), Commander Benjamin Cloud, had been aboard since August 11, 1972. The *Kitty Hawk* resembled the Great White Fleet of times past.[16] Of the 348 officers, only 5 were black, including XO Cloud. Of the 4,135 enlisted men, 297 were black (7.2 percent); none of the 219 officers in the Air Wing were black. "White crewmen worked in glamorous jobs such as aircraft maintenance, intelligence, and communications while blacks toiled mainly on the mess deck."[17] Moreover, by design and agreement, blacks and whites lived in segregated quarters, creating safe havens, but also institutionalized ghettos.[18]

The extension of deployment meant that there was no time for traditional liberty in Hong Kong or Japan. The only available location was Olongapo at Subic Bay, Philippines. Olongapo was a town of whorehouses, cheap bars, and tattoo parlors divided into a black zone, "the jungle," and a white zone, "the strip." Olongapo had a midnight curfew, strictly enforced by Philippine authorities. On October 10, 1972, the *Kitty Hawk*'s final evening in Subic Bay Naval Base before departing for the Tonkin Gulf, a fight broke out at a local bar that was hosting "soul night" for black sailors from the *Kitty Hawk*. There had been a series of minor fights between blacks and whites all evening, culminating in a brawl that required the marine riot squad to stop it. Several sailors from the *Kitty Hawk* were seen running back to their ship in order to avoid being arrested, arriving just before the midnight curfew.

The next day, *Kitty Hawk* resumed air operations against North Vietnam while also initiating an investigation into possible misconduct charges for what had transpired onshore the night before. A black sailor summoned for questioning brought nine other black enlisted men with him for support. The tagalong group was belligerent, leading the investigator to ban them from sitting in on the questioning. The sailor was read his rights, refused to make a statement, and was excused. Tensions from the night before continued to build between black and white sailors, and by nine fifteen that evening a group of black sailors had assaulted a white messman.

The group of nine had grown to over fifty and had assembled on the afterdeck, one of the two enlisted dining areas. The chief master at arms called for the ship's sixty-man marine detachment reaction force, whose onboard responsibilities usually involved guarding nuclear weapons, the brig, and other high-security areas. Captain Townsend had been asleep in his cabin when he received a call about a potential mutiny. "My main concern was the airplanes on the flight deck and the airplanes on the hangar deck. I put out the word to the master-at-arms to get out there and make sure we had people protecting the airplanes."[19] He then left the bridge to see for himself what was happening, hoping that his presence might help calm things down. Tensions ran high between the all-white marines and the group of black sailors. Commander Ben Cloud, who was of African American and Native American descent, tried defusing the situation, telling the sailors to go about their business and stop trying to make more trouble.

The group released by Commander Cloud from the mess deck returned via the hangar deck. The marines had been instructed to break up and disperse any group of three or more sailors who appeared on the aircraft decks. The marine guard was trained in riot control procedures. The group of black sailors refused to disperse, and the marines sought to contain them. The blacks began taunting the marines, raising their fists, overturning tables, and arming themselves with makeshift weapons. When a few blacks were handcuffed, all hell broke loose. Arming themselves with aircraft tie-down chains, black sailors marauded about the ship attacking whites, pulling many sleeping sailors from their berths and beating them with fists, chains, dogging wrenches, metal pipes, fire extinguisher nozzles, and broom handles. Many were heard shouting, "Kill the son-of-a-bitch! Kill the white trash! Kill, kill, kill! They are killing our brothers." The ship's dispensary was the scene of intense activity, the doctors and corpsmen working on the injured personnel. Alarmingly, another group of blacks attacked the dispensary.[20]

Cloud heard one sailor cry out, "They got the captain. They killed the captain. Oh my God."[21] Thinking that Captain Townsend was dead or wounded, Cloud made a shipwide announcement declaring an emergency, beseeching his "black brothers" to assemble at the mess deck. Marines were ordered to the forecastle. "I ask you, I implore you, I order you to stop what you are doing. All black brothers proceed immediately to the after mess deck. . . . This is an emergency."[22]

Townsend was not dead, but he was also not on the bridge, the place he should have been. Instead, he was with injured sailors in the infirmary. The captain did not understand why his executive officer was taking over the ship. He also did not want the marines to leave their areas. Townsend then also made a shipwide announcement, reporting that he was alive and that he was countermanding Cloud's previous order. "Cool it, everyone. Break up peacefully and proceed back to your spaces. The Marines will not use any weapons and will leave you alone. There will be no weapons used unless I call for it on this box. Those of you who have grievances I will meet with you right now on the forecastle. The rest of you, I want you to cool it. Knock off this senseless behavior before more of your shipmates are seriously injured. I know everybody is hot under the collar. I know you are disappointed about not going home as planned. So am I, but we've got to live with it, so cool it." Townsend later said, "Attica was always in the back of my mind. I did not want to ignite the situation any further by ordering a violent overthrow of the riot."[23]

When Townsend finally found Cloud in the ship's damage-control central, he said, "If anybody ever writes a book about this, this is going to be the most fucked up chapter."[24] Indeed, the conflict in command between Cloud and Townsend emboldened those bent on hurting people.[25] There were scores to settle for blacks against whites, but by now the white sailors were organizing themselves for retaliation against blacks. As Townsend's message was being broadcast, Cloud observed black sailors "indiscriminately beating all whites they encountered in passageways and berthing.[26] Cloud soon encountered an angry mob of about 150 men intent on tearing the ship apart. They threatened to throw him overboard because he was not a real brother. Someone yelled out, "Kill, kill, kill the motherfucker."[27] Cloud believed that had he not been black, the mob would have killed him. Invoking the words of Martin Luther

King and Gandhi, Cloud offered the black-power salute: "If you follow the practices of a Gandhi, and of Martin Luther King, Jr., you can live tomorrow and the next day in pride and respect, but if you continue to use the tactics that you are using here tonight, the only thing that you can guarantee is your death." Tearing off his shirt and grabbing a steel pipe, Cloud said that any man who doubted his sincerity could beat him. "If anyone in this crowd does not believe my sincerity, I hold this weapon and bare my back for you to take this weapon and beat me into submission right here."[28] The crowd went silent and began chanting, "He is a brother! Let's do it your way." Cloud returned another black-power salute. The sailors began throwing their makeshift weapons overboard and dispersed.

Meanwhile, Townsend had to deal with another war—air strikes against North Vietnam scheduled for the next morning. If the ship went to general quarters, the strikes would have to be canceled. He made the command decision not to go to GQ so that air operations could resume against North Vietnam the next morning.

On the same morning that air strikes resumed, Admiral Clarey, commander in chief of the Pacific Fleet (CINCPACFLT), stationed in Hawaii, called Bud to bring him up to speed on the *Kitty Hawk*. "You know that is the ship that we have a black XO on."[29] Clarey had been on the ship two weeks earlier and had seen no signs of trouble. His impression was that Cloud was "a really capable young man." Clarey thought that "the Marines sent in to restore order overdid it—that's too bad." Bud wondered why the racial tensions were getting more play than the war itself, saying, "It had been a bad week and I was very concerned about effects of long deployments."[30]

With order restored on the *Kitty Hawk*, Bud expected to devote attention to the war effort. Yet the biggest test lay just ahead, aboard the USS *Constellation*. "Connie was sort of a floating testimonial to the more occasional pertinence of Murphy's law: everything that can go wrong, probably will."[31]

The sixty-thousand-ton aircraft carrier *Constellation* had been deployed to Southeast Asia six times between May 1964 and June 1972, averaging almost nine months per cruise. She had been involved in the early Rolling Thunder bombing of North Vietnam and interdiction attacks in Laos, and in 1972 she joined the carriers at Yankee Station responding to the Easter Offensive.[32] In July, the *Connie* returned to San Diego for a two-month major overhaul, including a turnover of 1,300 people in a crew of 5,000. Repair work on the *Connie* was expedited so that she could return to the Seventh Fleet for the war effort.

By early October, the *Connie* was out for maneuvers and refresher training under the command of Captain J. D. Ward. Upon returning to port, the crew heard rumors about what had transpired on the *Kitty Hawk*. On the evening of October 17, five days after the riots aboard the *Kitty Hawk*, a group of approximately fifty black sailors on the *Connie* met in the barbershop area to support "the cause of their brothers on the *Kitty Hawk*."[33] The next day, a larger group met in the "sidewalk café," a safe haven on ships for black sailors, where sounds of Marvin Gaye and Motown filled the air. The sidewalk café doubled as a place for organizing and electing representatives under the new human resources system Bud had put in place for identifying examples of discrimination aboard ship. At this particular meeting, a long list of grievances was assembled, including allegations of biased quarterly marks in comparison to whites, assignments that were considered discriminatory, unequal general discharges, and unjust mast punishments (penalties ordered by officers without a court-martial) for blacks. When the meeting ended, a small group of black sailors, without any provocation, attacked a white cook and broke his jaw.[34]

Captain Ward's primary concern was not for perceptions of discrimination, but rather for training the air wing and maintaining a training schedule for Pacific Fleet rotations. The *Connie* returned to sea for maneuvers. His staff did not respond in a timely manner to the list of grievances. However, nothing in American naval history compared to what happened in San Diego during the first two weeks of November 1972.[35] On November 3, a group of 50 young black sailors staged a sit-in on the forward deck of the *Connie* and then reassembled at the after-mess deck. The group soon increased to 130, with 10 whites joining the majority of blacks demanding to see the commanding officer. Captain

Ward refused to meet with the group because night air operations were in progress. He would not leave the bridge for a meeting at the mess deck; instead, the captain said that representatives of the group were welcome to visit him at the bridge.

Two representatives carrying a list of their demands came to meet Ward. They also issued a threat. "If the Captain don't come down, his ship isn't going to be together much longer."[36] Others threatened "a blood bath worse than the *Kitty Hawk*."[37] The captain reiterated that he could not leave the deck, but said he was open to discussing all grievances. The representatives returned to the mess deck and told the larger group that Ward would not talk with them on the mess deck. Ward sent off a message to Admiral Clarey advising that a "cadre of blacks were working very hard to create a confrontation or racial incident and that twelve apparent ringleaders had been identified."[38] Fearing for the safety of his ship and his ability to conduct operations, Ward decided that the best strategy was to get the dissidents removed from the *Connie*. As they headed back to San Diego, the captain took special precautions, ordering his security detail to guard equipment from sabotage.

The *Connie* returned to San Diego early on the morning of November 4. The dissident group again presented Captain Ward with a petition, this one signed by eighty-two sailors, "from the oppressed people of this command, who after constant attempts to resolve our problems, request an immediate conference with the commanding officer."[39] Ward rejected the advice of NAVAIRSYSCOM (Naval Air Systems Command of the Pacific) that the group remain on board. He formed a shore detail and had the dissidents removed and moved into barracks at North Island Naval Air Station, where they now had access to the media, NAACP lawyers, and members of the congressional black caucus. This beach detachment was made up of 132 persons, 120 blacks and 12 whites. Most of those in the beach detachment were nonrated seamen; few had more than a high school education, and "most fell into the category III or IV of the AFQT."[40] More than two dozen had prior offenses. Also joining the group were senior supervisory personnel and members of the Human Resources Council (HRC).

. With the dissenters on shore duty, the *Connie* returned to sea. What

Ward had not anticipated was that, once the dissidents had been put ashore, the media would jump on the story as "a symbol for the Navy's racial problems."[41] Three thousand miles away, tensions in Washington were running high. Ward's primary concern had been to continue shipboard operations; he removed the sailors because he feared for the safety of his ship.[42] A flurry of top-secret exchanges among Bud and Secretary of the Navy John Warner in Washington and Admiral Clarey in the Pacific focused on devising a strategy for getting the situation cleared up fairly and quickly.[43] Warner and Zumwalt feared that the incident could affect the November 7 presidential election, in which President Nixon appeared to have an insurmountable lead over George McGovern.

Bud saw the *Kitty Hawk* and *Constellation* incidents as symptoms of the unevenness with which the navy had implemented "enlightened leadership" in race relations. This issue was best enunciated in a letter Bud received that week from Rear Admiral Draper L. Kauffman, commandant of the Ninth Naval District. Draper addressed the lack of understanding that "'we,' the white power structure of the navy, have for the problems, the beliefs, the aspirations of 'them,' the young blacks in the Navy. . . . One thing it took me a long time to understand is that race relations are vastly more important to the black sailor than to the white Admiral."[44]

Bud shared Draper's perspective but also realized that making the wrong move could jeopardize the navy's progress on the race front over the past two years. "I had to make my judgment in the light of my responsibility to the entire Navy, a responsibility that obligated me to find a way of combining the maintenance of discipline with the maintenance of progress in racial matters."[45] Bud thought Captain Ward was not part of the enlightened leadership program. "The picture back here is that he is operationally sound, the problem I think is he is a person who has not discovered the race problem and doesn't understand the depth and he has overlooked the human relations concern. There is some indication before the ship sailed there was some unrest."[46]

Bud wanted the sailors returned to the ship. He did not think the men should have been put ashore, and he felt it was essential they return quickly. He also thought there was little to be gained from taking a hard stance in getting them back, especially because they were involved in a nonviolent sit-in protest. To that end, he was willing to promise to create small focus groups to look into grievances and make recommendations for improving the situation aboard the *Connie*. He also promised to drop all charges. The only condition was that the sailors return to the ship. They were given one day to decide.

The problem was that the majority of dissenters had hardened their positions. They did not want to return unless their conditions were met. By this time, the NAACP was actively involved, demanding that the protesters have outside counsel. Captain Ward did not want a blanket amnesty under which all of the dissenters could come back to his ship; instead, he offered to interview each one of them to decide who was safe to allow back. This was rejected by the beach detachment. Ward then offered a compromise—he would allow any man who wanted to return to do so; those who did not return would receive a discharge or transfer to another ship. He also agreed to have each general discharge reviewed by an independent body.

Captain Ward soon received final instructions from Bud and Tom Moorer. They said he should meet with the protesters and present a final offer, giving them all twenty-four-hour liberty with orders that they must then return to the ship or be considered unauthorized absentees. Ward was advised to separate the sailors from the newsmen and lawyers and address them at the pier. He should review the steps being taken to resolve their grievances and urge them to return following liberty. Failure to return would mean they were absent without leave, and by violating repeated orders to return, they would also be in violation of direct orders. Ward went to the pier to carry out his instructions. When he asked for a show of hands on who was planning to return to the ship, in a photo broadcast worldwide, the sailors of the *Connie* raised clenched fists in a black power salute of protest. In interviews with the media shown on the evening news, sailors quoted Z-66 and demanded that their rights be honored by the navy.

This television and media coverage infuriated Nixon, partly because it was detracting from his election victory but also because he was old-school navy, believing that all orders needed to be obeyed and that no coddling or negotiating with dissenters could be tolerated. Nixon instructed Henry Kissinger to tell Bud that his commander in chief wanted the *Constellation* protesters to receive immediate dishonorable discharges, an order that Bud knew was in violation of the Uniform Code of Military Justice (UCMJ)—the military law of the land.

In a November 8 telephone conversation with Admiral Clarey, Bud said, "I like this better because it avoids the public appearance of a mutiny." Bud thought that Ward had handled this situation appropriately: "That permits the CO to say we have these discussions and made every effort to understand each other, we think we have the best possible insight into what is being done on the ship, what the circumstances are, we understand your problems better, this ends the period of discussions, you are all free to go on liberty which will expire at 0700 tomorrow morning onboard ship and we expect to see you there. They do not have to put it in the form of a threat. They can stress that there will be no disciplinary action for the fact that the people have gotten together up until this point."[47]

The next morning, the now dissolved beach detachment appeared at the pier, in uniform and with their gear. At 8:10 a.m., the duty officer from the *Constellation* came down to the dock and advised the group that liberty was over. Only five men returned to the ship, forcing Bud's hand. He first needed Secretary Warner's support. Warner now wanted the men returned to the barracks for a cooling-off period and wanted the *Constellation* out of San Diego because it had become a symbol. Bud vehemently disagreed. In a memorandum to Warner, Bud urged that the secretary rescind his order to detach the grievance group from the *Constellation*, to permit them to return to the barracks as a group, and to withhold action to take them into custody until he had a chance to implement a new plan. "It is my belief, shared by others . . . that we have reached the point where the basic structure of the Navy as a military organization will suffer grievously, throughout, if action is delayed longer. It is further our belief that delay increases the probability of physical violence with regard to the grievance group and weakens public confidence in the Navy."[48]

Warner ultimately approved the recommendation. The next day, the

Constellation departed San Diego. Of the protesters, 120 soon faced disciplinary action; 46 would receive discharges (36 honorable), and 74 would be transferred. Bud Zumwalt now had a target on his back.

Bud saw little loyalty from Warner. Especially infuriating was that as undersecretary of the navy, Warner "had apparently made numerous commitments to the old guard that when he became Secretary he would reverse the policies that John Chafee and I had articulated in the personnel field. He sought originally to do so by ordering me to stop the dissemination of any further Z-grams. I told him I would resign on this issue and he backed away."[49]

Bud considered Warner to be "one of the weakest characters with whom I associated . . . he has been described accurately in my opinion as a dilettante. He conducted his professional career somewhat along the lines of his marital career, that is, totally without integrity. He had the backbone of a jellyfish and great vanity which was stroked by those who thought to use him." Bud came to see Warner as "a man who relished the power and perquisites of his office while not facing up to the responsibilities that accompany that office."[50]

Admiral Clarey called Bud to say that the *Kitty Hawk* was back in Subic Bay and the judge advocate general (JAG) needed advice on whether the commanding officer needed permission to initiate trials of those who had requested civilian counsel. Clarey was also concerned that he was receiving no direction because of Warner's indecision. In a telephone conversation later that day, Warner told Bud, "We must keep flexible and watch our options."[51]

The buzzards were hunting for larger prey. "People are running around cutting up on you,"[52] warned Dan Henkin, the Pentagon's assistant secretary of defense for public affairs. These critics charged that Bud's programs had created the kind of permissiveness that led to a breakdown in order and discipline on these ships. *Zumwalt* and *mutiny* were used interchangeably. Bud told Henkin that he was aware that he had a few men in uniform who were acting up but he would be damned if he let them turn the clock back.

Bill Thompson, Henkin, and his deputy, Jerry Friedheim, were the public affairs personnel involved in damage control. Their collective counsel was that Bud should respond to the perceptions of permissiveness with a clear statement that outlined the navy's policy with respect to minority affairs. They agreed that the perfect setting for this type of speech was already on the schedule—an address to Washington-area flag officers. Bud decided to invite the press to the November 10 event, because "I wanted to put both the racists and those suffering from racism on public notice to follow orders."[53]

In a heartfelt speech to a group of eighty or ninety admirals, the CNO addressed core issues of race and change within the navy and implementation of over two hundred minority programs by the chain of command. Bud spoke of a recent retention study-group report that made clear to him that "the Navy has made unacceptable progress in the equal opportunity area. And that the reason for this failure was not the programs but the fact they were not being used." The recent racial incidents aboard ships "are not the cause of racial pressures; rather, they are the manifestations of pressures unrelieved."

Speaking without the usual jargon and exquisite politeness that traditionally characterized exchanges of this type, Bud made it clear that issues needed to be resolved within the framework of disciplined, efficient, orderly, and ethical military operations. "What we are talking about here is not a call for permissiveness, or a direction to coddle. Let me say again that discipline necessary for good order will and must be maintained." The issues involved discrimination and self-deception. Many in the chain of command gave only lip service to the words *equal opportunity*. "The most destructive influence on the resolution of racial problems is self-deception," said Bud. In short, the navy needed to do better in implementing equal opportunity programs.

Sounding like part social worker, part Hobbesian theorist, Bud spoke of the fallacy that began by thinking "that the Navy is made up of some separate species of man—that Naval personnel come to us fresh from some other place than our world—that they come untainted by the prejudices of the society which produced them. They do not. It is self-deception to consider all issues involving blacks and whites solely as racial in motivation. They are not. And, finally, it is self-deception to consider the Navy, or any military organization, as free-wheeling—'to each his own way'—civilian society. In fact, even a civilian society unbounded by military law and tradition can only exist within the system of law and custom. For a military society to fill its purpose every man must know his own role—and live within it. There must be no substitution of one prejudice for another. The prejudice against good order and discipline is as pernicious as the prejudice of race."

Bud ordered his admirals to punish any member of their commands who practiced discriminatory practices. He believed that while the incidents at sea were characterized as racial, that was because race was the only visible aspect. In fact, the causes were much more complex and numerous. Men at sea working seven days a week for months on end with aging equipment faced pressures almost inconceivable to those who had never known them.

"Saying equal means exactly that. Equal." Bud implored his commanders to understand that true leadership involved returning to the navy's oldest and most proven motto, "command by leadership." "You cannot run a Navy, or any large organization, if the top must provide all the solutions. Nor can you bring about any real change by obeying the letter and not the spirit of a program." Leadership that lacked commitment from "the heart" was essentially obstructionist. "It is my view that these current racial incidents are not the results of lowered standards, but are clearly due to failure of commands to implement those programs with a whole heart."

Bud's speech was the lead story in the next morning's *New York Times*, and "Equal means exactly that" was the paper's quotation of the day. Praise poured in from a wide range of sources. "Congratulations on your message to senior officers at the Pentagon last week," wrote Senator Hubert Humphrey. "I deeply appreciate your leadership and you can count on me for support."[54] Vernon Jordan, executive director of the National Urban League, wrote to "express my appreciation for the forthright way in which you expressed the Navy's official policy on equal opportunity. The remarks you made to senior officers last week were truly inspiring, and set an example of firm leadership every American should admire. You have the full support of the National Urban

League for your efforts to make democratic ideals come to life in today's Navy."[55]

Jordan's next syndicated column, "To Be Equal," paid tribute to Zumwalt. "Admiral Elmo Zumwalt, Chief of Naval Operations, put it on the line to an unprecedented assemblage of top Navy brass recently . . . shape up or ship out." Jordan thought Bud's remarks to admirals and commanders "ought to be required reading" for every government official who "thinks he's on top of the racial situation in his office or business." Pinpointing the destructive impact of self-deception that occurs when regulations are not enforced and command fails to embrace them with a whole heart, Jordan applauded Bud Zumwalt's courage in admitting that his directives had been sabotaged by lower-level officers. An officer's future in the navy was going to be tied to how well he embraced equal opportunity.[56]

Writing in reply, Bud noted the insidious and debilitating effects of racial disharmony and discrimination. "As I said in Z-gram 66 nearly two years ago, my goal is 'a navy family that recognizes no artificial barriers of race, color or religion . . . no black Navy, no white Navy, just one Navy—the United States Navy.' As we draw closer to that goal, I am delighted to know that you and your colleagues will continue to stand firmly and proudly by our side."[57]

Within an hour of Bud's speech, an Associated Press story spun it that Zumwalt had dressed down his admirals. "I have always thought that story was instigated and planted but could not prove it," writes Bill Thompson. "Even reading the speech today, it is hard to see how that speech could be viewed as castigating his admirals."[58] For the past few months Bob Salzer had been in a new position as Commander Amphibious Force, U.S. Pacific Fleet (PhibPac). "I read the newspaper account and I was absolutely amazed. I sent back for a copy to the text the next day, because I was going to the Marine Corps Birthday Ball and I wanted to see what CNO had really said—with my own eyes, verbatim—rather than trusting to a newspaper leak. What he actually said was considerably distorted in the press accounts."[59] A number of flag officers felt insulted and demeaned. They also resented that the CNO had allowed the press to attend the meeting. By the time Salzer got to the Marine Corps

Ball "the room was rife with insubordination." Salzer wanted to find out who leaked the press accounts: "It was that Goddamn black advisor of his along with his PAO, who was one of those idolators who thought Zumwalt did nothing wrong. I don't know what they could have been thinking about because it started a fire."[60]

In Key Biscayne, Richard Nixon was still fuming about the black-power salutes at the pier. Chief of Staff H. R. Haldeman recorded in his diary for November 10 that "he [President] called later this evening, furious by a Navy episode where Zumwalt had heralded on the blacks that refused to sail on the *Constellation*. He's told Henry to have all the men court-martialed and give them dishonorable discharges. Zumwalt, instead, gave them active shore stations and Coca-Colas and ice cream. You can't have a service without discipline, and he wanted E [Ehrlichman] to tell Zumwalt that the P [President] was terribly displeased; he said you know he's a McNamara man."[61]

From Key Biscayne, Henry Kissinger called Bud, who had just returned from an overnight trip visiting his ailing father, to say that from their perspective things did not look good. "Kissinger said that he had just finished talking to Mel Laird and that he thought that he should now call me so that I would fully understand how the President feels about the *Constellation* situation."[62] Nixon would not tolerate negotiating with people who did not carry out orders and still wanted them all dishonorably discharged. Bud tried to explain that "the President wants the most drastic actions taken" and that this was not what the Uniform Code of Military Justice called for. Commander Roberta Hazard was monitoring the call for Zumwalt from Kissinger and thought the latter "sounded like he was outside of his own skin he was so angry over this thing, threatening to have the Admiral fired."[63] Indeed, Kissinger had told Laird to fire Bud if he did not carry out the order. Laird ignored the order. Bud tried explaining that he needed to be "conscious of the need to get on with the problem of integration and this required dealing with the racial problem somewhat differently than one would deal with it if there were not racial overtones." Kissinger hit the roof, saying that racial policy and mutiny, "which has been occurring too frequently," were separate issues. "If he [Kissinger] had said, 'You're fired,'

I would have told him to have the commander-in-chief call me, because I wouldn't take it from him. But he didn't have the guts to say that."[64] Bud assured the hostile Kissinger that all those facing charges would receive appropriate punishment, but it was not possible to discharge all of them. "CNO also told him that he hopes HK and the President understand these things are taking place because of over deployments, under manning because of Congressional action."[65]

Bud and Henry Kissinger had gotten off to a good start. In their first meetings, Zumwalt found Kissinger to be charming and witty, making the new CNO feel as though "I was a person whose advice and assistance he uniquely sought."[66] Bud was more than willing to provide both types of resources. Here is how Bud described the honeymoon period. "Kissinger is a man of great talent and intelligence who is capable of great charm and persuasion. An initial experience with him to the uninitiated is like a delightful seduction. He exudes charm. The listener, particularly in one on one situations, is convinced that he is the most important person in the world to Henry."[67]

Correspondence from Zumwalt's CNO files reflects a burgeoning relationship. When Kissinger needed the CNO's barge for a Saturday cruise on the Potomac, Zumwalt was pleased to provide it.[68] The CNO gave Henry and Nancy Maginnes (his girlfriend at the time) tickets for the Army-Navy football game, as well as for numerous navy functions in town. Kissinger especially liked attending these navy galas. Kissinger also appreciated Bud's assistance in assigning Rear Admiral Rembrandt Robinson to his liaison office with the Joint Chiefs.[69] Zumwalt even approved the use of secure navy phone channels for communications that Kissinger did not want the State Department to get wind of. Kissinger liked navy channels because they never leaked, and Zumwalt's assistance was indispensable in the arrangements.[70]

As time went on, however, the seductive process became less charming. "This process of disenchantment becomes greater when one begins to perceive the disloyalty and lack of integrity of the man."[71] Bud knew Kissinger lacked line authority to tell him to do anything. Bud would not give dishonorable discharges without a written directive from the president. He was not going to carry out an unlawful order issued by the head of the National Security Council. "As you look through your lifetime, if there is any 30 seconds to pick out, of which you're proudest, that's the one I would pick out. When I refused to carry out that order. That's my highpoint."[72]

As soon as he got off the phone with Kissinger, Bud received a call from Rear Admiral Dan Murphy, Mel Laird's executive assistant, who confirmed that Kissinger had asked Laird to intercede but that Laird said he would not do so. Murphy reiterated his concern "that the picture the White House" had of Bud was that he was too soft. Murphy and Laird wondered why no one was taking a shot at John Warner. The whole thing was being blamed on Zumwalt. Murphy concurred that "there is simply no way to give them dishonorable discharges. Notwithstanding the direct signal from my commander in chief, I am proposing that we let this thing continue and we will see where the ice breaks."

The president did not like Bud's speech to the flag officers or the *New York Times*'s coverage of it. Bud worried that the president was holed up watching distorted news coverage; he wanted to get the message out that he was not coddling dissenters. Murphy told Bud that Nixon was "down on the CNO for permissiveness, mod navy and so forth, and he thinks the ice is getting awful thin." Bud wanted to send the full text of his speech to San Clemente, but Murphy said it would never get to him. "I really think people are out to cut you up." Murphy explained that in this administration it was impossible to get to the top man. Murphy advised Bud to act like a tough SOB, but Bud feared that "if we go too far there will be the biggest dissension ever because these guys already think the judicial system is stacked against them."

Bud was desperate for allies, and his best chance for survival was with Mel Laird, since John Warner was offering little protection.[73] Bud felt as if he were walking the plank. He admitted in a phone call with Admiral Clarey, "I am the guy under the gun."[74] He was especially concerned because Warner had decided that he and not Bud would be the spokesman for defending the navy's programs in the race area. Warner "will not make the case of what the Navy has been through—deployments, carrying brunt of things." In no uncertain terms, Bud told Clarey, "I would feel a lot better if Mel were doing it."

The next day, Bud placed a call to Laird in order to review the state of play and solicit advice from the man who had hired him and always stood by him. Laird was in complete agreement on the issue of blanket dishonorable discharges being in violation of UCMJ. The two men discussed having a military man like Alexander Haig explain this to Kissinger and Nixon. Bud sounded desperate, asking Laird if he thought he should fly to Key Biscayne to see the president personally. Laird stopped him in midsentence. "You have to keep cool . . . don't panic."

Throughout the crisis, Laird's support never wavered. Laird admired what he described as Bud's "matchless initiative, your great enthusiasm, and your determination."[75] He considered Bud to be "one of the finest officers of this era or any era" and believed the Z-grams "have brought good cheer and improved feelings about service life to countless thousands of navy men."[76] For Bud, Laird's support amounted to a life preserver. "You have no idea of how reassuring it is to know that your deep commitment to 'loyalty down' matches your proper expectation of 'loyalty up.' I pledge to you my continuing commitment to your objectives to the full spirit of the goals not merely the letter!"[77]

Bud was not lacking advice from former CNOs. Admiral Arleigh Burke understood that long deployments and congressional cuts had contributed to the problems at sea, but said that "unless people are given very severe punishments no matter what the basic reasons might be, you will have more trouble."[78] From reading press clippings, Burke thought Bud was blaming his senior officers; Burke did not think that was the right thing to do. Bud urged Burke to read the text of the speech and determine if it was a dressing down of flag officers. Bud implored Burke to remember that "we have only had trouble on three (of 596 ships)." Burke predicted that "it will spread." Years later Burke offered the following assessment: "I think he's a brilliant man—but I think he became impatient with other people who were not so brilliant. I think he got his personal ambitions confused with his obligations to the service, and you can never mix up your own personal ambitions with your service obligations."[79]

Admiral Robert Carney shared Burke's view. In a conversation with Bud, the former CNO recognized that the navy had reached "the elastic limits on what we are doing to our people" because of longer deployments and no access to good liberty ports. That said, Carney thought Bud's comments to flag officers "looked like an indictment against the whole command structure." Carney, who had read the text twice, advised Bud that "we can't have the point where these birds are demanding something from the commanding officer."[80]

Former CNO David McDonald wanted Bud to throw the book at the *Connie* protesters for being AWOL after he had been so good to them. "This is going to be hard—especially hard because of the preponderance of blacks. Problem is that for everyone black that thinks he is being discriminated against because of color, there are a lot of whites who believe we are giving them privileges because of color." Bud's only chance was to say, "Look by God, I meant it and I think it was for the Navy's good. I also meant it when I said I wasn't going to let any of these things lower discipline. . . . Whatever you do Bud, you are in for it. You can't please everybody."[81]

Meanwhile, a handful of retired admirals were working with key contacts in the media and on the Hill to undermine Bud's credibility. Their ringleader was Admiral George Anderson, who had succeeded the legendary Arleigh Burke, but after clashing with Robert McNamara during the Cuban Missile Crisis had not been reappointed for a second two-year term as CNO. At that time, Anderson did not seem to understand the nuances of the dangerous military-political game and highly resented Defense Department intervention into the deployment of ships. When Bud became CNO, Tom Moorer warned that Anderson would be a "dead hand on the tiller," meaning that he would always be second-guessing Bud and looking over his shoulder. As soon as the first Z-grams were out, Anderson started taking issue with them, speaking with anyone who would listen to his second-guessing.

Bud's first run-in with Anderson had been years earlier, when he commanded the *Isbell.* Bud had had to disembark a sailor on emergency leave in order to catch a plane. As he proceeded into port on the southern part of Taiwan, Bud saw Anderson's flagship. At that time Anderson was in command of the Formosa patrol force. Custom and tradition required that Bud give way and fall astern of his flagship in order to let him into port first. To do so, however, would mean that the sailor would miss the plane to the States. Bud sent a detailed message to Anderson, explaining the circumstances and requesting permission to proceed into port ahead of his flagship. No reply came, which meant either that Bud had permission or that Anderson had not received the message in time to veto it. Bud proceeded into port and disembarked the sailor. Shortly after Anderson's flagship arrived in port, Bud received a visit from an officer seeking proof that Bud had actually disembarked a sailor on emergency leave and had not shown Anderson up. "This lack of trust and basic sense or irritation in the circumstances, I found noteworthy," said Bud.[82]

Bud knew that Anderson was the "anonymous" White House source on most negative stories about the Z-grams and the rumors of Zumwalt's firing. Anderson was a member of the President's Foreign Intelligence Advisory Board and benefited enormously from the access and legitimacy of that position. "Admiral Anderson never had the guts to speak to me about his own views. . . . Admiral Anderson is by nature shallow, vain, egotistical," said Bud. "He is a back-biter, and quite two-faced, all in all, very well-suited to be a member of the Nixon White House staff."[83]

The Anderson Papers provide ample documentation of a personal vendetta against Bud and his policies. When James Holloway was nominated to succeed Bud as CNO, Anderson wrote directly to the nominee: "It is most satisfying to so many of us that the selection process for the office of Chief of Naval Operations has reverted back to the choice of an individual [who] by background and experience, is immediately qualified to take the helm."[84] He railed against Bud for "emasculating the list of senior flag officers" and that "the deliberate campaign of early retirement for experienced flag officers has been uniquely effective in eliminating experience in this regard." In a letter to Bud dated January 25, 1971, Anderson observed, "I had a different policy on girls [at the academy] on beards and mustaches when I had command of the Sixth Fleet."

During this period of crisis related to the *Kitty Hawk* and the *Constellation*, Anderson wrote Bud that "we on the retired list are always most anxious to be completely responsive to the requests and objectives of the Chief of Naval Operations. Because we believe that he is the one who is 'standing the watch on the bridge' in the stormy seas endeavoring to avoid the rocks and shoals which always seem to be threatening our wonderful Navy. It is most distressing when this support cannot be a hundred percent across the board and universal on the part of us."[85] That was an understatement. Bud was by now well on to Anderson's games, evidenced by his handwritten note at the bottom of a letter from Anderson apologizing that he could not attend Bud's speech at the D.C. chapter of the Naval Academy Alumni Association at the Army-Navy Club and regretted not being able to hear Bud's views on the needs of the navy. Bud wrote at the bottom of the letter, "and then the bunnie rabbits came out of the woods and fed carrots to the unicorns—B.S. . . . George."[86]

Dan Murphy warned Bud that Anderson had strong allies. "There were a lot of perturbed people that would use every chance they get to let you have it. You just don't hear much support any more."[87] Bud was puzzled because the navy "was carrying so much of the war and doing it well."

The activities of Bud's opponents were hardly stealthy, but he also had equally vocal backing. Retired master chief petty officer Chester A. Wright wrote to Secretary Warner, "It is with chagrin, shame and a sense of alarm that I found that a group of retired apostles of yester-year's vintage are attempting to scuttle the first breath of fresh air breathed into that Naval Leadership since 1919." Congratulating Bud for moving the navy into the modern era and facing the realities of the day, Wright understood that "what makes this attack doubly tragic is that if these retired armchair strategists had done their own homework while they were in the rough and tumble command arena, Elmo Zumwalt would not have half the racial troubles that he presently must deal with."[88]

Rear Admiral George H. Miller offered similar support. "Have been particularly concerned about reported second-guessing by some alumni who did little, if anything, to advance the cause . . . while they were in office. . . . Stick to your guns, and you'll win in the long run. As Smedberg once told me, if you can't lick 'em, outlast 'em. I'm with you 100 per cent." Bud wrote back that while Smedberg had the right idea, "I'd really like to lick 'em, too."[89]

The crisis also brought out the racists.[90] A sampling of the daily hate mail reveals an ugly underside of the 1970s. W. Joe Lowry of San Diego wrote, "Dear Admiral: Having nigger troubles? Give them an inch and they take a mile. Clobber them over the head with a club at the first sign of insurrection, and you may get by with them for a while until you clobber them again. But your navy loves niggers—make them officers and even Admiral. And you talk about equality! The nigger's equality is to dominate the whole picture and kick whitey in the rear. You will never give them enough to satisfy for it will always be more and more and more."[91] From a retired chief petty officer: "The idea of tongue lashing the upper echelon of the Navy makes me want to vomit. . . . Your command should be the U.S.S. African Queen named after the people you must represent and bow to."[92]

R. H. Rothley of Fairview Heights, Illinois, sent his letter directly to John Warner, urging that Zumwalt be fired. "It is evident that Zumwalt has the back bone of a jelly-fish, and is a bleeding heart do-gooder, unable to maintain order."[93] W. P. Ricks of Rocky Mount, North Carolina, suggested a new Z-gram, the shortest of all—"I quit."[94] From Yonkers, New York, Harlan Tucker wrote, "Booze, beer, broads, and beards on the ships of our navy, are a hell of a way to maintain discipline. Frankly, I think you are nuttier than that cockamamie name of yours: I doubt if you are in possession of the intestinal fortitude to resign; therefore, President Nixon should boot you out."[95] A. Miller called Zumwalt "a mush brained pseudo-intellectual dimwit."[96] One brave anonymous soul asked, "I wonder how many waves (navy girls, call girls) you have at your command. A different one each time you want a fuck." P.S: "That name sounds German."[97]

Recently released tapes reveal that President Nixon defined the situation in ways similar to the letter writers. On November 13, Kissinger tried redirecting a conversation to "that navy dude, that Zumwalt." Before Kissinger could complete his sentence, the president said, "About the race thing, calm down, quite frankly, if anything, now let's be quite candid about it, honestly, if anything the problem in the Navy and all the armed services is that they are determining too much for the blacks, giving them positions they aren't able to do." Nixon then asked, "It is mutiny, isn't it?"[98] On a December 20, 1972, tape, Kissinger questioned Bud's competence, telling the president that "his lack of intelligence has dropped on his lack of character."[99]

Bud had always enjoyed his working relationship with Congressman F. Edward Hébert, chair of the Armed Services Committee. Hébert had been predisposed to doing more for defense and was a strong supporter of navy budgets. He did not favor Bud's changes in the personnel area, but the two men managed to find a way to forge a productive working relationship. "Eddie did not think that long hair and beards were consistent with good discipline, did not favor women going to sea, and preferred to maintain the Navy's traditional policies towards minorities," wrote Bud.[100]

Bud's speech to the admirals was the tipping point for Hébert, who had attended the U.S. Marine Corps birthday party in New Orleans. Several senior members of the marine and navy old guard told Hébert that Bud had been way out of line giving that speech and speaking that way to flag officers. Urged by his close friends, like retired Admiral Anderson, who was leading the smear campaign, on Monday, November 13, Hébert called for an investigation and hearings into "alleged racial and disciplinary problems" in the navy.[101]

When Bud got word of the hearings, he called Hébert in New Orleans. The only good thing about this was that a friend was doing it to him, Bud told Hébert as he began a conversation with a man who had opposed the 1964 Civil Rights Act on the grounds that "I do not believe it is equitable. I do not believe it is valid."[102] Hébert wanted Bud to know that what really bothered him was chewing out his admirals. Commanding officers didn't "eat out their Admirals. You chewed out your own people on this side and didn't say anything about the other people . . . they got convicted in public without a trial." Bud held his ground. "All we said to the officers was get off your duffs and get this squared away." Hébert warned Bud that the navy was "headed for real trouble because of the so-called equality business and quota business of putting people in jobs they are not qualified for. . . . Gotta have a quota for blacks just because they are black—I don't buy that."[103]

Bud tried to explain that the offenders from the *Constellation* were being processed as rapidly as possible by lawful procedures, but Hébert sounded like Kissinger and Nixon. "You mean the CO is telling them not to be bad boys again. . . . They refused an order!" Bud said that the only order they had refused was to go back on ship, which is simply absence without leave. There were 593 other ships that hadn't had a problem and "are fighting a hell of a tough war while working 18 to 20 hours a day." Bud asked Hébert why the whole navy was being indicted.

Hébert wanted Bud to know that the backlash from his speech to the admirals had been horrible. Even the venerable Carl Vinson had called to say, "Get rid of Zumwalt." The now retired Vinson was a founding parent of the two-ocean navy and was often referred to as Mr. Armed Services or Mr. Navy.[104] Hébert appointed Democrat Floyd Hicks of Washington to chair the subcommittee investigation. Hicks was flanked by lame-duck Republican congressman Alexander Pirnie from New York, a man Bud knew "believed in busting heads to maintain discipline." The other member was Wilbur Clarence "Dan" Daniel, a Democrat from Virginia whom Bud regarded as "very pro-defense and very people oriented, but definitely segregationist." Bud understood that Hébert had assigned Daniel to the committee "to ride shotgun, as it were, on Hicks, and report back every evening on how the chairman had done that day."

Admiral Bill Thompson saw the hearings as a charade by "southern congressmen making a move to show what they thought of Zumwalt's programs relative to minorities—blacks, Hispanics, Filipinos and women—also their dislike for beards and women as aviators and serving aboard ship."[105] Bud's former flag aide Lew Glenn wrote directly to Hébert, registering astonishment at the scurrilous attacks on his boss: "During the past few weeks I have read with increasing dismay and disbelief the stories concerning permissiveness in the Navy. Having worked closely with Admiral Zumwalt in Vietnam when he was Commander, Naval Forces, VN and having served on board ship after he became Chief of Naval Operations, I have seen his firm leadership in the close combat environment and felt his direction over the entire Navy. Having served under his leadership in these capacities, it amazes me that Admiral Zumwalt's total dedication to law and order and sound discipline is not apparent to everyone."[106]

Bud began to see the Hicks hearings as an opportunity to make his last stand against those who sought to pin the label of permissiveness on the navy's personnel policies and those who argued that integration had created permissiveness. Bud asked aloud why these types of problems had not yet occurred in the Atlantic Fleet or even on Atlantic Fleet ships operating in the Pacific.[107] There could be only one answer—extended deployments, not the Z-grams. He was not going down without a fight. He would be fired, ordered to slow down equal opportunity, or win. His own political antennae told him that he would have the support of all the liberals in Congress who had opposed his defense budgets. When Bud turned to John Warner for support, he was told it was his mess to clean up. "The fascinating thing then was to try to carry out these hearings with a Secretary of the Navy who was hostile to what I was trying to do."

Bud would need guile, skill, and luck over the next six weeks. "I have never been in a nastier fight or one that was more important to win."[108] He and his staff went into crisis mode, a "total preoccupation with survival,"[109] working day and night, putting out fires, building support, rallying political pressure. Bud felt under siege, much as Richard Nixon must have felt, except "I didn't feel like a crook."[110] Bud also began to see the bigger playing field—a "right wing racist reaction" organized by Anderson and others, who was working in the Nixon White House as a presidential assistant with good contacts among the press and the retired-officers community. He also took notice of people taking sides, like Admiral Isaac Kidd, whose deviousness would never be forgotten. Anderson kept in close contact with Kidd, who was providing minute-by-minute advice from the inside as to how best to proceed while keeping his hands very well covered. Bud was told by his friends on the Hill that Kidd was also using Admiral Rickover to promote Kidd as the man who could straighten the navy out. Making matters worse was the perception that John Warner "was working against Bud with the Admirals," said Roberta Hazard.[111]

John Warner was growing weary of Bud's insistence that he be allowed to hit the airwaves. Warner did not think that using the rationale of extended deployments would have any traction with the general public or in Congress. In a memorandum for the record, Bud noted, "My decision is to use it but make it one of the last factors in my briefing."[112] J. Fred Buzhardt, Jr., general counsel for the Department of Defense, was now closely involved as one of Bud's confidants and strategists. Buzhardt agreed with Warner in this instance. He was "very reluctant to see me put too much stress on the deployments because this turns the press off." Warner warned that there could be no distance between the two men in public statements. Deputy Assistant Secretary of Defense for Public Affairs Jerry Friedheim urged Bud to understand that "we have to make it down the road for the good of what you and John will do next year."[113]

The Hicks hearings were scheduled to begin on November 20. Bud's sources on the Hill had already made clear that the hearings were going to focus on whether or not there was permissiveness in the navy. Meanwhile Bud was fuming about the inspector general's recently released report on the *Connie*. That report identified Z-grams as a contributing factor in the breakdown in intraship command and control. "I went through all Z-Grams and I'll be damned if I can find anything that could be interpreted that way," Bud told Clarey. "Z-Grams speak to sailors and by-pass the chain of command," replied a frank Clarey.[114] Clarey wanted to improve the middle-management guys. "The black CPO [chief petty officer] and first class [petty officers] are looked at as being Uncle Toms—we have to educate these people—tell them to get with it or get out."

With the hearings approaching, Fred Buzhardt thought the time had arrived for Bud to reach out to Carl Vinson through retired rear admiral Robert H. Hare. Hare had spent part of his younger years growing up in South Carolina, had graduated from law school there, and had retired to South Carolina after serving with distinction as deputy judge advocate general. He was good friends with both Carl Vinson and Eddie

Hébert. The plan was for Bud to ask Hare to arrange a meeting with Vinson.[115] Bud began the conversation with Hare by saying, "I'm in a bit of trouble," adding that he could use some help. Bud told Hare that he had just mailed him a copy of his speech to flag officers as well as the latest Z-grams on the importance of good order and discipline. He asked Hare to "find it in his conscience" to speak with Vinson, because Vinson could not possibly be getting the full story. Bud explained that he had "been fighting my heart out for about a year," because navy personnel had been stretched to the breaking point. Six rather than three carriers were deployed in Southeast Asia at the same time as the stand-down in Vietnam and congressional budget cuts. Certainly problems existed, but Vinson could not possibly be getting the full story.

Hare didn't think Vinson was likely to change his mind, as Bud recounted. "Once he has made it up he has never known him to be willing to change it easily. . . . Vinson has his thinking pretty well in concrete but if he finds out about this point he may have a tendency to water down his venom." Hare offered to help, but warned Bud that "the haircuts have been bugging people ever since it came out." Bud reminded Hare that Admiral Moorer had actually endorsed this policy before Bud even became CNO!

Bud Zumwalt and John Warner were soon on a plane to Milledgeville, Georgia, for lunch at a local Holiday Inn with Eddie Hébert and the eighty-nine-year-old Carl Vinson. Vinson wanted Zumwalt to get a "good scrubbing" from the committee.[116] Bud tried breaking the ice by recalling something that happened when Vinson had chaired the House Armed Services Committee. Bud was working for Paul Nitze. Vinson had been invited to give the graduation address at the Naval Academy. Bud and Nitze drove there with Vinson, and during the ride, Vinson told them that when he really wanted to do a job on people, he would put Mendel Rivers in charge; when he was really mad at someone, he would appoint Porter Hardy of Virginia to the committee; and when he wanted to make certain that the person would get really scarred up, he would add Eddie Hébert to the unholy three. Vinson loved the story, repeated it, and winked at Hébert.

Prior to lunch, Vinson asked many questions, making clear that what had really gotten him riled up and made him call Eddie Hébert about holding a hearing was the black-power salutes on the pier beside the *Connie*. "He said as he saw that he couldn't help remembering the [original] *Constellation* as the first ship in our history to score a victory and he resented the comparison."[117] While in Milledgeville, Fred Buzhardt informed Bud that a group of retired admirals had been calling reporters to say that "you are being fired and Chick Clarey is replacing you."

Bud could sense the fat hitting the fire. Following his meeting, Bud wrote Vinson, "As I attempted to describe to you during our conversations, order and discipline do in fact prevail within the naval establishment at large. Although our record is by no means perfect, the overwhelming majority of Navy men and women are indeed dedicated to their country and its purpose and have discharged their individual responsibilities with honor, valor and dignity."[118] Bud enclosed a signed copy of a book, *Riverine Warfare: Vietnam*, and two documents—one a report on Project Sixty and the other a copy of his letter to flag officers at the two-year point of his tour. Bud confided that he had the feeling of "being shot at from both the extreme right and from the liberal groups for exactly opposite reasons. A lot of concern on the liberal side as to why in these episodes the blacks were charged and, on the other hand, you get the very right side that is concerned as to why more decisive action wasn't taken more quickly in the case of the *Connie*. Just a symptom of the tough times we are going through."[119]

Bud was scheduled to be the first witness at the hearings. A day before the hearings, his testimony was being rewritten so that the phrase *discipline and order* always preceded the word *race*. Daniel Z. Henkin, assistant secretary of defense for public affairs, counseled Bud not to be defensive and to stay on script. Bud called Hébert in order to review the main points of his testimony—the importance of combat effectiveness, good order and discipline, and human-relations goals. Hébert warned Bud that "we should not emphasize trying to bring in any particular ethnic group—we want effectiveness—we want the best man for the job." Bud joked that "I feel like the man [who] said to Abraham Lincoln when he was being ridden out of town on a rail—if it weren't for the honor of this thing, I'd just as soon walk." Hébert said he had faith in

Hicks, who was neither right nor left, and a fair person. "The whole thing was we cannot emphasize we want the ethnic groups or the minority groups to take over," said Hébert.

As Bud Zumwalt left home to testify before the Hicks committee, his daughter Ann left a note in his briefcase: "Dearest Daddy, I just wanted to wish you very good luck today. What words of easement can one speak of? I only hope those who dramatically oppose you will eventually see the role and importance your thoughts and you are, and what your and mommy's vision is doing as a unifying force to the service of man."

Chairman Hicks opened the hearings by saying the focus would be on "whether or not these incidents indicated a permissiveness which had led to a breakdown of discipline."[120] This sounded like little more than a plan to place blame on the CNO for increasing the number of blacks through reverse discrimination, lowering standards, and reducing the quality of recruits. "We cannot overlook the possibility that there may exist at this time an environment of—for lack of a better word—permissiveness, wherein all that is needed is a catalyst. Perhaps perceptions of racial relations in the cases provided the spark," said Hicks in his opening statement.

Bud appeared twice before the subcommittee, the first time during its opening session on November 20 and later in mid-December when the committee returned from California, after visiting the *Constellation* and holding hearings there. In the first session, Bud tried putting the incidents in perspective, emphasizing how proud he was of the navy's performance since the massive Easter invasion when he had been required to commit unprecedented forces from a greatly reduced base. The navy had over 650 ships, many operating under wartime conditions, "unmatched elsewhere in our society for rigor and danger. . . . Their hours of work have been extreme, their separation from home and family long, and time with loved ones short."

Bud reasoned that discipline and order were the sine qua non of an effective fighting force and that conversely the effectiveness of the force is the essential measure of its good order and discipline. If the question was how best to ensure that discipline and maximum effectiveness are maintained, the answer was obedience within the context of an evolving society dedicated to humane goals and equal opportunity. An environment with no draft, an all-volunteer force, the greatest combat stress and longest extended deployments since World War II, and a base of shrinking resources required that commanders instill "intelligent obedience" in military personnel. Black sailors entering the navy from a society that discriminates expected to find the same in the navy. The navy had to make every effort to ensure a state of good discipline by eradicating all vestiges of prejudice.

Two other people spoke in defense of Bud's programs, Admiral Bernard Clarey, commander in chief of the Pacific Fleet, and retired admiral U. S. Grant Sharp, Jr., commander in chief, Pacific, 1964–1968. "These equal opportunity programs have to go on and they have to be implemented," Sharp told the committee. Hicks told Clarey that what had transpired on board the *Constellation* was "deliberate mutiny" and wanted to know whether Clarey agreed. "Yes sir, certainly that was collective insubordination in my view." Hicks was incredulous at that type of nuanced language. "Well that might be a nicer way of saying mutiny. They didn't exactly try to take over the ship, but they decided it was going to run the way they wanted it to run." Clarey rejected the very premise. "I see no evidence of a breakdown of good order and discipline. On the contrary I see a great deal of evidence that the overall level of discipline in the fleet is as high as at any point in history."

In the second session with Bud, the questioning quickly became nasty, and Bud became much more vigorous in defending his personnel policies. He was not going to deny that there was racism in the navy. Hicks then used letters and examples to mock the abolition of Mickey Mouse regulations, citing "sloppiness and unbuttoned jackets." A petty officer had told him, "If I can't tell a man to get a haircut, I can't tell him to do anything." Representative Daniel accused Bud of putting his head in the sand: "You absolutely refuse to face reality. We went aboard the *Constellation* and it looked like a pigpen. It was absolutely filthy. . . . You are letting small minorities disrupt this great record in the public mind of those people performing so well."[121] Throughout the session, all three committee members kept referring to Bud's November "berating" of the admirals. One committee member saw a link between recent cases of sabotage aboard certain ships and the racial incidents, alleging destruction of government property by minorities running rampant.

In California, the committee did everything possible to maneuver Captain Ward into admitting that the human-services officer on the *Constellation* was responsible for the breakdown in discipline. Ward refused to bite, insisting that his officer played an important and constructive role, giving the example of black sailors being addressed as "boy" by petty officers. Chairman Hicks insisted that the term *boy* had "nothing to do with color, race, or anything else." Congressman Pirnie told Ward that blacks should be grateful for even being allowed in the navy. They were "lucky," because standards had been lowered and they now had better jobs than they might otherwise have had. Ward interjected by saying that his Human Relations Panel would never say such a thing, because it implied blacks were inferior. "If he doesn't pass the test, isn't he?" asked the congressman.[122]

Commander Cloud's testimony was especially important. He was black; the committee, entirely white. Cloud came out swinging, reminding the committee that the navy was the last armed service to comply with executive orders banning segregation and that many of the sailors had fathers who had served in World War II, when the navy was known as a place where blacks could rise no farther than steward's mate. Their fathers held the view that "the Navy is the most biased, segregated branch of the service there is, so if you have to go into a service, don't go into the United States Navy." Cloud reminded the committee that Truman's 1948 mandate that the armed services integrate had had only limited effect on the navy. He testified that at the time he enlisted in 1952, he could not use base facilities at many installations, including Pensacola. "The black, no matter what you say, feels the pressures of segregation more in his attempts to secure housing and education for his children than he does anyplace else. The Navy, up until recently, has condoned, basically, segregated housing, [and] medical as well as school facilities."[123]

The pressure on Bud during this period was most obvious to those closest to him. In late November, Mel Laird sent a birthday greeting: "I find it hard to believe that the 49-year old Admiral I swore in as CNO in 1970 is now 52. The time has passed fast. . . . Of course, when you have as many 'balls in the air' as you do, you are bound to drop one or two. Sometimes those that are dropped get all the emphasis, and the successes are somewhat overlooked. On this important day in your life, your 52nd birthday, I want you to know that I measure your performance as CNO by the countless successes. Keep those 'balls in the air,' Bud. You are a fine leader. You have my full confidence and backing."[124] Laird then went public with a strong statement of support, insisting that Bud would finish his term. Dan Murphy saw this as the best evidence that "Laird and his people are all behind you."[125]

Some of Bud's support came from the unlikeliest quarters. In a handwritten note of appreciation for visiting him in the hospital during the pressure-packed days of the hearings, Admiral Rickover said, "Please accept my best wishes during this troublesome period. Nothing is actually as bad as it seems at the time. I believe the worst is over and that the Navy will be the better off to have its problems aired."[126]

Bud also received a call from former secretary of defense Robert McNamara offering a strong possible defense. "Tell them McNamara told him some time ago he regretted he had not enacted in advance the movement to get racial equality—say it!"[127] Bud refused to put any of the load on McNamara, telling him that "this was an effort to turn the clock back." He informed McNamara that he had a group of active admirals "whose sense of law and order has permitted them to go in revolt." McNamara agreed that "we can't turn it back."[128]

Sometime during the Hicks hearings, two officers visiting the Pentagon presented Bud with a poster depicting a black kitten clinging desperately to a narrow rod with the caption, "Hang in there, baby."[129] Bud told his family, "I just keep leaning forward."

Bud's attention was diverted from the hearings to Vietnam on the morning of November 30, when he attended a JCS meeting at which the president and Henry Kissinger planned to brief the chiefs on the draft treaty being negotiated by Kissinger and Le Duc Tho in Paris.[130] The president began the meeting by saying that on May 8 he had laid down three conditions for peace; one, a cease-fire; two, return of American prisoners of war and an accounting of the missing in action; and three, assurance that the people of South Vietnam will have the right to determine their future without the imposition of a communist government or communist coalition. "The proposal made by Hanoi on October 8 meets these requirements but now Saigon and some in the U.S. say this is not enough. The facts are, however, that if the American people knew all the details of what has been offered, they would never continue to support a prolongation of the war."

Bud was one of the first to recognize the depths of Nixon's deception. "The President's discussion of the status of the cease-fire increased my sense of being on a strange planet. I could not help reflecting that the crews of *Kitty Hawk*, *Hassayampa*, and *Constellation* had not been able to deal with the Vietnam War from armchairs in a quiet study. It was perfectly obvious to all of us at the time that the promise of massive American assistance to South Vietnam and of prompt U.S. retaliation to serious truce violations were the critical elements in securing the cease-fire and that the fulfillment of these promises would be the critical element in maintaining the cease-fire. Yet the administration never really let the American people—or Congress—in on this non-secret, apparently on the assumption that the critical element in persuading Americans to accept the terms of the cease-fire was to allow them to believe that it meant the end of any kind of American involvement in Vietnam no matter what happened there after the cease-fire was agreed to. Not even the JCS were informed that written commitments were made to Thieu [President of South Vietnam Nguyen Van Thieu]. There are at least two words no one can use to characterize the outcome of that two-faced policy. One is 'peace.' The other is 'honor.'"[131]

The Hicks committee completed its work in December, and the chairman pledged to release a final report in early January. Rumors that Bud was being fired because of violence and unrest in the navy pervaded the news. Under the headline REPORT SAYS ZUMWALT IS ON THE WAY OUT, the December 17, 1972, *Washington Star* quoted Congressman William

L. Clay (D-Missouri) that President Nixon had decided to reassign Zumwalt. "You can quote me as saying Kissinger told Zumwalt he was going to be fired." *Human Events* predicted that Bud's "knuckles would be rapped" for going too far too fast. The story sourced Representative Harold R. Gross (R-Iowa), ranking member of the Foreign Affairs committee. "If Admiral Zumwalt is so liberal and soft that he cannot enforce discipline in the Navy, he should get out or be tossed out."[132] Bud lamented that "the pros are coming out with favorable articles and they are hurting me because they are too liberal and the antis are coming out with the other stuff in rebuttal."[133]

Conservative columnist Max Rafferty took aim at Zumwalt in a column titled "Wave of Tolerance Swamping Navy": "A Navy that is rotten with rebellion, palsied with permissiveness and disintegrating with disobedience is far more dangerous to our national security than having no Navy at all." The column was sent to Zumwalt with a letter: "If you were any kind of man you would resign. You don't belong in the Nixon camp anyway. You're a Johnson type rinky-dink."[134]

Bud's supporters also surfaced, in force. Retired lieutenant R. P. Leavitt of Hartsdale, New York, wrote, "If the President replaces you, at least you will have the satisfaction of knowing that you have been a man just a little ahead of his time at a time when 'time is out of joint.' If the rumor proves to be true and you lose the battle, it will be just one more instance in which later times recognize a man who, if heeded, could have greatly eased the transition from past to clearly predictable future. I feel strongly that you personally embody the whole future of the volunteer Navy in these changing times."[135]

Lieutenant General Robert E. Pursley, who had been outside the door when Bud first arrived in Washington for his CNO interview and was now commander of U.S. forces in Japan, offered support. In a "Dear Bud" letter on December 18, 1972, he wrote, "Times are never easy for those who are too innovative, imaginative, and energetic. . . . I know there has been turbulence during your tenure as CNO. I want you to know that at least one of your most avid supporters, this old 'fish' here, views the activity in an affirmative context. Recognizing that taking hard knocks is not the most pleasant pursuit imaginable, I hope you will continue to press on and will continue to be the uniformed leader who commands our highest respect. You have mine, Bud."[136]

While waiting for the Hicks report, Bud received a note from Howard Kerr. Four years earlier, they had arrived together on day one in Vietnam. "I believe that the position you have taken is nothing less than Lincolnesque. What is at stake is nothing less than the quality of our commitment to our service, our country and our fellow man—And these are not divisible. You are running with the tide of history. I am proud to know you and serve you."[137] Another former Vietnam hand, Bob Powers, wrote that he had heard on Armed Forces Radio that the critics were having a field day. "What I really want to say is that there are many who haven't lost the faith. If there is anything I can do, I would be eager to respond."[138]

A *Los Angeles Times* editorial titled "Zumwalt and the New Navy" took a similar position: "Hicks is wrong. The ultimate question was not 'permissiveness' but rather failure of the command structure to carry out Zumwalt's orders. The ultimate question here is whether there will be a modern Navy, a Navy responsive to the principles of human equality. Only that kind of Navy will attract the best men and women. Nothing is more important to the mission of the Navy than the quality of its personnel. That is why Adm. Zumwalt must succeed."[139]

The Hicks report was released on January 2, 1973.[140] In some ways, the report was as harsh an indictment as could have been issued within the norms of congressional propriety. After fifty-six witnesses and 2,500 pages of executive-session testimony, including two contentious sessions with the CNO, to no one's surprise, the committee's final report identified the "permissiveness" of recent reforms as the cause of unrest. The committee's investigation revealed prejudgment and preconception from the outset. "The United States Navy is now confronted with pressures, both from within and without, which, if not controlled, will surely destroy its enviable tradition of discipline." The committee warned that the "environment of leniency, appeasement and permissiveness" enhanced chances for mutinous acts and brought about "an alarming frequency of successful acts of sabotage and apparent sabotage on a wide variety of ships and stations within the Navy. . . . If discipline collapses, a military force becomes a leaderless, uniformed mob, capable only of accomplishing its own destruction."

From the very beginning, the hearings had been structured to obtain this foreordained result. An unedited version of the transcript showed Chairman Hicks saying, "We are going to do the very best we can to make a fair conclusion based on the evidence, as it appears to us." And so they did.[141] The issue of disciplinary laxity throughout the navy became the primary focus, even though the investigation was triggered by racial incidents on only 3 of the navy's 650 ships. The committee had its tentative conclusions and then went in search of the facts to support them. On the other hand, the navy was asked to prove that permissiveness had not caused the riots. Witness after witness, through a tangle of badgering and leading questions, was directed toward the committee's conclusions—discipline was deteriorating; permissiveness prevailed; authority had been diluted; the chain of command had been ignored; blacks wanted special treatment; and there was no racial discrimination.

The conclusions of the report were often cloaked in innuendo. For example, the committee report noted that the *Constellation* and *Kitty Hawk* incidents "might be characterized as mutinies." They might also be characterized as riots, assaults, or disobedience of orders. In its first so-called finding of fact, the committee stated that permissiveness existed in the navy and could be servicewide. The committee reached its conclusion about blacks not perceiving racial discrimination on the *Kitty Hawk* without interviewing a single black crew member. With respect to the *Constellation*, perception of discrimination was conceded, but the committee concluded that several of these perceptions were not accurate. The committee concluded that the human relations councils manned solely by minority personnel were used to bypass the chain of command and were disruptive to good order and discipline. This racist point somehow suggested that if whites had been on the councils, they would have been improved.

Astonishingly, even the White House ended up seeing the matter more clearly than the committee. In a "No Eyes" memo to President

Nixon, William Timmons, the White House staffer tasked with the evaluation, observed,

> The early fear that the Subcommittee was a "hanging jury" seems to have been justified. Committee staffers report that it was "adopted" and fed by a number of retired senior officers who have been broadly opposed to Zumwalt and his reforms on "beards, beer and broads." The report reads as if it were guided by a priori conclusions. The sweeping generalizations drawn from the two cases are simply not supported by the evidence. . . . To say as the report does that there is no racial discrimination in the Navy is obviously untrue. Zumwalt has gone far in removing its institutionalized aspects but habit and attitude die hard and discrimination persists at the personal level.
>
> The report completely ignores a critical dimension to race problems in the services. Young recruits bring with them the trends and attitudes ruling in society, and black separatism has come to the navy just as it has to high schools and universities. This polarization permits individual grievances to escalate quickly into group grievances, and hence troubles. The Navy, however, seems to be dealing with this phenomenon far more effectively than civilian institutions.[142]

Timmons noted that the report totally ignored the effects of long deployments in Southeast Asia. "Without building a case, the report concludes that 'permissiveness' could be found Navy-wide, yet elsewhere makes the finding that the navy has performed with outstanding combat effectiveness in SEA [Southeast Asia]. Other members of the Armed Services Committee point out that the Zumwalt reforms averted a manpower disaster in the Navy. Yet the report does not even mention the dramatic reversal of reenlistment trends under Zumwalt (10% in 1970, 23% in 1972) and the great improvement of overall morale among both officers and enlisted."

———

The period was a "watershed" in many ways. "When I survived, then the right wingers went in under a rock, and were just kind of sullen, but not mutants for the rest of my watch." The committee made no recommendation of rebuke for Bud Zumwalt. "I had a nasty month or two, but the giant torpedo that I had feared would blow my policies clear out of the water turned out to be a damp little squib."[143]

With the crisis behind him and the Nixon presidency about to be consumed by Watergate-related issues, Bud penned a handwritten note to Bob McNamara: "Now that the shot and shell of reaction to Navy's racial episodes have been left astern and with the failure of the effort to convert the word 'integration' into a synonym of 'permissiveness,' I wanted to let you know again how much I appreciated your thoughtful telephone message of support during the height of the action. I have thought of it, often."[144]

CHAPTER 12

THE ZUMWALT INTELLIGENCE SERVICE

If it was a movie, you wouldn't believe it.
—RICHARD NIXON, WHITE HOUSE TAPE,
DECEMBER 22, 1971[1]

At the outset of his term as chief of naval operations, Bud had been enthusiastic about the opportunity to be in close consultation with President Richard Nixon and Special Assistant for National Security Affairs
Henry Kissinger, the principal actors who formulated and conducted
American foreign policy. As naval advisor to the president, Bud anticipated communicating directly with his commander in chief, which
Nixon assured him at their first meeting would be standard operating
procedure. Within a matter of months, however, Bud realized that he
was working for an administration whose style of governance emphasized secrecy, distrust, and back-channel communications. This design
represented "the deliberate, systematic and, unfortunately, extremely
successful efforts of the President, Henry Kissinger, and a few subordinate members of their inner circle to conceal, sometimes by simple silence, more often by articulate deceit, their real policies about the most
critical matters of national security."[2] Primary among these issues were
strategic-arms negotiations with the Soviet Union, termination of the
war in Vietnam, and the opening of diplomatic relations with China.

Bud came to see the modus operandi of the Nixon administration
as "divide and conquer the bureaucracy by selectively withholding information."[3] The back-channel style derived from Kissinger and Nixon's
concern with leaks emanating from the bureaucracy. The intent of the
back channel was therefore to create a separate channel for communications independent of the Pentagon or State Department. "Essentially,
a backchannel is a communication system that seeks to circumvent

normal procedures," wrote Kissinger.[4] This mind-set was captured perfectly in two Oval Office conversations. The first was on June 13, 1971, when Kissinger reminded Nixon that "[Secretary of State John Foster] Dulles always used to say that he had to operate alone because he couldn't trust his own bureaucracy." Nixon quickly agreed, saying, "I just wish we operated without the bureaucracy." While Kissinger laughed, Nixon caught the absurdity of his own comment, interjecting, "We do! Yeah we do, we do, we do." Kissinger closed by saying, "All the good things that are being done [are accomplished by back channel]."[5] In a conversation on December 22, 1971, Nixon told H. R. Haldeman, "I tell you whenever there's anything important you don't tell anybody. We don't tell Rogers, Laird, anybody. We just don't tell any son-of-a-bitch at all."[6]

As the private channels started paying dividends, documents stamped Top Secret or Eyes Only became the norm. The first private channel had its roots in relations with the Soviet Union, a top priority for Nixon personally. The president instructed Kissinger to open a private channel with Soviet ambassador Anatoly Dobrynin, below the radar of Secretary of State William Rogers. Dobrynin's daily and routine official contact would be with Rogers, but anything important would be transmitted through Kissinger. In his memoir, Dobrynin noted that this "two-tiered" method of diplomacy seemed strange, especially to his boss, Foreign Minister Andrei Gromyko, who did not know which loop Secretary of State Rogers was in, although it was apparent that Rogers was not in the most important one. The private channel paid important dividends in U.S.-Soviet relations by reducing tensions and facilitating agreements over Berlin, Cuba, and the Middle East. President Nixon also utilized the channel to ask for Soviet assistance in pressuring the North Vietnamese in their negotiations to end the war in Vietnam, eventually leading to private talks in Paris between Kissinger and Le Duc Tho.

The historic 1971 diplomatic breakthrough with China evolved when Pakistan's president Yahya Khan, architect of the Bangladesh genocide of 1971, offered to serve as Nixon's secret channel. In November 1970, Khan personally delivered a message to Chinese premier Zhou Enlai relating Nixon's desire to send an American emissary to China. On July 1, 1971, Kissinger embarked on an elaborately deceptive journey orchestrated by President Khan and code-named Marco Polo in honor of another Western traveler who had journeyed to China. Flying first to Vietnam for consultations in South Vietnam and then to Pakistan for a well-publicized visit, Kissinger developed a "stomachache" at dinner and excused himself. The feigned illness would keep Kissinger in bed for a few days and out of sight of the press. Meanwhile, Kissinger was driven to an airfield in the middle of the night, where a Pakistani jet flew him to China. The world's press, the U.S. embassy staff, and members of Nixon's cabinet were kept in the dark about Kissinger's whereabouts in Beijing.[7] For two days, no one, not even the president, knew whether the mission was a success. Only when Kissinger sneaked back into Islamabad was he able to cable the code word "Eureka" to his assistant, Alexander Haig, who called Nixon with the single-word message of success.

The Pakistani back channel paid such great dividends that "from then on the contacts were kept secret from the Department of State, and indeed from NSC staff members who had no need to know."[8] The bureaucracy would be used for technical expertise but not strategy development. This modus operandi required keeping multiple sets of briefing books. Winston Lord, Kissinger's special assistant for the China trip, recalled that there were actually three sets of briefing books: "One set for those who were going on to China. . . . Another set were for those knowing that people were going into China, but were part of the cover team staying in Pakistan. . . . And then the third briefing book was for those who didn't even know there was going to be a China leg."[9]

There were times when the house of cards almost collapsed. At one NSC staff meeting, for example, Melvin Laird's deputy, David Packard, was sitting next to JCS chairman Admiral Thomas Moorer. He noticed that Moorer's briefing book contained a document on White House letterhead that Packard knew Laird had never seen. This was because Kissinger didn't want Laird to see it. The secretary of defense was on the "no need to know" distribution list for that particular White House policy paper. To counter this type of inadvertent error, Kissinger's staff was instructed that henceforth the White House letterhead was to be removed from documents going to Moorer before photocopying.[10] This way, the next time Packard or Laird looked at Moorer's briefing book, they would not catch any discrepancies. "You can't have people looking over your shoulder," said Lord. "You've got to hand out the right book to the right person."[11]

Laird, Moorer, and Bud Zumwalt developed strategies for overcoming the back-channel routine, which was leaving them in the dark on critical issues of national importance. For Bud, this began with the practice of what he called Kissingerology, premised on the assumption that he could not believe anything that Kissinger told him and that Kissinger's assent did not mean that he agreed or intended to do what he said, because it "could very well be a mere ploy to defuse or dismiss me."[12] By the end of his second year as CNO, Bud concluded that if he wanted a voice at the table to advocate for the navy's political-military policies, he needed an early-warning system to learn what Kissinger was up to. This meant that his relationship with Kissinger was "almost entirely [that of an] adversary."[13]

The practice of Kissingerology was like putting pieces of a jigsaw puzzle together, taking each piece of human intelligence and analyzing it, just as Sovietologists did in assessing the motives of their adversary. It involved finding out what members of Kissinger's staff and extended network were up to. A case in point was Seymour Weiss at the State Department, a close confidant of Kissinger. Weiss was a defense and foreign-policy expert known as a hard-liner, who warned against Soviet intentions and cautioned against détente and arms reduction. Bud nurtured a relationship with Weiss and appointed him to a CNO executive panel. Weiss spilled few beans, but when he and others explained administration rationale, "sometimes one learned a little from observing which beans they took the most care not to spill."[14]

SALT was all back channel, and Kissinger "was wholly and unquestionably in command."[15] The back channel "was so far back that neither the members of the U.S. SALT delegation nor even those senior officials who sat on the Verification Panel knew about it."[16] Moorer, Laird, and

Zumwalt were on the periphery, "hence, we could not affect those negotiations, which is precisely what Kissinger and Dobrynin intended."[17]
Laird told Moorer that "he was really mad as hops that Kissinger had
been by-passing him on things."[18]

Bud could always identify the day his relationship with Henry Kissinger
started to unwind—Saturday, November 28, 1970, during a train ride
returning from the Army-Navy football game. At Bud's invitation, Kissinger and Nancy Maginnes joined him and Mouza in their special box
for the game. It had been a delightful day until the ride home. Bud's
handwritten notes of the conversation offer his account: "K. does not
agree with the President that American people can be turned around.
He states strongly that the President misjudges the people. K feels that
the U.S. has passed its historic high point like so many other civiliza-
tions. He believes that the U.S. is on downhill and cannot be roused by
political challenge. He states that his job is to persuade the Russians to
give us the best deal we can get, recognizing that the historical forces
favor them. He says that he realizes that in the light of history he will
be recognized as one of those who negotiated terms favorable to the
Soviets, but that the American people have only themselves to blame
because they lack stamina to stay the course against Russians who are
'Sparta' to our 'Athens.'"

During the ride, Bud challenged Kissinger's position, saying he
could not accept giving the Soviets superior capability in either strategic or conventional fields, believing that the issue needed to be presented to the public. Once the people understood the risks, Bud believed
they would approve additional spending to guarantee their survival.
Kissinger disagreed. "You don't get re-elected to the Presidency on a
platform that admits you got behind. You talk instead about the great
partnership for peace achieved in your term." When Bud challenged
this premise, Kissinger threatened: "You should take care lest your
words result in a reduction in the Navy budget. There are subtle retri-
butions available."[19] It was at this moment that Bud Zumwalt "opted not
to be Kissinger's whore."[20]

The conversation over, Bud went through the train to find Mouza.

He told her, "I have seen Henry's view of the future and I don't like it."[21] Every instinct and his totality of life experience led Bud to reject a theory that accepted the inevitability of America's decline. Thinking back to the time following World War II when the United States was scrapping its military strength while the Soviets were on the move in eastern and central Europe, Bud recalled General Marshall's words at Pinehurst, "Don't ever sell the American people short. They have vast reserves of hidden strength, ready to use when the crisis is clear." Bud would not accept Kissinger's view of American decay. "It was then, I think, that the Kissinger-Zumwalt mutual admiration society began to become unglued."[22]

In retrospect, their friction was based on the simple fact that Kissinger felt the issues involved in strategic-arms reduction were too important to be made public, while Bud felt they were too critical to be decided in secret. Bud wanted to sound a loud trumpet blast to warn the American people of the danger so they would be convinced to support larger military budgets. Kissinger felt that the people would never agree to support an effort to achieve superiority, so it had to be done in secret. Moreover, by going public, the Russians would know the United States was in trouble. Kissinger believed he would lose negotiating leverage. Bud thought that Kissinger's view sold out the American people by not trusting them and not believing in them. The topic was so important that it needed an open national debate; Bud was willing to trust the judgment of the people to make the right decision. "Kissinger has operated in what I consider to be an immoral fashion in the conduct of his foreign policy, that is, he has not leveled with Congress," said Bud. "He has deceived important members of the executive branch. I believe that no one, not even the President, is aware of the extent of his commitment to the Russians."[23]

David Halperin, who worked for both men, keeping Kissinger's back-channel files while always remaining loyal to Bud Zumwalt, observed that "part of the reason that Kissinger and Zumwalt don't get along is that they're both—in some respects they're quite alike, they're powerful figures that cannot be had and Kissinger was in a position where he was going to have Zumwalt and you know Zumwalt is never had."[24] The animosity and distrust between the two men transcended their years in government service and defined much of Zumwalt's post-CNO public service career. Bud thought that "Kissinger basically had a Spenglerian pessimism of a man who escaped from a fascist system, went right to the Ivy league, and didn't understand America."[25]

Bud Zumwalt, Tom Moorer, and Mel Laird fought back against the back-channel system with a "three tiered spying operation" targeted at Kissinger and aimed at ascertaining what was being withheld.[26] Secretary of Defense Laird was especially taken aback upon learning that he had not even been told about the start of the first secret meeting in Paris on August 4, 1969, between Kissinger and North Vietnamese representative Xuan Thuy. The president did not want either his secretary of defense or the Joint Chiefs to know about the meeting.

Unbeknownst to Kissinger, Laird decided to make use of the top-secret National Security Agency (NSA) and Defense Intelligence Agency (DIA) to monitor White House back-channel activity. The NSA is the Department of Defense's communications espionage agency; the DIA is its all-source military intelligence agency. Laird told each of the military men he selected to spy on the White House that "they'd better be loyal to me." He promised each four stars after four years and kept his promise.[27] From this deal, Laird got all the information he needed about Kissinger's contacts with foreign governments, his secure telephone conversations, and the back-channel messages on his secret conversations in Paris with Le Duc Tho and on the SALT negotiations.[28] "Henry was very smart, but Mel was smarter," observed former secretary of state George Shultz. "Mel enjoyed the bureaucratic in-fighting and was a match for Henry."[29]

NSA operatives intercepted all back-channel messages, and the U.S. Signal Corps tracked White House conversations. The Special Air Missions branch of the air force controlled the planes that Kissinger used for his secret trips.[30] Laird often knew Kissinger's destination before Henry did, which is probably the reason Laird reportedly had advance knowledge of Kissinger's secret trip to Beijing. Laird never let Kissinger in on the fact that he had this information.

Another tier in the counter-Kissinger surveillance program was controlled by Bud Zumwalt through a network of loyal navy aides strategically located within Kissinger's office. Bud took pride in creating his own intelligence network, which provided him with details on just about every activity of the Nixon and Ford administrations. This human intelligence operation provided the rawest types of data. Kissinger's assistant Peter Rodman thought that Bud's operation was "perhaps even more productive" than Laird's because Bud had a private window into Kissinger's front office in the White House.

When their relationship started, Bud persuaded Kissinger that he needed more navy personnel on the NSC, and this started a regular yearly rotation of Bud Zumwalt special assistants. Located in a cubicle between Kissinger's office and Haig's, these individuals "had an unequaled vantage point on everything going on—discussions of the most sensitive diplomacy, helping prepare the most secret plans, listening in on Kissinger's telephone conversations with Nixon, Rogers, Dobrynin, and others."[31]

Bud told JCS historians in 1990 that "assigning carefully chosen lieutenants to serve as Kissinger's aides" was how he got his best information. Seymour Hersh described this as the Zumwalt Employment Service.[32] The archival records document the accuracy of the information, providing the equivalent of an opponent's game plan days before kickoff. "I had my own spies" working for Kissinger.[33]

Kissinger liked having navy men on the NSC because of their administrative efficiency, since the job involved scheduling, making personal arrangements, taking notes at meetings, and taking care of Kissinger's administrative and organizational needs. Three people occupied the position—David Halperin, Coleman Hicks, and Richard Campbell. Each performed superbly in his responsibilities; each provided terrific staff work for Kissinger and personal loyalty to Zumwalt. "I want to add that David was a key member of my NSC staff during one of the most complex and difficult periods in which I held that position," wrote Kissinger in support of Halperin's membership application to the Council on Foreign Relations. "He was an outstanding assistant, with a keen and sensitive grasp of strategic issues and a rare ability to get things done."[34]

All of the men Bud placed into Kissinger's shop were navy, and there was very little they did not share with Bud. In return, Bud protected them and nurtured their career advancement. In a 1988 letter to Paul Nitze, David Halperin noted that "I believe you were in very large measure the man who shaped the life and career of Bud—who remains, for me, the greatest man I have ever known."[35] Coleman Hicks replaced Dave Halperin. Hicks had been a roommate of Bill Bradley at Princeton and a graduate of Yale Law School. He joined the Judge Advocate General's Corps in the navy in 1969 as part of a four-year naval officers program. By November 1971, he was an instructor at the Naval Justice School in Newport, Rhode Island, but he left after a few weeks to become personal assistant to Kissinger, based on the recommendation of Halperin, who had consulted Zumwalt prior to making the recommendation to Henry.

When it was time for Hicks to move on, Bud recommended Richard Campbell, describing him as "a very outstanding young naval officer" in the same mold as Hicks and Halperin. Kissinger replied that Haig had already put Campbell on the radar screen, and "he was the number one choice."[36] In a telephone conversation the same day, Kissinger told Bud, "You have never steered me wrong."

Richard Campbell had gone to Vietnam in 1968 "to develop strategy for the Brown Water Navy." Afterward, he went to the Center for Naval Analyses in 1970, and in the spring of 1972, he went to work as Kissinger's executive assistant, in which post he remained until 1974. "That has got to be the most intense period around," said Campbell of the years in which Kissinger negotiated arms treaties with the Soviets, made overtures to China, orchestrated the end of the Vietnam War, and won the Nobel Peace Prize.[37]

What Kissinger did not know was that Campbell was a regular "line out" to Zumwalt on just about everything that went on inside the White House. Bud was meticulous about dictating whatever Campbell told him. These tapes were then transcribed, and they provide a hitherto unavailable view into the Nixon-Kissinger-Haig nexus. Zumwalt used "DC" to tag Campbell's comments and debriefings, rarely referring to him by name except in one key crisis memo during the Hicks Committee hearings of November 1972. In this document, "Special Contacts," dated November 29, 1972, and stamped Sensitive—Eyes Only, Bud mentioned "Charlie DiBona to Dick Campbell (good for feed out—no good for feed in)," meaning that Campbell's position could not be compromised.

A sampling of DC's input offers tantalizing views into Kissinger's thinking, his office rivalries and personal jealousies, and the effects of Watergate. In early June 1972, DC reported that "Kissinger was distressed concerning Laird's SALT testimony—Kissinger didn't look at it hard enough—he was unhappy with the Laird add-ons to the budget in the name of SALT. There is a problem concerning the SALT agreement—the interpretation of it—Kissinger won't take a firm position on anything until all the uncertainties are sorted out."[38] He reported, "Trade is the carrot that is being used—with regard to SALT Phase II the Soviets want FBS [forward-based (missile) systems close to the Soviet Union] included." When Zumwalt asked if the PRC (People's Republic of China) was in the SALT loop, DC replied, "Soviet matters are never discussed with the PRC."

DC informed Bud that the president was concerned about the legislative position that Zumwalt had outlined for the Joint Chiefs of Staff because he was worried about the McGovern forces. "He doesn't want to be dollar limited in Defense and wants to put more on research and development. . . . Haig sees the President a lot while Kissinger is away—during the Moscow trip, Haig was in regularly cutting Kissinger down. He made the point that more analysis of Kissinger's position is needed. . . . It is clear that Haig is going to stay on."

DC also told Bud, "The President is against General Abrams because Secretary Laird is so strong for him . . . Laird was fighting for command and control changes . . . so that Kissinger could not continue his end-runs and use the JCS as he was doing."

DC later briefed Bud on how Kissinger was trying to maneuver the Soviets and PRC to make it difficult for Hanoi not to listen in the Paris negotiation. "The President plans to carry on the signal sending for about three more months and then to really pull the plug and blast Hanoi . . . go balls out—trying to hold down civilian casualties." DC also reported that Dave Young was working on security things for Ehrlichman. "P [president] may be ready for K [Kissinger] to leave; maybe John Connally as replacement."

In the summer of 1972, DC reported that "Kissinger is away, first in Paris and then off to Saigon where he is focused on negotiations and getting Thieu on board with the deal. Kissinger is getting lots of pressure from the President and is fluctuating wildly—he has been very dovish lately. . . . The President is getting miffed with Kissinger—his egocentrism is getting on the President's nerves. The President is playing it very rough."

DC provided an example of Kissinger's behavior. He was having breakfast with William Buckley when the president called. Kissinger told the president that he was busy and asked if it was really necessary to interrupt his breakfast. "The President was furious. . . . Kissinger lies to the President at least on minor things," such as "Kissinger is telling the President that he has told Laird to do certain things but not necessarily telling Laird."

In the same debrief, DC discussed the Haig-Kissinger and Zumwalt-Haig relationship. Bud viewed Haig as a man of professional competence but never really trusted him. "Both [Haig and Kissinger] were masters of deceit and disingenuousness. Each was capable of running the other down. Each has done so to me."[39] They had first met during the Cuban Missile Crisis, when Haig was a lieutenant colonel working as a special assistant to Joe Califano, who was special assistant to Cyrus Vance, secretary of the army. Haig's great knack was the ability to ingratiate himself with his superiors and to move rapidly up the ladder of assignments. He rose from the rank of colonel to four-star general during a single tour in the office of the president. This was a phenomenal rise without any command experience during senior grade.[40]

DC wanted Bud to know that Haig was being promoted to vice chief of staff for the army. Haig "sensed Zumwalt was opposed" and considered this "very dangerous," but Kissinger had told him not to worry because Zumwalt "would never become Chair of the Joint Chiefs of

Staff." DC also informed Bud that "Haig's replacement is likely to be Gen. [Brent] Scowcroft." He then provided Kissinger's precise schedule for the next few weeks—Camp David, Key Biscayne, the West Coast via Chicago, then after Labor Day, to Moscow for SALT II, back to D.C. on September 14, 1972, and then in October to China.

By late December 1972, DC reported that the president wanted Haig to go with Kissinger to the Paris negotiations. "The President, whose mind is being poisoned by Haig on the subject, feels that Kissinger is screwing the negotiations up—Haig told the President that Kissinger would get yo-yoed by the North Vietnamese—Haig called Kissinger to relay this as the President's judgment—Kissinger was furious." Kissinger was acting like a defeated man. Haig forced the B-52 issue over Kissinger's preference for tactical air, a "less bloody image." Haig told DC, "Henry is trying to promote me out of the White House." In Key Biscayne, Kissinger told the president that they needed to get Haig back into the army "because nobody was watching Abrams from within the Army." DC concluded that "Haig is playing the sinking ship routine very skillfully—he plays both sides working Henry and telephoning the President."

DC also reported that "Kissinger discussed Zumwalt with DC—admitting that Zumwalt had a difficult problem in trying to turn the Navy's racism around, but he isn't sure that he can survive the Congressional concern about integration of the Navy." When DC tried responding, Kissinger said, "You're defending the Navy because you're a Zumwalt man." Kissinger tried making vicious remarks about Bud to see if he could get a rise out of DC. "Kissinger said that what really bothered him about Zumwalt was that he stood up to Kissinger all the time."

DC concluded this debrief with some Watergate intel: "David Young is in trouble with regard to Watergate phone tapping, etc. "Kissinger wants to be sure that all his papers are saved for his memoirs and has had Rodman trucking them out to the Rockefeller estate in New York, all of the most sensitive papers." DC urged Bud to try to heal his relationship with Haig—he felt that as a result of the strong rift and jealousy between Kissinger and Haig and with the president's hatred of

Kissinger, that Haig will get even closer to the President . . . almost all contact between the White House and Laird has ceased." DC pointed out that Kissinger's job is on the line and if he fails to get a deal, will likely quit.

In another detailed report, DC let Zumwalt know that the president asked Kissinger for the names of "philosophers for dinner." DC put Paul Nitze on the list but "Kissinger struck him saying President hates him." DC also reported that Press Secretary Ron Ziegler called Kissinger to say that "two unimpeachable sources" say that Kissinger called Zumwalt "and told him the President wanted him to quit." Ike Kidd was going to take over the Navy.

In March 1973, DC told Zumwalt about an NSC meeting that Paul Nitze attended when his name was not on the official staff participation list. This "irritated President" and "Kissinger said we've got to get rid of Paul Nitze." DC said, "Kissinger is paranoid over Haig (Haig has had a secure phone installed in his office as Vice Chief of Staff for Army by special arrangement and talks to the President regularly and this drives Kissinger wild)." Moreover, "Haig sent his aide to work on DC to inform on Kissinger's schedule" because he wanted to know whom Kissinger was seeing. "All of the staff working for Kissinger are paranoid—they debrief DC and cut him in on key items." DC has been encouraging Kissinger to reopen dialogue with Zumwalt as a way of checking Haig's influence, but "Kissinger doesn't want to see too much of Zumwalt because he believes that Zumwalt won't work in Kissinger's best interests."

Later in March 1973, "DC says that Kissinger feels more secure with President now that Haig has left."[41] Bud was also receiving regular updates on the Watergate investigation. A June 1973 debriefing is almost totally related to the Oval Office taping system. "Haig knew about the taps, was in bed with Haldeman and Ehrlichman and was aware of the Plumber operation. Kissinger was not aware of the Plumbers taps, but did request that his own staff be tapped. The Plumbers felt that the SALT leak source was the Pentagon." In news that must have delighted Bud, DC reported that "George Anderson's stock is going down, he tried so hard to get Zumwalt fired that he over-exposed himself." DC ended by informing Bud that Kissinger will be secretary of state and that the president was considering releasing a study of other presidents' wiretapping.[42] The Watergate theme continued in the July 25, 1973, debriefing—"Dean is singing" and Kissinger was "treating DC like a King, he's so worried." "Haig and Kissinger have formed an alliance for survival." "The President is optimistic about Watergate—Buzhardt is afraid to tell him how bad it is." DC told Zumwalt, "Haig is working on you again because he wants to be Chairman [of the Joint Chiefs of Staff]."

Bud believed that the system created by Nixon and Kissinger could not have produced any result other than "leaks and spying and all-around paranoia. Indeed, they had created a system in which these were everyday and essential elements."[43]

The final tier in the counter-Kissinger strategy was supplied by the liaison office between the chairman of the Joint Chiefs of Staff and the National Security Council. When Admiral Moorer replaced General Earle Wheeler as chairman, there was already in place an established liaison officer to act as a go-between for the NSC staff and the JCS chairman, to ensure that there was a flow of information pertaining to policy, as well as to the activities of both. In essence, this person was on the staff of the chairman of the Joint Chiefs, responsible to Admiral Moorer, and also on the staff of the National Security Council, responsible to Kissinger.

This was a demanding job—to keep information flowing from the NSC to the JCS so that the chairman was briefed and prepared for meetings. The liaison officer was given the title assistant for national security affairs to the chairman of the Joint Chiefs of Staff. Daily duties entailed keeping Admiral Moorer and the Joint Chiefs advised of matters under consideration within the National Security Council system, making recommendations on staff positions and on responses relating to those queries, and providing a means for the Joint Chiefs of Staff to make their military advice directly available to the National Security Council staff, as required by statute.

With regard to National Security Council duties, the liaison officer was a senior staff member of the NSC with an office in the old Executive Office Building. The liaison officer had a special White House pass permitting access throughout the Executive Office Building. The liaison office was located with other NSC staff offices, administratively integrated with them and subject to the same staff administrative procedures, except for communications to and from Kissinger and Haig dealing with military matters. These were transmitted by hand delivery outside usual staff channels.

The first liaison officer in the Nixon administration was Rear Admiral Rembrandt Robinson, a man of unique talent and professional competence, able to "suck up information like a sponge," who took on vast amounts of work and responsibility. As a member of the NSC staff, Rem was on regular, authorized distribution lists of copies of daily cablegrams received by the White House, intelligence reports, study directives, study responses, decision memorandums, incoming and outgoing correspondence with other branches of the executive, intrastaff memorandums and papers, and a wide variety of other material. Much of this information was highly classified and extremely sensitive.

Kissinger and Bud treasured Rem Robinson. In a November 6, 1970, meeting, Kissinger told Bud that "Rem Robinson had his full confidence and was authorized to tell Admiral Moorer and me everything K was doing."[44] When Rem departed for his next rotational assignment, Bud wrote a personal note that his detachment from the Washington area brought mixed emotions. "Joy on account of the honor of breaking his flag at sea, but also remorse, because your splendid performance in a particularly delicate assignment will be sorely missed. Words are inadequate to describe how very much you have assisted me during my first year in office."[45] Upon his untimely death in a helicopter crash in Vietnam, Bud wrote to Rem's widow, Joan, "I came to feel personally so close to Rem that I feel part of me is gone."[46]

It was apparent to Rem that, notwithstanding Kissinger's professed desire to keep the JCS chairman fully informed, there were many things he was not sharing with the chairman and the secretary of defense. Zumwalt started keeping his own record of these omissions, prepared by executive assistant Burt Shepherd. "My impression from my discussions with Henry Kissinger and with other members of his staff is that he and the President wanted Admiral Moorer to feel that he was getting all the information but they in fact did not intend him to get all information. And that they had their own little plot within plots that was designed to exclude him."[47]

But there was another counterplot within the counterplot, involving navy yeoman Charles Radford.

Radford had joined the navy in Reno, Nevada, on July 8, 1963. He found the navy to be a rewarding and meaningful experience and went to great lengths to please his superior officers and serve his country. On September 18, 1970, he reported for duty on the staff of the chairman of the Joint Chiefs of Staff, working for Rear Admiral Robinson.

Robinson told Radford that he was in a position of great trust. He would be seeing things that, in some cases, Robinson would not be seeing and therefore needed to keep an eye out for materials he thought Robinson might be interested in. It would take a while for Radford to learn what types of things Robinson might need to see, but Rem made it clear that he wanted copies of such materials. "He stated that if I was able to make a copy, to do so and give it to him. He made it clear that my loyalty was to him, that he expected my loyalty, and that I wasn't to speak outside of the office about what I did in the office. He further stated that he worked directly for the Chairman and that it was his responsibility to keep the Chairman informed and that I was to help him do this. He also stated that I would find some people who would deliberately leak things to him to prove their own points, and that I should be careful of any information that was deliberately given to me, to assess the person's character and actions at the time I received the information and communicate these things to him, in order that he could properly assess the value of the information he received."[48] Radford's role was to be "a shadow of him, or an extension of himself, which I don't mind being, as I admired the man."

Radford soon proved himself up to the task, impressing Robinson enough that he arranged for Radford to accompany Haig on a fact-finding mission to Vietnam. Radford was an exceptional stenographer, and he served as Haig's stenographer on the trip. The assignment meant that Robinson would be privy to all cable traffic and messages between Haig and Kissinger. "The Admiral stated that this was an opportunity to do a job for the Joint Chiefs, and that it would be of quite a bit of value; that he would like me to keep my eyes open for any and all information that might be useful to the Joint Chiefs." Intelligence mastermind Vice Admiral Earl Rectanus knew exactly what was going on, telling Paul Stillwell, "Someday you might want to keep in the back of your mind— I'm not going to go into it, but Rembrandt Robinson and his days in the White House with Henry Kissinger, the opening of China and the relationship of Naval Intelligence, with that whole thing."[49]

Robinson was specific with regard to the types of information he needed—troop strength and projections of reductions in Vietnam, the details of any agreement the White House might be discussing with President Thieu of South Vietnam, and any information Radford might see bearing on General Lon Nol in Phnom Penh. Radford returned from his trip with Haig on December 17, 1970, giving Robinson copies of materials he had collected. Robinson seemed quite pleased, as were members of the JCS staff. "One time I remember specifically that Captain Train walked through the outer office from Admiral Robinson's office and said to me, 'Radford, you do good work.' He didn't say any more than that and kept walking through the office, smoking his pipe. Since I had just returned from the second trip with Brigadier General Haig and had given Admiral Robinson much information, I knew what the captain was talking about."

In the first week of June 1971, Rear Admiral Robinson was transferred to the command of Cruiser-Destroyer Flotilla 11, homeported in San Diego. He was replaced by Rear Admiral Robert O. Welander. Bud considered Welander a good staff officer but not in the same league as Robinson. Welander soon approached Radford about going on a trip with Kissinger in July 1971. "Admiral Welander told me that he would be interested in anything I could get my hands on." By this he meant diplomatic dealings with China and anything else. "He cautioned me to be careful and don't get caught. He said, 'Don't take any chances.'" Eager to please his new boss and having gained the trust of Haig and Kissinger, Radford expanded his role from transmitting information to stealing it, being proactive about getting it.

The Haig trips had provided useful information, but the assignment to accompany Kissinger on his July 1971 worldwide trip, with the secret detour to China, was where Radford hit the mother lode. Radford was in the briefing book three category, no need to know, but he found out about the trip by rummaging through Kissinger's personal papers, burn bags, and briefcase in Pakistan. Sneaking into Kissinger's room, the yeoman rifled through Kissinger's personal belongings and discovered that Kissinger was heading to Beijing.[50] He had so many documents by the time he got to New Delhi that he asked a friend to mail them back to him at his White House office. Radford coded the envelopes so he would know if they had been tampered with.

Radford wanted to make sure that Moorer had the information Kissinger was withholding. Most of these documents had been obtained by going through Kissinger's briefcase, "reading or duplicating whatever papers he could get his hands on, and sometimes retaining discarded carbon copies of sensitive documents that were intended to be disposed of in the 'burn bag.' "[51] He brought back copies of everything he could. Moorer would soon be seeing materials he clearly was not intended to get pertaining to Kissinger's first meeting with Zhou Enlai. "The Chiefs thus had an extra measure of knowledge about Kissinger, much as Henry did about Secretary of Defense Melvin Laird thanks to the wiretap the FBI had placed on Laird's military aide."[52] Welander complimented Radford on a "great piece of work" and warned him to "never tell anybody that I had done it."

Radford returned from the trip with Kissinger by way of San Clemente. He telephoned Welander in Washington, and in the course of that conversation, "He [Welander] asked me to get a copy of the agenda items of the meeting that was going to be held in San Clemente that he and Admiral Moorer were going to fly out for at the end of the week." Radford made an exact copy of the agenda book being used and personally gave it to Welander in El Toro, where his plane had landed. Welander later told him, "I had no idea how helpful it was for the Chairman of the Joint Chiefs to walk into a meeting and to know what is going to be said."[53]

Two months later, in mid-September 1971, Welander told Radford that he was needed once again to accompany General Haig to Saigon. "I told him that I would go and he told me that I would be doing the same sort of things as before. He further told me that a lot of things were going on, and that he wanted information relating to the peace settlement in Saigon, or relating to South Vietnam, or relating to any messages that were sent by Henry Kissinger or Brigadier General Haig. He stated that information relating to these items would be of great interest to the Chairman, Admiral Moorer."[54] Radford was able to provide field reports by rifling through Haig's briefcase as he was returning from South Vietnam. By his estimation, Radford took over five thousand documents in his fifteen months on Kissinger's staff.[55] He knew it was wrong, but "I was loyal to the 'cause'—the Navy. . . . The government stank. The JCS weren't getting all the information that they wanted and were forced to steal their information."[56]

It remained a highly efficient, productive, and clandestine method for obtaining information until December 1971, when a series of Jack Anderson's syndicated *Washington Merry-Go-Round* columns appeared. The period of March to December 1971 is usually referred to as the Tilt, a time during which the Nixon administration abandoned its public pronouncement of neutrality to favor Pakistan in its war with India.[57] Created in 1947 by a partition of India, Pakistan, a nation predominantly of Muslims with a minority of Hindus, was divided into East Pakistan and West Pakistan. The Punjabi elite controlled the central government in West Pakistan, and the Bengali minority lived in East Pakistan. India, a democracy under the leadership of Indira Gandhi, was predominantly Hindu.

In 1971, Pakistan's third president, General Agha Mohammad Yahya Khan, decided to squash an independence movement in East Pakistan by dispatching his army from West Pakistan to target Hindus living in East Pakistan. Somewhere between 500,000 and 1 million Bengalis were killed in a genocidal purge, triggering the largest mass exodus in history. Hundreds of thousands of Hindus escaped into India. When Nixon remained silent on the genocide, Prime Minister Indira Gandhi called for the independence of East Pakistan. In the White House, Nixon called Gandhi "an old witch" and "that bitch, that whore."[58] He feared that Gandhi's actions would undermine his China initiative.

Despite the administration's public line of remaining neutral in the conflict, privately Nixon was leaning toward Pakistan because of its role in brokering an opening with the People's Republic of China. Then, on December 3, 1971, Pakistan attacked India. At a Washington Special Action Group (WSAG) meeting that same day, Kissinger moaned, "I am getting hell every half hour from the President that we are not being tough on India. He has just called me again. He does not believe we are carrying out his wishes. He wants to tilt in favor of Pakistan." These classified meeting notes were released verbatim in Anderson's columns. He later released the complete texts of the minutes from other WSAG meetings of December 4, 6, and 8.

The conflict between India and Pakistan became another proxy battle in the Cold War. China backed Pakistan; the Soviets backed India. Ever the Realpolitik pragmatist, Nixon cashed in with Pakistan, needing a wedge for his opening with China. Nixon could not afford to do anything that could be construed as an affront to Yahya Khan, because Pakistan served as Nixon's gateway to China. "We don't really have any choice," Kissinger told Nixon. "We can't allow a friend of ours and China's to get screwed in a conflict with a friend of Russia's."[59]

Laird, Moorer, and Zumwalt had reservations about the tilt. At a December 6, 1971, top-level senior staff meeting, Bud told Laird, "I am disturbed. The United States will take a lot of lumps siding with the Pakistanis."[60] Bud believed that East Pakistan would inevitably gain its independence and the Soviets would end up on the winning side. Bud soon had another reason for concern. On December 10, without consulting either Laird or Moorer, the president ordered a naval task force from the Seventh Fleet to be assembled with the aim of preventing the disintegration of Pakistan. Nixon feared that Gandhi would use the crisis as a pretext to destroy West Pakistan.

Bud was ordered to assemble a task force from the Seventh Fleet, then located off the coast of Vietnam. Neither Moorer nor Bud was given a mission for the ships to carry out. The administration's official public explanation stated that the flotilla was being sent to evacuate fifty Americans from Dhaka in East Pakistan. The Indian government and just about everyone in the State Department scoffed at the transparency, noting that the task force included "the world's largest nuclear aircraft carrier (USS *Enterprise*), the amphibious assault ship *Tripoli* with a marine battalion, assault helicopters, and a nuclear attack submarine."[61]

Zumwalt and Moorer feared that a naval task force being sent into a war zone without orders could lead to unintended consequences. Moreover, the Indian Ocean Soviet navy deployments outnumbered the task force, and Bud worried about a confrontation at sea. "In the short term, the military balance in the Indian ocean area will go against us," the CNO warned.[62] The counsel fell on deaf ears, because for Nixon this was a power play to support Pakistan. On December 13, the flotilla (Task Force 74) was ordered into the Indian Ocean in daylight so that it could be easily detected by India and the Soviets.[63] Inside the Oval Office, Kissinger said to Nixon, "You're putting your chips into the pot again." Nixon felt that the Chinese, Indians, and Soviets needed to know that "the man in the White House was tough."[64]

The White House was in a rage over the leaks to Anderson, especially the details of policy discussions from the Washington Special Actions Group. Within twenty-four hours of the first Anderson column quoting from the classified meetings, Kissinger persuaded Nixon "to launch a full-scale investigation of Anderson's penetration."[65] Ehrlichman knew the perfect person for the assignment—David Young and his Plumbers unit, the secret White House group created to stop leaks, named after Young's father, who had been a plumber. Young had previously worked as Kissinger's personal assistant but was now on Ehrlichman's payroll. The Plumbers had been relatively inactive since breaking into Daniel Ellsberg's psychiatrist's office over Labor Day weekend in 1971, looking for evidence that might discredit the Pentagon Papers leaker.

Meanwhile, Admiral Welander was stunned when reading the first of Anderson's columns. He saw a reference to "guided missile destroyers *Parsons*, *Decatur*, and Tartar SAM." Weeks earlier, Kissinger had asked Welander to prepare a list of ships that were to accompany the USS *Enterprise* into the Indian Ocean. Welander recalled asking Radford to type the document for his signature. When reviewing the list, Welander caught the misuse of a word in identifying a ship in the destroyer category as a Tartar SAM (Tartar being the surface-to-air missile that the *Parsons* and *Decatur* carried). Welander personally corrected the error and made only five copies of the document. Welander realized that the leak must have come from his files, meaning that the leaker was someone who had access to them. This had to be Radford, but Welander was in a bind, because he knew Radford had been stealing for the Joint Chiefs, but apparently he was also in bed with Jack Anderson. Welander had no option other than dutifully informing Haig and Laird that he suspected Radford as the source of the leak to Anderson, never mentioning that he was also a spy for Moorer. Welander had the uneasy sense that his own neck was on the line.

On December 16, navy yeoman Charles Radford was ordered to the Pentagon for questioning by Young and W. Donald Stewart, the Pentagon's chief investigator whose office was in the Defense Investigative Service. The investigators were relentless in their focus on Radford's relationship with Anderson, demanding a detailed accounting of his activities over the past weeks. Stewart cursed at Radford, threatening that he and Anderson would both end up in federal prison for violating the Espionage Act. "He used some of the worst profanity I ever heard, most of it directed toward Mr. Jack Anderson and 'his kind' and toward 'traitors.' At one point I thought him on the verge of hysteria," recalled Radford, who was so shaken that he began sobbing.[66]

Radford denied giving any classified materials to Anderson, but four polygraphs indicated otherwise. According to Haldeman, the "polygraph makes clear that he did it."[67] Stewart was certain that he had found "the son of a bitch that's giving everything out to Anderson."[68] Radford disclosed that he and Anderson were friends; in fact, they had dined together on December 12, the night before Anderson's first India-Pakistan story appeared in the papers. Radford said it was coincidence. In an earlier assignment, Radford had been stationed in India, where he'd met Anderson's parents in New Delhi's Mormon church. When Anderson's parents had a problem with their visas, Radford came to their assistance. As a way of thanking him, when Radford returned to the States, the Andersons invited him and his wife to dinner at Jack's home. They became fast friends. They were members of the same Mormon church, and their wives enjoyed each other's company. Radford had Indian students staying with him at his home. He was pro-India and, of course, the White House position was anti-India. Radford felt that Kissinger and Nixon were "irrational" when it came to India; he had an "animus" that the secret tilt was "very hypocritical, very two-faced."[69]

To investigators Young and Stewart, it was a slam dunk. Still, Radford refused to confess anything related to Anderson, swearing time and again that he had not passed classified materials to Anderson. Following his initial interrogation, Pentagon security agents arrived at Radford's office to change all the locks on his file cabinets. Radford's access to the building was revoked as well. Radford was certain he was going to prison. "I felt like I'd have a knock at the door and be whisked away in the middle of the night and I wouldn't see my family again." In desperation, Radford's wife, Tonne, called the Andersons, who soon arrived at the Radford home. Jack warned that it was not safe to speak because the government probably had the home under surveillance. Anderson recommended taking a drive in his car, but not before the two men, with flashlights in hand, thoroughly searched Anderson's automobile for bugs. "Under the seats, under the dashboard, behind the steering wheel, in the trunk, underneath the car and around the gas tank," recalled Radford.[70]

Driving around Washington, Anderson reassured Radford that only two people knew how Anderson got the classified materials. "I won't tell them anything, so the only way they will know is if you tell them," said Anderson.[71] There were no other witnesses. Nothing could be proven. Radford agreed that the only way to save his career was to deny everything, to say he could not recall, to stand shoulder to shoulder with his friend Jack Anderson. "You are going to hurt a lot of other people if you don't come clean and tell us why you gave these papers to Anderson," warned Agent Stewart. Radford never changed his story about Anderson, who went to his grave insisting that his source was not Radford.

If Radford had not given classified materials to Anderson, then what did he do with them? Here Radford offered up a bombshell, admitting to stealing documents from Kissinger and Haig and giving them to Welander, who slipped them to Moorer.[72] "When we broke Radford that night, that's where I got the *Seven Days in May* idea," recalled Agent Stewart. "I said Jesus Christ, here's the military actually spying on the President of the United States . . . this is a hanging offense."[73] In the end, Radford was "a reverse agent," not really working for Anderson, although he appears to have been the source, but for the Joint Chiefs. Nixon and Kissinger had their back channel to cut out Laird, "and here they find that Moorer's been double-dealing them."[74]

At 6:07 p.m. on December 21, 1971, ten minutes after arriving at the White House by helicopter, President Nixon met in the Oval Office with Attorney General John Mitchell and presidential assistants H. R. Haldeman and John Ehrlichman.[75] Haldeman wrote in his diary that the meeting had been called because the investigation of recent Jack Anderson columns "had uncovered the fact that a yeoman in the NSC shop, assigned to liaison with the Joint Chiefs, was the almost certain source of not only the leaks, but also the absconding of information from Henry's and Haig's and other people's briefcases, which were turned over to the Joint Chiefs of Staff. The President was quite shocked, naturally, by the whole situation and agreed that very strong action had to be taken, but very carefully, since we don't want to blow up the whole relationship with the Joint Chiefs of Staff."

During this personal White House briefing, Ehrlichman told the president that Young and Stewart's investigation yielded only one place the leak could have originated, "right here in the Joint Chiefs of Staff liaison office" inside the White House.[76] "Jesus Christ!" exclaimed Nixon. Ehrlichman explained that Radford had "access to everything from State, the Pentagon, National Security Council. Everything else. And he just Xeroxed this material for Anderson. There's no question."

Nixon wanted to know, "How in the name of God do we have a yeoman having access to documents of that type?" Ehrlichman explained that he'd been traveling with Kissinger to India and Vietnam with Haig. Nixon grasped that Radford had "been right at the precipice." Ehrlichman pointed out that Radford typed all the documents for "contingency plans, political agreements, troop movements, behind-the-scenes politics, security conferences going on between our government and foreign governments. . . . This sailor is a veritable storehouse of information."[77] Nixon wondered aloud, "If you can't trust a yeoman in the navy, I don't know goddamn who you can trust." Attorney General Mitchell reminded Nixon that "the yeoman served in India. He married his wife in India." That was enough for Nixon. "Oh, he's pro-Indian? Well, then, he did it." The president was aghast that the information came from Kissinger's briefcase: "Oh my God. Can him. Can him. Can him. Get him the hell out of here."[78]

When Nixon first heard the news, he blurted that the Joint Chiefs had committed "a federal offense of the highest order."[79] John Mitchell thought "Zumwalt was involved in this." Nixon repeated, "Zumwalt." Ehrlichman said, "I'll hazard that this was basically a Navy operation with Moorer, Zumwalt and a Yeoman involved." Nixon agreed.[80] The president felt that Bud Zumwalt was involved because of his strong objections to sending the Seventh Fleet task force into the India-Pakistan front, endangering his men. The JCS are "a bunch of shits," said Attorney General Mitchell, and Zumwalt is "the biggest shit of all."[81]

Nixon thought Al Haig must have also been involved. The Pentagon's general in the White House "must have known," said Nixon. Haldeman and Nixon discussed wiretapping Haig. "It seems unlikely he wouldn't, he wouldn't have known," repeated Nixon. Mitchell agreed that a wiretap made sense.

Welander was summoned to the White House by John Ehrlichman, where he confessed that he had been aware of Radford's actions "as a thief in the employ of the nation's military commanders."[82] He admitted to ordering Radford to look specifically for materials bearing on withdrawal of troops from Vietnam and notes from Kissinger's secret meetings with Zhou Enlai and Le Duc Tho. Welander also acknowledged taking these directly to Admiral Moorer. When Ehrlichman presented him with a prepared statement on White House stationery for his signature, Welander balked at signing a "statement [that] would have had me admit to the wildest possible, totally false charges of 'political spying' on the White House."

Nixon was soon comparing Radford with Alger Hiss and Daniel Ellsberg. "He's another Ellsberg. That's the thing that concerns me."

Haldeman agreed, "except that he probably knows a hell of a lot more than Ellsberg." Nixon agreed. "Yeah, he really knows more . . . because he's been in on the hard-core things. . . . That Radford, the culprit who turned this crap over to Anderson . . . goddammit, leaking it, that son of a bitch should be shot. He has to be shot!"[83]

Ehrlichman suggested sending in a high-ranking Mormon from the Pentagon to break Radford, but Nixon rejected the idea, saying, "Those Mormons are really turning out to be a bunch of scabs." Nixon obsessed about how Anderson got Radford to turn over classified documents. He needed to find out what was in it for Radford. There had to be something besides his love of the navy. Nixon did not think it was bribery, so it had to be homosexuality. On December 21, Nixon asked, "Is [the] yeoman deviate [sic]?" Ehrlichman wrote the query down in his calendar for the next day's Plumbers' assignment. Two days later, in an Oval Office meeting, Nixon said that "after sleeping on it" he had instructed investigators to see if the relationship between Radford and Anderson was "sexual up the ass."[84] Nixon saw an analogy with Alger Hiss and Whittaker Chambers. "Hiss and Chambers, you know, nobody knows that, but that's a fact how that began. They were both that way. That relationship sometimes poisons a lot of these things."

Kissinger was furious upon learning about Radford and wanted the yeoman and Moorer in jail. Nixon explained to Kissinger that none of them could afford to have the truth come out that the Joint Chiefs were spying on their president. "You see, Henry, if you were to throw Moorer out now, the shit's gonna hit the fan. And that's gonna hurt us. Nobody else. We get blamed for it." The presidential election was less than a year away. "The main thing is to keep it under as close control as we can. . . . We've really just got to keep the lid on it." Kissinger, the man who had authorized taps on his own staff, could not believe Nixon's response. "They can spy on him and spy on me and betray us and he won't fire them."

Instead of prosecution, Mitchell urged that the JCS liaison office be closed immediately and that those involved in the theft be transferred. This would amount to a de facto admission of guilt on the part of the Pentagon. "The important thing, in my way of thinking, is to stop this

Joint Chiefs of Staff operation, and the fuck-up of security over here. And if Moorer has to order Welander off to Kokomo or wherever it is— what to do with Robinson I don't know—then they have taken recognition of this. And they, in effect, are admitting to this operation."[85]

Mitchell warned the president "as to what this would lead to if you pursued it by way of prosecution or even a public confrontation. You would have the Joint Chiefs allied on that side directly against you. What has been done has been done and I think the important thing is to paper this thing over. First of all, get that liaison office the hell out of the NSC and put it back in the Pentagon."[86]

In the end, all the parties entered into a high-stakes confidentiality agreement.[87] Haldeman described it in his diary as "a monumental hush-up all the way around." In his memoirs, Nixon acknowledged that Radford could not be prosecuted because he was a potential time bomb and state secrets could be released—"it was too dangerous to prosecute the yeoman."[88] Radford, Anderson, Welander, and Moorer "could each expose the military spy operation if Nixon tried to go after them," later concluded journalist and professor Mark Feldstein in *Poisoning the Press*. "The President in turn could prosecute each of the men if they dared go public about the scandal. The end result was an uneasy balance of mutually assured destruction."[89]

The president telephoned John Mitchell at 5:33 p.m. on December 24 to convey this very message. "I think the main thing is to keep it under as close control as we can. But I—We cannot move to do anything to discredit the uniform. That's what I'm convinced of. . . . Our best interests are served by not, you know, raising holy hell."[90] In an election year, Nixon could not afford to have a story about a military establishment gone berserk. Ehrlichman thought Moorer was now "preshrunk," meaning that when Nixon said, "Jump!" the chairman would ask, "How high?"

With the decision made to cover up the episode and retain Moorer as chairman of the JCS, Nixon had to decide what to do about Radford and Welander. If Radford was not going to be prosecuted, they would have him transferred.

On January 4, 1972, Acting Secretary of the Navy John Warner called Bud, saying he had orders from the White House to get the first-class yeoman assigned to Rear Admiral Bob Welander out of town. Warner directed Bud, in the name of the president, "to get the orders written in such a way that no one other than me would 'know what was going on.'" When Bud questioned this move, Warner told him not to challenge a direct order from his commander in chief.[91]

The CNO dutifully contacted the senior naval officer in the Bureau of Naval Personnel, Rear Admiral Jim Watkins, "and ordered him not to ask any questions but write a set of orders assigning Radford to the Thirteenth Naval District, whose headquarters were in Seattle, and to get it done that day."[92] Zumwalt and Moorer believed that Radford deserved, out of fairness and due process, to be charged and that the system should deal with him. The White House rejected this course of action. Bud went to Warner and recommended that because Radford was a navy man, formal charges be brought. Warner refused to do this and ordered Bud to cease his inquiry. Shortly thereafter, Warner called to change Radford's orders from Seattle to Portland, Oregon, because the White House did not want Radford going to John Ehrlichman's home state.

The next day, January 5, 1972, at ten p.m., Radford's household goods were picked up, and he was out of town, heading to Salem, Oregon, for assignment at a naval reserve training center. Wiretaps were installed in Radford's home in Salem and stayed in place until June 20, 1972. The taps, authorized by Mitchell, yielded only one congratulatory call to Anderson after the columnist won the Pulitzer Prize for the December 1971 stories on the secret White House tilt.

Meanwhile, two secret investigations of the spy ring were initiated. The first was conducted from the White House by John Ehrlichman. Once Laird learned about the White House investigation, he assigned his general counsel of the Department of Defense, Fred Buzhardt, responsibility to supervise his own Pentagon investigation of the unauthorized disclosures as well as of all communication channels between the Joint Chiefs and the National Security Council staff.

Young's report concluded that the military had "trained" Radford

"how to steal" and that "he stole" for the Pentagon and then for Anderson. Buzhardt's report was delivered to Laird on January 10, 1972. It documented "a morass of mistrust and spying—the inevitable offspring of
Nixon's own secretive and manipulative style that infected every office
from the White House on down." Buzhardt's investigation concluded
that Welander and Moorer knew how Radford was getting the documents and did not ask him to stop.[93]

After reading Buzhardt's report, Laird called Moorer into his office
to say that "it was the greatest personal disappointment that I'd had."[94]
Moorer said he had done it for the good of the Pentagon "and for your
good too, Mr. Secretary." Laird understood but also told the chairman, "Well, I just want it to stop."[95] Copies of Buzhardt's report have
disappeared.

In Bud's view, "Henry's duplicity left booby traps everywhere."[96] In
this one instance, Nixon seems to have agreed with his CNO. "The real
culprit is Henry," Nixon told Haldeman. "We'd all like to find someone
else to blame—the goddamn state department or the defense department. But Henry could never see anything wrong with his own staff."[97]

On July 1, 1972, Admiral Thomas Moorer began his second term as
chairman of the Joint Chiefs of Staff, the principal military advisors to
the president, the National Security Council, and the secretary of defense. Overshadowed by the Watergate scandal and the investigation
of the Plumbers, the story about internal espionage remained buried
until January 1974. After the story broke in the papers, Admiral Moorer
assembled the JCS for a briefing on January 11, 1974, to offer his side.
He began by saying that Ehrlichman and the other Plumbers, who had
recently been indicted for the Los Angeles break-in at Ellsberg's psychiatrist's office, were trying to justify their actions by reference to national
security leaks. In his tape transcripts, Bud stated that this was the first
time he learned that Welander had improperly provided information
from Kissinger's NSC to the Joint Chiefs.[98]

Moorer told his colleagues that, in his opinion, Radford had changed
his story, saying he had been ordered to steal for the JCS, because he had
lawyered up and was trying to deflect the investigation from him and

Anderson to Moorer. Moorer swore that he did not know that the documents supplied by Radford to Welander had been purloined. In another wheel within wheels, "Ehrlichman was trying to discredit Kissinger at the time of the investigation as part of the power struggle within the White House, and Laird was trying hard to insulate Kissinger from the Pentagon. And this is what led to two separate investigations, neither of which Admiral Moorer had ever seen." When Moorer asked Laird why Radford had not been court-martialed, he answered that he was told not to, on orders of the president.[99] "Ten years from now," Moorer told the chiefs, "no one would remember that Kissinger hated Laird and that Laird hated Ehrlichman and Haldeman, but the important thing was not to let the JCS get sullied by this."[100]

In February 1974, the Senate Armed Services Committee, chaired by Senator John Stennis, held hearings that resulted in the report *Transmittal of Documents from the National Security Council to the Chairman of the Joint Chiefs of Staff.*[101] Appearing before the committee were most of the principals—Radford, Welander, Kissinger, and Buzhardt. After four days of testimony, Stennis later told Seymour Hersh, he realized that if he dug any deeper, he would "destroy the Pentagon."

When the story broke in 1974, there was a new secretary of defense in place, James Schlesinger, who wanted to speak with Bud about the spy ring. Bud contacted Welander, who had just returned from a private meeting with Senator Stennis. The conversation between Zumwalt and Welander focused on whether or not Welander knew that Radford had stolen from burn bags as well as Kissinger's briefcase. "Was I correct [in telling Schlesinger] that the only one you really knew had not been [a] legitimate acquisition was the burn bag?" asked Zumwalt. "Yes sir, that's correct," replied Welander. "Is it correct that Senator Stennis has the appreciation that Chairman [Moorer] was not aware of any illegitimate acquisition?" asked Zumwalt. "Yes, that's correct," said Welander. "I just need to be sure of that because that's the way I understand it and relayed it when asked. So you don't feel that anything that Senator Stennis has been given changes the parliamentary situation that concerns you and the Chairman?" asked Zumwalt. "Not at all, sir," replied Welander.[102]

A few days later, Bud asked his staff to prepare an Eyes Only—
Sensitive talking-points memo titled, "Principal Points of CNO Recollection Re: Radford/Welander Matter."[103] The memorandum provides a
fascinating chronology of events from the perspective of the CNO: On
January 4, 1972, Acting Secretary of the Navy John Warner called Bud
with orders to have Radford reassigned. "CNO was told he could not be
told the reason or other particulars." Warner told Bud to do so in such
a manner that "contacts in Navy should be kept to an absolute minimum." Only after the story broke in the newspapers did Warner tell
Bud that his orders concerning the reassignment of Radford "had come
from Mr. Laird and Mr. Buzhardt" and not from the president. After
receiving this order, Bud went directly to Admiral Moorer "in an effort
to find out the reason in order to insure Radford's rights and interests
were protected." Moorer said that "there was evidence that Radford had
passed classified information to columnist Jack Anderson." Bud thought
that proceedings should be initiated against Radford, leading to a possible court-martial. Moorer agreed, but told Zumwalt that "Mr. Laird
had said that the President had said not to court martial Radford. He
should just be transferred." Warner wanted Radford sent to Oregon, to
any area except Portland, and he selected the Naval Reserve Training
Center in Salem.

The White House liaison office had been closed, and Welander was
now available for reassignment. Moorer advised Bud that "Welander
had been caught in a three-way squeeze. That Mr. Laird had considered the possible leak by Radford to Anderson as cause to move to close
the liaison office. Mr. Laird viewed the office as cutting him out of the
communications pattern between Kissinger and the JCS." In his own
discussions with Welander, Bud learned that Welander had "blown
the whistle on Radford for giving information to Jack Anderson." Even
though Zumwalt, Kissinger, and Moorer were aware that Welander
had encouraged the pilfering of documents, "CNO was advised by both
CJCS [Moorer] and Mr. Kissinger that RADM [Rear Admiral] Welander
had done a fine job and should be given a good assignment." They
all had the goods on one another. Welander was assigned to Cruiser-
Destroyer Flotilla Six, and in 1973 he was reassigned to duty as deputy

chief of naval operations for plans and policy, in which post he continued to support Zumwalt directly.

On January 15, 1974, Secretary Warner told Zumwalt that with respect to the Welander/Radford probe, he was "not able to remember anything concerning his involvement in this matter."[104]

By February 1974, Welander learned that he was going to be called before the Senate Armed Services Committee to testify under oath about the leaked papers. He called Bud to warn him that, if asked, he would have to say that Captain Burt Shepherd, who had been Bud's executive assistant, once asked, after Radford returned from Vietnam, what type of information he had brought back. Bud wanted to know how Shepherd would have known that Radford had brought something back. Welander said that Radford must have told him; it was the only way he could have known. So for Zumwalt the issue was trying to know if he had seen information that Kissinger wanted him to have or information he was not supposed to have, supplied by Radford. There were so many papers going back and forth, Bud had no idea which was which from the "Machiavellian nature of Kissinger's operation."[105] Bud wanted to see Shepherd's detailed handwritten notebooks from all of Bud's meetings. Don Pringle, Shepherd's relief as executive assistant, called Shepherd to say that the CNO "wanted to see all of his notebooks that related to the spy ring investigation." Shepherd dutifully sent them all in, which was the final time he or anyone else has ever seen the notebooks.[106]

The circumstantial evidence puts Bud's claim that he did not know anything about the Moorer/JCS bugging of Kissinger and the White House on shaky ground. "He wanted to know everything that was going to help the Navy," recalled Shepherd. "He was always looking for people to 'spy,' but he was seeking information—they were told to keep him informed." Bud's NSC liaison officers had direct orders to brief him on Kissinger's activities. Rex Rectanus received his third star and became Bud's director of naval intelligence (DNI) in 1973. In that role, Rex most certainly knew what was going on with the bugging, especially Radford's spying on Kissinger, Nixon, and the NSC. He also must have known about Laird's using NSA and Army Signal Corps intelligence to do the same. Bud survived four years as CNO because he always had his antennae up. He was a savvy Washington player and it is difficult to accept the idea that he was ignorant of all this spying going on while he was CNO, especially with Rex as his DNI. How else can we interpret Rex's subtle comment at the close of a lengthy oral-history interview when asked if there was anything else he wanted to say: "Someday you might want to keep in the back of your mind—I'm not going to go into it, but Rembrandt Robinson and his days in the White House with Henry Kissinger, the opening of China and the relationship of Naval Intelligence, with that whole thing."[107]

In a moment of rare self-reflection, Richard Nixon realized it was all on him. "Damn, you know, I created this whole situation—this, this lesion," Nixon said in an Oval Office conversation. "It's just unbelievable. Unbelievable. There have been more back channel games played in this administration than any in history 'cause we couldn't trust the goddamn bureaucracy."[108]

Bud Zumwalt wondered how it was possible for rational men to act the way they did, concluding, "Maybe they weren't rational."[109]

CHAPTER 13

RUFFLES AND FLOURISHES

Schlesinger talked about the fact that the worst period of his whole stewardship as Secretary of Defense was the period from the 27th of June to the 30th of June, when I was in my final hours. He reported that the instructions from Haig and Kissinger had been to fire me with only two days left in my watch.[1]

Bud Zumwalt's last eighteen months as a sailor were not what he had hoped for. Instead of steaming serenely toward the horizon, he was "awash in controversy and the target of recriminations that were not only acrimonious but personal."[2] Prior to Richard Nixon's 1972 election, Secretary of Defense Melvin Laird had announced that he would not continue for another term. Nixon nominated Elliot L. Richardson to succeed him, but Richardson, sworn into office on January 30, 1973, served less than four months before being nominated to be attorney general. It was understood that Richardson would guide the administration's handling of the Watergate investigation, which had reached a critical stage.

To replace Richardson, President Nixon chose Dr. James Rodney Schlesinger, then serving as director of the CIA.[3] Bud had known Schlesinger only briefly from his time at the Bureau of the Budget (BOB), and he asked Captain William A. Cockell, Jr., to provide him with a detailed analysis of Schlesinger's writings "with emphasis on his interests and perceptions of strategic issues."[4] The initial backgrounder might have led Bud to think he was reading about himself: "Tough, hardheaded and conservative, a mission oriented executive, capable of intense concentration, holds things close to his vest, is always pushing for the right answers, strives to excel in whatever he does."[5] At the bottom of Cockell's report, Bud scribbled "very good" and asked that Cockell speak with Rand Corporation president and Bud's longtime friend Henry Rowen "for his evaluation of Schlesinger." Cockell soon provided an Eyes Only personal and sensitive assessment from Rowen: "I am confident that the best way to deal with him on issues is on an incisive, factual, analytical, no b.s. way. Clearly he is capable of understanding arguments that are subtle and complex, but arm waving, ideology, vagueness, etc, should be avoided." Rowen wanted Bud to understand that Schlesinger lacked experience and background in "dealing with allies and a certain lack of sensitivity to their problems and concerns. I suspect a lot needs to be done to educate him in that area."[6]

Ever since 1972, when Nixon and Kissinger had rushed the SALT I accords—the Anti-Ballistic Missile Treaty and the Interim Agreement on Strategic Offensive Arms—through both houses of Congress as an executive agreement, Bud knew that the Kremlin had outmaneuvered Kissinger and Nixon, who wanted a deal before the election. The source of Bud's angst could be traced to an error made by Kissinger at the key stage in the negotiations. As discussed in Sherry Sontag's and Christopher Drew's *Blind Man's Bluff: The Untold Story of American Submarine Espionage*, Kissinger had offhandedly agreed "not to ask for limits on the Soviets' 'massive efforts' to build the Deltas, a new class of submarines that would far surpass the Yankees and carry ballistic missiles with ranges of 4,000 miles. Zumwalt was furious."[7] Bud believed that this opened the door to giving the Soviets a great advantage in submarine-based missile attacks.

To undo his negotiating error, Kissinger went to Dobrynin and executed, without informing anyone in the chain of command, an agreed-upon "clarification." Executed in secret and without any arms-control agency providing scrutiny, this clarification created a new loophole by defining a modern submarine-launched ballistic missile (SLBM) as one "deployed on a nuclear-powered submarine commissioned since 1965." This faulty wording made it possible for the Soviets to build any number of diesel submarines and install new nuclear missiles on them. Bud and Paul Nitze were the first to pick up this error, but in 1973 the rest of the government discovered the "secret covenant" when the Soviets told the SALT negotiating team about it.[8]

Dismissing State Department officials as "bed wetters" for advising him not to challenge Kissinger, Bud did all that was possible to force Kissinger to pay for the mistake by seeking approval of a more powerful new class of missile subs—the Tridents. "It was a battle he would win."[9] In the summer of 1973, Congress approved the funding of the Trident submarine, seen as the key component of the U.S. nuclear deterrence program.[10] Zumwalt could not have accomplished this without the strong support of his new secretary of defense. "Though he had nothing of Laird's flair for coaxing his projects through the bureaucracy and the Congress," Bud wrote in his memoirs, "he had the equally important quality, which Mel Laird lacked, of formulating issues precisely and clearly and thus forcing the bureaucracy and Congress into making conscious decisions."[11]

A significant portion of Bud's final eighteen months as CNO was taken up with the deep division within the chain of command on arms control. Bud had no doubt that Kissinger was the source of leaks portraying him as Dr. Strangelove, the unhinged general who orders a first-strike nuclear attack on the Soviet Union. In the film of that name, the president and his advisors must then try to recall the bombers to prevent a nuclear apocalypse. The U.S. SALT II negotiating team was approaching the next round of talks from the concept of "equality in capabilities," meaning that, according to team member Paul Nitze, "our principal goal was to secure Soviet acceptance of the concept of 'essential equivalence' as set forth in the Jackson amendment."[12] The language of Senator Henry Jackson's amendment, which passed Congress by a vote of 56 to 35 on September 11, 1972, expressed the consensus that Congress "urges and requests the President to seek a future treaty that, inter alia, would not limit the United States to levels of intercontinental strategic forces inferior to the limits provided for the Soviet Union."[13] Equality in capabilities did not mean that everything had to be exact, but rather that the strategic nuclear capabilities of the two sides should be essentially equal to each other. Going into the negotiations, the Soviets had not accepted this concept of "essential equivalence." Instead, their position was premised on the concept of "equal security taking into account geographic and other considerations."[14]

The SALT talks had resumed in Geneva in March 1973, just as domestic political considerations related to the Watergate investigation began to preoccupy Washington, a situation the Soviets were carefully monitoring. Recognizing that the outcome of these negotiations was likely to have a decisive effect on strategic military forces and the national security of the United States and its allies, Secretary Schlesinger created a Department of Defense SALT Task Force so that Defense could play a vigorous and constructive role in the negotiations. Schlesinger was keenly aware that Mel Laird had been cut out of the SALT I negotiations, and he appointed Dr. N. Fred Wikner as director of the task force, with "the responsibility of the coordination of Defense SALT policy recommendation."[15] In another important selection that delighted Bud, Paul Nitze was appointed special assistant to the secretary of defense for SALT.

For Henry Kissinger, the success of U.S. foreign policy depended on détente with the Soviet Union. The new relationship with China, the peace agreement with the Vietnamese, and an improved workable relationship with allies had all been accomplished because of détente and Soviet assistance. All of this threatened to unwind on October 6, 1973, when Egyptian and Syrian forces attacked Israel on the Yom Kippur holiday. The attack came as a complete surprise to Israel. The Soviets, who had advance knowledge of the attack, chose not to share this information with the United States. "Now it was evident that their commitment to détente was nil," wrote Nitze.[16]

Kissinger chose to define the situation differently, seeing instead an opportunity to broker a lasting Middle East peace agreement, so long as the Soviets exercised restraint. Kissinger was looking at the bigger stakes—a long-term peace in the Middle East and the effects of adverse American intervention on détente. He therefore favored a "partial" Israeli victory that would be accomplished without direct intervention and provide the seeds for a lasting settlement. "The best thing that could happen to us," Kissinger wrote to Secretary Schlesinger on October 7, "is for the Israelis to come out ahead but get bloodied in the process."[17] Kissinger told Israel's foreign minister Abba Eban that American foreign policy depended on creating a military position that would constitute "an incentive for a cease-fire."[18]

With Nixon distracted by the Watergate crisis and the need to look for a new vice president, day-to-day management of the crisis fell to Kissinger, who understood that Israel was an important ally, but who also weighed such issues as a potential oil boycott, the importance of not antagonizing the moderate Arab states, and most of all, détente with the Soviet Union. He seemed not to care that, while the U.S. and Israeli intelligence agencies had been caught off guard, the Soviets knew of the invasion beforehand. "In reconstructing American policy during October 1973, it becomes evident that Kissinger persuaded Nixon that, once having broken out, the Arab-Israeli war should be manipulated so as to create the conditions he considered necessary for successful long-range peace diplomacy by the United States," wrote Tad Szulc. "The rewards of such a policy, Kissinger reasoned, would be the disappearance of the threat of an Arab oil embargo, the establishment of a new relationship between the United States and the Arab world, and ultimately, the elimination of Soviet influence in the Middle East. But for this policy to work, it was necessary that neither the Arabs nor Israel win the war militarily. What Kissinger needed was a stalemate among the three bloodied and exhausted combatants so that, at the proper time, they would turn to the United States in search of peace. To Kissinger, therefore, the war itself was but a cruel sideshow, serving the larger interests of American policy as he perceived them."[19]

To accomplish this objective, Kissinger made the decision to delay providing Israel with equipment and supplies. This policy was based on two premises: that Israel would quickly defeat its foes and that the United States should maintain a low profile. He made it clear that Military Airlift Command (MAC) should not deliver supplies. As Israeli casualties mounted in the Sinai counteroffensive, Defense Minister Moshe Dayan advised Prime Minister Golda Meir that the numerical superiority and technological sophistication of Arab arms and supplies meant that there was no alternative other than "a supreme effort to secure planes and tanks as quickly as possible from America, and perhaps try to get tanks from Europe too."

Kissinger was awakened by a call from Israeli ambassador Simcha Dinitz with an urgent plea for help. Israel had suffered enormous losses in the counteroffensive, including five hundred tanks. Recognizing that they could not win a war of attrition, the Israelis requested a massive airlift of supplies and equipment. So desperate was their situation, Prime Minister Golda Meir offered to fly incognito to see Richard Nixon, which Kissinger rejected as looking like "hysteria or blackmail." For two days, Kissinger let Israel bleed, stalling on the resupply.

By now, the Soviets had reassessed the situation. Thinking the Arabs had a chance to win, on October 10, the Soviet Union began a massive airlift to Egyptian and Syrian forces, a clear violation of their understanding with the United States that both superpowers would stay out. The Soviets also began pushing for a settlement "in place," one that would allow the Egyptians and Syrians to hold on to their current gains on the battlefield. On the same day of the Soviet airlift, JCS chairman Tom Moorer informed the chiefs in a Pentagon meeting that the secretary of defense's "guidance in general about the way to respond to Israeli requests for supplies is that we are to be overtly niggardly and covertly forthcoming." Bud immediately saw this as Schlesinger's way of working around Kissinger.

In his memoirs, Kissinger insists that the Pentagon, particularly Schlesinger, was dragging its heels on the question of resupply. He presents himself as Israel's great protector. Schlesinger always denied this and claimed that his shoes were nailed to the floor by national policy. Bud was a strong proponent of resupplying Israel rapidly and was disturbed by Kissinger's tactic of blaming Defense. Indicting Schlesinger was preposterous, "just a tale, and an extraordinarily disingenuous one at that," wrote Bud, who knew that Kissinger, in the name of the president, had ordered Schlesinger to stall. "This is one of Henry Kissinger's great lies. Incidentally, in 1978, when Golda Meir came to this country, she asked to see me. She said, 'I want to hear it from you. Is it true that my friend, Henry, withheld the supplies?' I had to say, 'Yes.' She was terribly crestfallen."[20]

Scoop Jackson was receiving daily communications on Kissinger's fraud from both Schlesinger and Bud. On October 11, Bud took a step that by his admission he would never have taken if he'd been sure that "Richard Nixon rather than the unelected, unaccountable Henry Kissinger was making national policy about the war." The CNO went directly to Jackson, pleading with one of Israel's strongest Senate advocates to save Israel from defeat. Bud told Jackson that without resupply, Israel would lose the war. "Jackson met with President Nixon that day, exhorting him to act."[21] Nixon needed little convincing. When he asked Kissinger what had taken so long, his secretary blamed the Pentagon. Preoccupied and already drinking heavily, Nixon ordered Kissinger and Schlesinger to make certain that Israel got everything it needed. "Goddamn it, use every one [of the supply planes] we have. Tell them to send everything that can fly," instructed Nixon. On October 18, Jackson cosponsored a resolution that sailed through the Senate with overwhelming support, calling for "decisive action to assure that essential military equipment be transferred to Israel on a time scale and in whatever quantities are required to enable Israel to repel Syrian and Egyptian aggression."[22]

When the airlift began to show results, the Soviets threatened to send troops into the war unless the United States stopped Israel from destroying the elite Egyptian Third Army Division. Nixon accepted the compromise, because the Soviet navy outnumbered the U.S. Sixth Fleet by three to two and could have brought overwhelming air power to bear. In an October 24, 1973, letter to Nixon, General Secretary Leonid Brezhnev made clear "that unless the Israelis were forced to end their encirclement of the 3rd Army, the Soviets would go in and free them." British prime minister Edward Heath's office called the White House just before eight p.m. to ask to speak with Nixon. "Can we tell them no?" Kissinger asked his assistant, Brent Scowcroft, who had told him of the urgent request. "When I talked to the president, he was loaded."[23] Nixon was asleep and too drunk to be awakened, so Kissinger took the lead as part of a WSAG group, which placed two million U.S. troops on the high state of nuclear alert called DEFCON 3, leading the Soviets to back down in return for U.S. support in enforcing the cease-fire on Israel.[24]

During the winter and spring of 1974, the JCS had written a number of memorandums dealing with SALT, multiple independently targetable reentry vehicles (MIRVs), and the Threshold Test Ban (TTB), registering grave concerns. In almost all cases, the chiefs had requested that Schlesinger present their views to the president, but the bureaucratic process was not bringing the Pentagon's professional opinions to him. Lieutenant General Edward Rowny was the JCS representative to the SALT delegation. During a dinner party at Rowny's home in the fall of 1973, General Alexander Haig invited Dr. Wikner, the director of the Defense Department SALT Task Force, to call on him whenever he thought a matter required so. On May 26, Wikner, who had a home in Key Biscayne, somehow managed to get through security and visit with Haig at the general's villa at the Key Biscayne Hotel. It was a remarkably candid meeting, one that resulted in Wikner being fired.

The director informed Haig that Defense and the JCS were not having their views aired with the president. Kissinger was keeping the president in the dark. Haig told Wikner that "Henry has mentioned to me that there have been differences between himself and Jim (Schlesinger) but he has talked with Jim about this and has reported back that the two of them settled their differences." Wikner told Haig, "This is a very inaccurate report. All due respect to the Secretary of State, but he is acting as a very selective filter in the reporting of information to you and the President on Defense views." Wikner handed Haig a report that had been written by Paul Nitze, saying that it represented the views of the chiefs and that, while the secretary agreed, "Jim thinks the letter is very good but feels that he cannot forward it without attaching his resignation."

After spending close to ten minutes reading the letter, Haig commented, "There is no reason why Jim shouldn't have sent this letter." There had been dozens like this. "As you can see," said Wikner, "there are really fundamental differences between the Department of Defense and Kissinger; both on technical matters and on negotiations . . . it is clear we have substantially different opinions on how negotiations in SALT should be conducted." Haig must have reported this secret meeting to either Kissinger or Nixon, because within weeks Wikner was out of a job. "In the wake of this episode, I became increasingly convinced that my talents and expertise could be put to better use elsewhere than on the delegation," wrote Nitze.[25] Nitze went to Schlesinger to say he wanted to resign, but "Jim said his particular concern was Haig, whom he regarded as unpredictable and fully capable of trying to organize a military coup d'état to keep Nixon in office."[26]

Schlesinger wanted Nitze to remain in service of the government and offered him a position he had previously held, assistant secretary of defense for international security affairs. This time, the intent was to head off Kissinger, who Schlesinger feared would offer too many concessions in the upcoming SALT negotiations. Bud urged his mentor to accept. "Paul, you know it's essential that somebody take over ISA and really run it and that somebody is you." A reluctant Nitze agreed, but Senator Barry Goldwater, who eleven years earlier had voted against Nitze's nomination as Secretary of the Navy, informed the White House that if Nitze's appointment went forward, he could not be counted on as a sure vote if impeachment proceedings reached that point.

It was just as well. No longer able to function under the dark cloud of Watergate, Nitze felt he had no option other than to resign in protest. "I believed that in an effort to rescue his administration and to rally the public behind him before impeachment proceedings began, he would make imprudent concessions to the Soviet Union on arms control to strike a deal."[27] On May 28, Nitze notified Schlesinger and the president that he was resigning, effective May 31. He enclosed a statement that made his position perfectly clear. "For the last five years I have devoted all of my energies to supporting the objective of negotiating SALT agreements which would be balanced and which would enhance the security of the United States, and also of the Soviet Union, by maintaining crisis stability and providing a basis for lessening the strategic arms competition between them. Under the circumstances existing at the present time, however, I see little prospect of negotiating measures which will enhance movement toward those objectives. Arms control policy is integral to the national security and foreign policy of this nation and they, in turn, are closely intertwined with domestic affairs.

In my view, it would be illusory to attempt to ignore or wish away the depressing reality of the traumatic events now unfolding in our nation's capital and of the implications of those events in the international arena. Until the office of the presidency has been restored to its principal function of upholding the Constitution and taking care of the fair execution of the laws, and thus be able to function effectively at home and abroad, I see no real prospect for reversing certain unfortunate trends in the evolving situation."

The Zumwalt intelligence service was in high gear, providing a steady flow of information on Nixon's emotional fragility and the jockeying for influence between Kissinger and Haig. One report from Captain Cockell noted that "mention of Nitze provoked strong reaction from Haig: 'That God damned guy—he and his overbearing ways did the country great damage under McNamara. It's just as well that he's going.'"[28] Years later David Halperin took the time to write Nitze. "It has been some years since we met—I worked for several years while in the Navy for Bud Zumwalt and he was kind enough to bring me along to your house one fourth of July. That must have been in 1968 or 1969, shortly before I went to work for Henry Kissinger as his Personal Assistant. In that year I recall that we occasionally met, on one instance (which was as confusing to me then as it is to me now), when I had the responsibility of editing certain records of the SALT negotiations which Henry was prepared to have you read, in the West Wing, without others present. Why the record provided to you was incomplete I did not know then (or now) understand."[29] Halperin went on to say, "I don't believe there exists a man with the country's interest more at heart and with the intellect and the years of involvement with the Soviets to be able to make judgments of such awesome importance. . . . By your example, you affected Bud powerfully; by his example, he has similarly affected me and countless others. In this very attenuated way, I feel you have touched my life. . . . It is of great comfort to me to know that you are there, watching over the nation's future in these years when critical decisions will be made affecting the nation's destiny."

———

In order to have an orderly transition, Bud advised Secretary John Warner that it was important to choose Bud's successor at least six months before July 1, 1974. Warner assured him that they would work together to assemble a list for Schlesinger, but Warner kept stalling. Warner had his reasons, which Bud only accidentally discovered when Warner asked that they meet in Schlesinger's office. Once Bud arrived, Warner remarked that the timing was perfect because they were meeting about Bud's successor. Bud sensed a trap, thinking that Warner was trying to "make it difficult for me to put my thoughts together." Warner began by saying that there was just one viable candidate, Ike Kidd.[30]

Kidd was a very strong threat to all the personnel programs Bud had set in motion. He was also someone not committed to the high-low mix in Project Sixty or strong enough to take on Rickover. Moreover, Bud knew that Kidd had been part of the Anderson cabal during the Hicks hearings period, when his name was usually the first cited as replacing a sacked Zumwalt. "His friends had made him the man whose appointment as CNO would be perceived by the fleet as heralding a return to the status quo ante. I therefore regretfully concluded that Ike could not be on my list of CNO candidates."[31] Bud also believed that Kidd lacked the character to keep the navy "from being suffocated in the political miasma that was enveloping ever more closely the Nixon-Kissinger-Haig White House."[32]

John Warner and his top aide, Deputy Secretary of Defense Bill Clements, favored Kidd for the very same reasons Bud was opposed to him. They believed the navy personnel system needed to be returned to its pre-1970s style. "Both of them were anxious to slow down the rate of integration. Both of them objected to the provision of equal opportunity to women, to the extent that the Z-gram had provided it. Both of them felt that Ike Kidd would work with them to accomplish their turn-back objectives."[33] Bud had refused to do so in November 1972, and he was certainly not going to do so in June 1974. Conversely, John Warner had failed to accomplish his objective in 1972, but saw a new opportunity in 1974. "John Warner had been unhelpful to me in just about every way a

Secretary of the Navy could be unhelpful to the Chief of Naval Operations," said Bud.

Bud's ambivalence regarding Kidd was a personally troubling one. Kidd's father was killed aboard the battleship *Arizona* at Pearl Harbor, and in memory of his father's name, Ike became an object of veneration. But as time went on, Bud saw him as a man who developed very early an overriding ambition with no ethical limits, although he had high regard for his professional competence. Their first close association came when Bud was working as Nitze's executive assistant and Kidd was executive assistant to CNO Dave McDonald. "In that role, Ike's relationship to me, although he was senior, was one of absolute unctuousness. He appears to maintain that attitude toward all those who are in positions of authority, or in positions to help him. It almost defies the imagination to watch him come into a room, and speak to his seniors as 'you giants' and men of gigantic intellect, you gentlemen who are so much smarter than I am, etc. Yet, the impact of this flattery seems to have helped more than it has hurt him. . . . Over time, I came to the conclusion that he was a very devious man. But I retained a tremendous admiration for his ability, as a professional man."[34]

By the time Bud became CNO, Kidd was a three-star admiral in command of the First Fleet, with orders to next command the Sixth Fleet in the Mediterranean. John Chafee overrode Admiral Moorer on this appointment because Kidd was not an aviator. Bud promoted Ike to his fourth star and made him chief of naval material. He had to convince Chafee, Laird, and Packard that Ike was right for the job, hoping that Ike had outgrown the earlier characterizations. Before long, however, "Kidd began the practice of almost routine end-running the system."[35]

As early as December 17, 1973, Bill Cockell had passed on a No Eyes—Sensitive message from Paul Nitze that "he did not think Ike was the right man," but "the one person really working for the job was Ike. He had much support from the retired flag officers and had sold Clements [William P. Clements, Deputy Secretary of Defense] on himself." Nitze reported that in a recent meeting, Secretary of Defense Schlesinger made clear that he was not high on Ike but did mention Worth Bagley. Nitze said "he thought the world of him." Schlesinger also asked about four other flag officers as possible replacements for Bud—Noel Gayler, James L. Holloway III, Maurice "Mickey" Weisner, and Stansfield Turner. Nitze told Schlesinger that Gayler, commander in chief of the United States Pacific Command, was fifty-nine, hence would require a special congressional act to extend his active service. Nitze thought more of Weisner than Holloway. "He is tough, capable of making decisions and fighting for them." He was "not sure Holloway has the strength of character to do so." Nitze did not know Turner very well, "but from all he had heard, he was a competent officer—perhaps more in the analytical line than in command, however."

Schlesinger had gotten to know Kidd when they were both at the Naval War College—Schlesinger an instructor and Kidd a student. Schlesinger shared Bud's assessment of Ike and confided to Bud that for him, Kidd was a nonstarter. Still, no one knew what Nixon would do.

Schlesinger held Bud in such high regard that he offered his CNO an unprecedented two-year extension, promising to secure congressional authorization. Bud rejected the offer based on the fact that during his own tenure he had enforced the daisy chain, that is, the policy that flag officers should move "up or out" in order to provide promotional opportunities and incentives for those who were more junior. Bud knew that he had no chance of serving as chairman of the JCS, because it was on a rotation basis and Moorer was navy. Bud also turned down an offer to serve as head of the Veterans Administration. The offer was made by Alexander Haig on May 14, 1974. "While I am honored that you would consider me for the position, I cannot accept," Bud wrote in a letter to Haig. "I deeply regret having to decline since I sincerely believe that the Veterans Administration must provide assistance to those who have served our nation so honorably and so well. However, I must base my decision upon your expressed views of the Veterans Administration's future and do not feel that I could perform a useful function within the limited possibilities outlined during our meeting."[36] The Zumwalt intelligence service reported that "Haig was furious when Zumwalt leaked turning down the offer." After having lunch with Zumwalt to feel him out on future plans, Haig returned to Kissinger and said, "I've just had lunch with Bud Zumwalt, the most hypocritical man in Washington."[37]

Schlesinger instructed John Warner to poll the admirals for their opinions as to who should be the next CNO. Warner told Bud that he should find out whom Rickover favored and not share the complete poll results with Schlesinger. Rickover refused to get involved, saying it was none of his business. "I never had any confidence in John Warner's word and found many occasions in which he had actually lied to me," said Zumwalt. Warner soon forwarded three names for vetting—Mickey Weisner, James Holloway, and Worth Bagley. "Each of these candidates would have made a superlative CNO," wrote Bud, who felt a special affinity for longtime friend Worth Bagley. Bud had brought Worth to Washington for Project Sixty and promoted him to deputy CNO for program planning, a three-star job. In 1973 Worth got his fourth star when he was sent to be commander in chief of U.S. Naval Forces in Europe and served superbly during the Yom Kippur war. He came from a distinguished family, the son of Vice Admiral David Bagley; his brother David served as Bud's chief of naval personnel.

James Holloway III was likewise from a family of admirals and was the youngest man in Bud's own academy class. They had known one another as classmates but were not close friends. During the Battle of Leyte Gulf, both men had been serving on destroyers—Bud aboard the *Robinson* and Holloway as a gunnery officer on the USS *Bennion*. Both were selected a year early for captain. On that occasion, Admiral James Holloway, Jr., Bud's former mentor, wrote in a personal letter to Bud that "it's a special delight to me to discover that my son was the junior man on the selection list and that you were the junior man on the alpha list." They were together again at the National War College in 1961 and 1962, which is where they became good friends. When Bud went to Nitze's office, Holloway went off for a period of indentured service with Rickover to command the nuclear carrier *Enterprise*. He was promoted to rear admiral one year after Bud and James Calvert, the second increment of their class to make admiral.

Holloway was heading off to command a carrier division at sea when Bud entered as CNO. He performed so well with the Sixth Fleet during the Jordan Crisis of 1970 that Bud decided to groom him as a potential successor. After the tour as carrier division commander, Bud nominated

Holloway to be the deputy commander in chief of the Atlantic Fleet under Admiral Charles Duncan, wherein he did so well that Duncan recommended him as his own relief. Bud wanted to reduce the criticism leveled at him for not having sufficient operational experience in high command in both oceans, so Bud sent Holloway to be commander of the Seventh Fleet in the Western Pacific. He did a great job and was one of the eight or nine in the daisy chain about whom Bud threatened to resign if Warner did not honor the promotion list. Holloway got his fourth star as vice chief of naval operations.

Like Holloway, Maurice Weisner was an aviator. Weisner commanded the USS *Coral Seas* during the early years of the Vietnam War. As a vice admiral Weisner was appointed commander of the United States Seventh Fleet and held that position until June 1971. Bud then brought him to Washington as deputy CNO for air and soon promoted him to a fourth star to serve as his vice CNO. He was then relieved by Holloway and next took command of the Pacific Fleet in Hawaii.

By March 1974, the nomination list with these three names was forwarded to the White House, two months later than Bud desired. Preoccupied with Watergate, Nixon sat on the nomination, refusing to make a decision. Each time Bud asked Haig about it, he was told that the president was still thinking. In truth, Ike Kidd was lobbying for himself directly with the White House. A team of retired admirals was trying to make their voice heard, but many of them had already been discredited by their 1972 anti-Zumwalt campaign. Support for Kidd was seen as a vote against Zumwalt.

On March 16, in a telephone conversation with Haig, Bud asked if there was any news on his relief. Haig leaked the selection by saying, "Yes, but don't let me put any time on it. He doesn't know this fellow. He did know his father and doesn't seem to have any problem with the nomination since all the principals concurred. However, he does want to review the whole spectrum with Secretary Schlesinger." Bud was pleased that the choice was James Holloway and not Ike Kidd. "Jim Schlesinger told me as late as the day before the White House made its final decision that President Nixon was still deliberating as between Jimmy Holloway and Ike Kidd, believing that he really wanted someone who would turn the clock back," said Bud. "But apparently, he finally concluded that he should support the Secretary of Defense's choice. Thus, by this slender thread, the Navy's programs would survive into a new era."[38]

When the selection was announced, Holloway was described by associates as an "enlightened traditionalist," not likely to turn back many of the Zumwalt reforms.[39] Writing in the *Chicago Tribune* under the title "The New Navy Chief: A New Big Z," Bill Anderson thought that Holloway's appointment was likely to reassure junior people who were ready to leave the navy had Ike Kidd been selected.[40]

Bud immediately sent his successor a handwritten note: "Dear Jimmy, I am honored and proud to be succeeded by one so wise and capable. I will leave with absolutely no concern for the future of the navy. . . . I pledge you my total cooperation in the turn over and in the years ahead."[41] Meeting shortly afterward with his preferred choice, Worth Bagley, Bud offered the following candid assessment: "Worth, I will tell you what my feelings are. My feelings are that Jimmy doesn't think as big as you and I do. You are going to have to help him enlarge his horizons, there is going to be a certain amount of reluctance on his part because he tends to think in terms of the narrow interpretation, more like Admiral McDonald's approach. . . . I found his counsel to be wise and mature. And the way in which he has handled Ike I think has been adept. I think he feels the system is one that he can let work in his favor than to stress in the way I have."[42]

The announcement of Holloway's nomination brought heightened awareness that Bud's navy career was almost over. Arleigh Burke wanted Bud to know that retirement was likely to bring mixed emotions. "Memories come flooding in of associates and associations precious from Midshipmen days on. You can't help missing the day to day dispatches and the day to day knowledge of naval matters. The mantle will fall on other shoulders and you will be both relieved by the sudden lack of pressure and saddened by a certain remoteness from the line of scrimmage. But a whole new world will be opening up for you to challenge. May you and Mouza have a happy civilian life and thoroughly enjoy the opportunities that have not been available in Navy life."[43]

One of the most sensitive issues during the CNO transition centered on housing for the CNO's staff. Shortly after Gerald Ford became vice president, Secretary Warner, without consulting anyone in the navy, hatched a plan to take the CNO residence at the Naval Observatory from the navy and designate it the official home of the vice president. Ford liked the idea and used his extensive contacts on the Hill to have legislation introduced. Upon learning about this, Bud called his friend John Marsh in the vice president's office to say that as CNO he planned to oppose the loss of this house. Marsh reassured Bud that Ford would understand and there would be no hard feelings.

Throughout the hearings and in the process of reconciling House and Senate versions of the bills, Ford was always civil and straightforward. The Senate version called for the house to become the permanent home of the vice president with staffing support provided by the navy. The House version called for it to be temporarily the home of the vice president with staffing support by the General Services Administration. Bud decided to visit Ford to suggest a compromise whereby the navy would provide the staffing support but the house would be statutorily described as the temporary home of the vice president. Ford readily agreed to the deal. "Had I been dealing with the President, Kissinger or Haig on a similar issue there would have been threats, nasty calls to my superiors and all kinds of paranoid expressions," wrote Bud.[44]

In his first transition meeting with Holloway, the issues of housing and the promotion process were at the top of the agenda.[45] Worth Bagley was to be the vice CNO. Bud did not feel Worth should live on the Navy Yard because he needed a home to accommodate his family and young children. The perfect home for Worth was currently occupied by none other than Vice Admiral Ray Peet. "If we bump Peet of course that is going to be immediately seen as harassment because Peet doesn't play ball with us but I think it has to be done," Bud told Holloway, who concurred in the recommendation.

Peet's years of hostility toward Bud quickly surfaced. He refused to move into his new assigned quarters, a much smaller home than the one he currently occupied. When Bud ordered Peet to move, Peet refused to acknowledge the order. Bud threatened to charge Peet with disobeying a lawful order. On June 11, Bud told Peet, "You know it's this sort of thing, Ray, that I think is really, really going to cause an awful lot of concern within the Navy for your personal MO." In a conversation the next day with Deputy Secretary of Defense William Clements, Jr., Bud explained that Holloway had made a decision that his new vice chief would move into Peet's place and Peet would move to a smaller home. "Ray, when he got these orders from a Vice Admiral, refused to carry them out and said he would have to consult with Jim [Schlesinger]." Clements replied that he was "100% in accord with you," and warned Bud that "you've got to keep your guys under control and all I can say is I am sympathetic to you and you do what the hell you have to with that guy." Bud told Clements and Schlesinger that he would initiate charges against Peet. "We just can't run a building in which everybody decides whether or not they are going to carry out orders. That's right before a coup."[46]

It all came to a head on June 20. Peet had received a direct written order from Bud, but still refused to move. "I consider that disobedience of an order," said Bud, who accused Peet of playing "Jim Schlesinger against us." Peet was defiant. "Bud you have done some pretty raw things in your time—you know it and I know it." "None that I am ashamed of," replied Bud. "You write all the written directives you want, Bud," replied Peet, who insisted he would not relocate.

In the end, Peet did move, insisting that he had behaved the way he did just to jerk Bud Zumwalt's chain. Knowing that Bud was a lame duck and aware just how far he could string out the charade, Peet had some fun at Bud's expense, giving little thought to how his behavior reflected on himself.

On June 5, 1974, the president, Mrs. Nixon, and Alexander Haig joined Bud for the graduation ceremony at Annapolis. This was the first time since the Kennedy administration that a president had attended graduation ceremonies at the academy. It was an especially poignant occasion for Bud, because in the audience was Admiral Chon for the commissioning of his son, Truc.

Bud was allotted three minutes to speak, introducing the president. In his remarks, which had not been cleared by the White House, Bud took a minor detour, noting the great pride he felt in looking out at the class of graduates representing the first wave of diversity. By 1976, the CNO predicted, 131 years after the Naval Academy was built and 200 years after the nation's founding, "this Naval Academy will be truly representative of the nation at large." Bud next went to the precipice. "Mr. President, this is the class which has had the watch as those barebones defense budgets which you have submitted to Congress, described by you as such, have been cut. . . . This class observed defense budget cuts of two billion, three billion, five billion, and three and a half billion. And as a result, you and I have had to see this Navy reduced from 976 ships to the lowest figure since 1939—508."

Nixon privately seethed, but he also had important points to make in his own remarks that were certain to antagonize Bud and his ally Scoop Jackson, who had recently sponsored legislation that would deny favored-nation trading status to the Soviet Union unless it lifted restrictions on Jewish emigration. Declaring that his policy of détente was aimed at reducing the chances of nuclear war, Nixon charged that critics were foolish to insist that the United States press the Soviets on domestic matters. More than three hundred members of the House and seventy-nine senators had endorsed linking U.S. concessions with free emigration. The president warned that there were three alternatives to détente—a runaway nuclear arms race, a return to constant confrontation, or a shattering setback to the hopes of building a new structure for world peace.[47]

The next day, Jim Schlesinger told Bud, "You are in the dog house again. . . . They are so paranoid over there it is just unbelievable." Schlesinger told Bud that "he had gotten his ass chewed out several times from Kissinger and Haig," and the president had even accused him of "disloyalty and not keeping the Pentagon in line and so forth."[48]

In a meeting that same day, Tom Moorer confided to Bud, "I think that Schlesinger is glad that you are leaving."[49] Moorer confided that he had gone up for breakfast with Bill Clements and learned that Schlesinger was supposed to have breakfast with Kissinger but Kissinger canceled it because of an article in a newspaper in which Schlesinger said he had no idea what Kissinger had decided on a deal with the Israelis. "So Henry apparently told him he should watch his ethics and he told Henry if there was anyone that could give him a job—or a lesson on the lack of ethics Henry was the man—after that he hung up and he came in there and said we take no more guidance from the State Department." Bud predicted that George Brown, the incoming JCS chairman, is going "to find himself right in the middle of the dogfight you used to ride out." Moorer replied, "Except worse," because Laird circumvents, unlike Schlesinger, who confronts, and "they are basically two different people." The two admirals then joked that even though Brown was bringing in a new team, the same old problems would show up, and then "they are going to say, What do we do now, Momma?" joked Moorer. Bud replied, "Right and we can go fishing."[50]

The limitation of underground nuclear weapon tests, also known as the Threshold Test Ban, sought to establish a nuclear "threshold" by prohibiting tests having a yield exceeding 150 kilotons (equivalent to 150,000 tons of TNT). The negotiating strategy adopted by Kissinger forced Bud to take a position that antagonized the White House. The Russians had proposed it for inclusion in an agreement that would be discussed in the June 1974 meetings in Moscow. Bud worried that Kissinger would bite on the bait for a threshold or upper limit on the size of nuclear devices that could be tested. It seemed unobjectionable and would easily pass Congress and be approved by the people. But there was one big catch—the Soviets had already deployed dozens of missiles with larger warheads and tested many more warheads than the U.S. The proposal was a ploy to ensure Soviet superiority.

When the chiefs opposed it, Kissinger railed against them for their skepticism about Soviet intentions. On June 17, 1974, with little to lose, Bud wrote a lengthy memorandum to the president, headed "Strategic Arms Limitations." Knowing that if he sent it directly to the president, the White House staff would bury it, Bud routed the memo via the secretary of defense, increasing the chance that it might get to Nixon. Before sending the letter, Bud wanted to express his views personally to the president, but Schlesinger told him it would be like shooting off your foot. "Every time you go in there and say that to the President,

Haig goes violent. Haig worries the Army budgets will be cut and the Navy's will go up. Kissinger goes violent, his worry is that it will bungle up the appreciation people have for détente, to understand that really what's happening is the Russians are gaining military superiority, and the President goes wild."[51] Schlesinger wanted Bud to accept the fact that "you're having zero impact and all you're really doing is just outraging them." Schlesinger warned Bud that "they are all paranoid over there, they will resent very strongly the fact that you are making these arguments, they don't want to admit that these circumstances are coming to pass and they can't afford the political risks."[52]

In the memorandum, Bud maintained that no subject was of greater importance to the long-term security interests of the United States than SALT. As CNO he reminded the president that he had a statutory responsibility as his naval advisor "to provide you with my military judgment on the current state of the SALT." SALT I had shifted the strategic balance to the disadvantage of the United States rather than equivalence in strategic capabilities. "The reasonable conclusion to be drawn from the Soviets' behavior thus far is that they are not now disposed to negotiate a comprehensive permanent agreement on terms compatible with the national security of the United States." Bud advocated true equivalence as the only acceptable negotiating outcome. Anything else would be politically unacceptable because "the U.S. public will not willingly accept a position of inferiority, with all the military risk and loss of international influence that entails. The Soviets should be made to understand that their failure to agree to strategic equivalence will drive the U.S. in the direction of expanded strategic programs, which will inevitably destroy the atmosphere and domestic political support essential for a policy of détente. In my judgment, failure of the United States to convey this fundamental fact to the Soviets runs the risk of producing both an unsatisfactory SALT outcome and the ultimate destruction of détente."

Turning to a familiar theme, Bud told the president that "it is absolutely essential that we be totally forthright with the American public about the true state of affairs and what is required to attain an equitable agreement. If the public is accurately informed, I am confident that it will appropriately respond. The signal we must convincingly convey to the Soviets is that the US people will unhesitatingly support whatever programs are necessary to ensure that the Soviets do not gain permanent superiority in strategic capabilities." Bud's position was that détente should be approached on a businesslike basis whereby the Soviets got what they needed from the United States in the form of trade and technology in return for being reasonable about parity in the strategic field.

Taking a parting shot at the Kissinger back channel, Bud observed that the U.S. negotiating positions had been developed without any real coordination with key elements of the government. "I think it essential that our procedure ensure that you receive in clear and undiluted fashion the judgments of both your political and your military advisors before reaching key decisions on U.S. positions. From my observation, the system as presently operated fails to assure you of such balance in the consideration of major SALT issues, hence runs the risk that positions potentially detrimental to the country's long-term security may be adopted. To rectify this situation, I would recommend strongly that you periodically confer directly with the Secretary of Defense and the Joint Chiefs of Staff and solicit their advice on these subjects of such far-reaching national importance."

A few days later, Bud wrote a memo to Schlesinger: "Negotiating Leverage for Salt II." He urged that "the appropriate leadership from both parties in both bodies of Congress be briefed by the President on the status of SALT II negotiations *prior* to the President's departure for Moscow, and that at that time, he submit a request to them for a budget supplement providing for significantly increased strategic expenditures. I strongly recommend that the President ask that these leaders obtain a joint resolution from the House and Senate prior to his departure, expressing support in principle for these increased expenditures as an alternative to be examined if the Soviets do not accept a satisfactory SALT II."[53]

Bud had by now learned from one of his well-placed sources that at the NSC meeting, "The President shook his head and indicated in certain words that he was not too happy with Zumwalt."[54] Nixon believed that his CNO was not being realistic and that he, the president, was entering negotiations from the perspective that "we must go over with something between what the Soviets want and what we want." He had to compromise and wanted the chiefs and his CNO especially to know that further expressions of doubt would be counterproductive. "The President stated that we cannot look at the problem academically; we must consider what we *can* do, not what we want to do." Nixon took aim at Bud, predicting that "the CNO doesn't want this type of agreement and will probably make this known. However, he, the President, has the responsibility."

Tom Moorer had been at the meeting and afterward sought out Bud, informing him what the president had said about his letter. "I am sure that he had been told that there was a disastrous letter in from Genghis Khan," joked Bud. "Bud, I told them you did not concur with many aspects of this approach and that you so stated in a letter to the boss on the 17th. He kind of shook his head and indicated he knew all about the letter and was not too happy about it. That was about the thrust of it."[55] Moorer told Bud that Nixon simply asked, "Is that the same guy that took a cheap shot at me at the Academy?"[56]

Bud soon received another report on the NSC meeting from Admiral Bob Welander. "The President indicated that he was upset with the CNO's letter of June 17, asking, 'What in hell does he want?'" The president "expressed the view that the CNO had taken a cheap shot at him to which he could later refer. The president further stated that he had noticed at Annapolis that the CNO was antagonistic toward him, implying that he had not supported the Navy."[57]

Bud still had a few more bullets to fire and arranged a meeting with Holloway and Bagley.[58] "The reason that I am going into this, Jimmy, is because I really think it is kind of the first major brushing you are going to get and that Congress is really going to be looking you over when you have to go in there and testify on whatever they come home with from Moscow." Bud warned Holloway that the president had not seen the last thirteen Joint Chiefs of Staff memorandums on SALT, because Kissinger made sure they never reached him. Kissinger had confirmed at the NSC meeting that there had been interim MIRV discussions between both sides. "That's how they learned that the Soviets had already rejected an interim MIRV agreement like the one Schlesinger presented at NSC."

With Watergate unraveling the presidency, Nixon and Kissinger went to Moscow for the SALT II talks. Meanwhile, Bud had been invited to speak about SALT on *Meet the Press* the day after his change of command, scheduled for June 30, 1974. For obvious reasons, Nixon and Kissinger did not want Bud to appear on the show. Kissinger warned Schlesinger that if Zumwalt went on TV, the navy budget would be cut. "Communiques were filling the airways with orders and threats, including court-martial for the admiral if he spoke out," recalled Howard Kerr, by now working in the White House for Vice President Ford, but providing regular intel for his boss. Nixon and Kissinger "saw it as a plot." They called Schlesinger and told him to tell Zumwalt, "If he goes on, we'll destroy him." Nixon was so angry that he instructed Schlesinger not to present Bud with the Distinguished Service Medal that had already been approved. Schlesinger was also ordered not to speak at the change-of-command ceremony. Schlesinger advised Bud to cancel his appearance, something Bud refused to do without a written order from the president. Schlesinger pleaded with Bud that if he went on the show, he could not speak about SALT; that he must defer and dodge any questions on the subject.

By now, Bud's mood was generally one of "gloom and impatience." He could not wait to be away from "a wrecked President and an unprincipled Secretary of State." In the midst of a constitutional crisis, Nixon and Kissinger were negotiating a new arms-reduction treaty with the Soviet Union. Moreover, Nixon's preoccupation with political survival had left Kissinger with an even freer reign. "Both Kissinger's low opinion of the good sense and the resolve of the American people, and the obvious ego-gratification he derived from engineering foreign-policy spectaculars made me distrust him absolutely as a negotiator."[59]

Throughout Zumwalt's final week in office, tributes arrived from a broad range of friends. California governor Ronald Reagan expressed "gratitude for your many contributions to the defense of your beloved country. Your sincere interest and devotion to the welfare of the individual sailor and the recognition of his importance within the military establishment will remain an inspiration to all who follow you."[60] Rear Admiral William Crowe noted that "the U.S. Navy has been extraordinarily fortunate to have a leader with your vision, energy and courage during these rather trying times. The Navy is not an easy institution to move, even when it's for its own good, but you have literally succeeded in bringing it into tune with the times and giving it an up-to-date sense of purpose. . . . I will always be proud of the fact that I was in the Navy during this dynamic and exciting period when we turned the corner to face the future squarely."[61]

One of the most poignant expressions was a letter signed by sixty-seven officers and crew members of the USS *Benjamin Franklin*, expressing gratitude for Zumwalt's commitment to sailors. "We are grateful for your constant efforts to recognize our importance and dignity as individuals. As you leave the service those of us who remain promise that we will continue to strive to be worthy of the trust and respect you have shown us. Smooth sailing, Admiral. The Navy will miss you."[62] Bud quickly replied, "My heartfelt thanks. . . . No other letter that I have received has made me so regretful to be leaving such a magnificent Navy team. None other has done more to reconfirm that our efforts over the past four years were truly worthwhile."[63]

From Vice Admiral Robert Baldwin, commander of the Naval Air Force, U.S. Pacific Fleet: "First and foremost, you've left an indelible mark. You have been many things to many people, but for an integrated whole of intellect, determination, vision, action and compassion, nobody can hold a candle to you."[64] Tom Moorer joked that by the end of the week, they would both be contributing to the unemployment rate. "I want to express to you my appreciation for your full support and professional assistance during four turbulent years. I believe it is fair to say that we have been through the worst of it—World War II not excepted! In any event, let us look forward—and not backward—and do what we can in the future to contribute to the stability in this shaky land."[65]

Bud had one final note to write before leaving office, reminding his successor, "Jimmy," to never forget that "little people" need to see the chain of command as interested in their welfare.[66] Bud attached a letter from Missile Technician Second Class (Submarine) Joseph J. Campisi, Polaris Missile Facility, Atlantic Fleet. "In the last four years I've watched you bring the navy out of the dark ages into the 21st century. My only hope is that your successor will continue the work started under your leadership." Campisi wanted his departing CNO to know that even though he had emphasized *people*, the "SSBN [nuclear-powered ballistic missile submarine] Fleet didn't get the word." Campisi was leaving the navy because of "irreparable damage." For those who came next, he advised, "Treat your people as you would like them to treat you. Don't ever treat your men like so many pieces of furniture, because even the best furniture will wear out before its time if you abuse it. . . . I only hope you are not the first and last CNO to gain the admiration of your men, the way you have."

Bud was delighted when Holloway wrote back to say, "These are as important as the letters from certain flag officers, probably more so." Things would change under Holloway's leadership of the navy, but Bud refused to be a "dead hand on the tiller." Holloway's vice chief, Admiral Harold E. Shear, put it this way: "What he [Zumwalt] did with regard to racism, to blacks, and to women and so forth was absolutely positive. All the other things he did were not good for the Navy. Something that Holloway and I had to do was quietly and firmly get the Navy back to battery. We never put out anything that said that we were getting the Navy back to battery, but we just slowly and calmly just took a round turn, just took a round turn. Squeezed the ratchet a little bit, and it became obvious in a matter of months that we were getting the Navy back where it ought to be. . . . It wasn't any big publicity that, 'By golly, we're dropping Zumwalt, we're changing things in a different way.' "[67] Shear's use of nautical phraseology as in "taking a round turn" refers to putting two loops of line over a bitt or bollard, thus restraining the ship at the other end of the line. In the figurative sense, Shear meant that he and Holloway were seeking to restrain unwarranted personal behavior with respect to uniforms, haircuts, and beards.

———

Contemplating the day's catch at Hockett Meadows, 1937. This was Bud's father's favorite fishing spot in the Sierras. The following summer Bud would depart for Rutherford Preparatory School in Long Beach. *(Photograph VA047627, Admiral Elmo R. Zumwalt, Jr., Collection, The Vietnam Center and Archive, Texas Tech University.)*

Bud as a midshipman at the Naval Academy. He graduated summa cum laude with one of the most demanding extracurricular loads of any midshipman. *(Courtesy of Zumwalt family.)*

Lieutenant Elmo Zumwalt on the deck of the USS *Robinson* in the South Pacific just before the end of World War II. He received the Bronze Star for meritorious conduct in action in the Battle of Surigao Straits, during which the *Robinson* and other destroyers launched a torpedo attack against Japanese battleships. *(Photograph VA047591, Admiral Elmo R. Zumwalt, Jr., Collection, The Vietnam Center and Archive, Texas Tech University.)*

As prize crew officer of HIJMS *Ataka*, Lieutenant Elmo Zumwalt is seen with Japanese prisoners on a dock in Shanghai, China, October 9, 1945. He worked closely with Rear Admiral Milton E. "Mary" Miles, the commander of U.S. naval forces in China disarming the Japanese. *(Photograph VA047641, Admiral Elmo R. Zumwalt, Jr., Collection, The Vietnam Center and Archive, Texas Tech University.)*

Celebrating with the *Robinson* crew, Bud Zumwalt *(seated third from the right)* as a lieutenant in Shanghai, 1945. *(Photograph VA014785, Admiral Elmo R. Zumwalt, Jr., Collection, The Vietnam Center and Archive, Texas Tech University.)*

The Russian Orthodox wedding of Bud Zumwalt and Mouza Coutelais-du-Roché in Shanghai, China, October 22, 1945. The *Robinson* was scheduled to depart the next morning. *(Courtesy of Zumwalt family.)*

Wedding photograph. *(Courtesy of James G. Zumwalt and Ann Zumwalt Coppola.)*

Bud's visit to Pinehurst, North Carolina, and conversation with General George Catlett Marshall changed the course of his life by keeping him in the navy. Second from left is the iconic Christina Wright, whom everyone called "Aunt Tina." She was close friends with Mrs. Marshall and arranged for Bud and Mouza (*far right*) to meet the Marshalls. Young Elmo is in front of General and Mrs. Marshall. *(Courtesy of James G. Zumwalt.)*

The family in 1953: newborn Ann, Elmo (born 1946), and Jim (born 1948) on his father's lap. Mouzetta would join the family in 1958. *(Photograph VA020755, Admiral Elmo R. Zumwalt, Jr., Collection, The Vietnam Center and Archive, Texas Tech University.)*

As naval aide to Paul Nitze during the Kennedy administration, Bud was able to expand his policy portfolio. His apprenticeship under Nitze provided him the equivalent of a PhD in foreign policy. *(Photograph VA047640, Admiral Elmo R. Zumwalt, Jr., Collection, The Vietnam Center and Archive, Texas Tech University.)*

"To Mouza, My beloved wife, with whom life is never dull. Love, Bud. July 10, 1961." *(Photograph VA047639, Admiral Elmo R. Zumwalt, Jr., Collection, The Vietnam Center and Archive, Texas Tech University.)*

Paul Nitze and daughter Ann shoulder board the new rear admiral in 1965. Bud then took command of Cruiser-Destroyer Flotilla Seven with the flagship USS *Canberra* in July. *(Photograph VA017484, Admiral Elmo R. Zumwalt, Jr., Collection, The Vietnam Center and Archive, Texas Tech University.)*

With Captain Arthur W. Price, commander, River Patrol Force (CTF 116). Until Bud arrived in Vietnam, the Brown Water Navy got little respect. Bud had a new aggressive strategy, and his helicopter became a frequent sight in the field. *(Photograph VA046910, Admiral Elmo R. Zumwalt, Jr., Collection, The Vietnam Center and Archive, Texas Tech University.)*

"For the men upon whom we all depend" was Bud's way of recognizing the bravery of his sailors. A commander's responsibility was a commitment to the welfare of those who served under him. Sailors never forgot what their admiral was doing for them. *(Photograph VA046917, Admiral Elmo R. Zumwalt, Jr., Collection, The Vietnam Center and Archive, Texas Tech University.)*

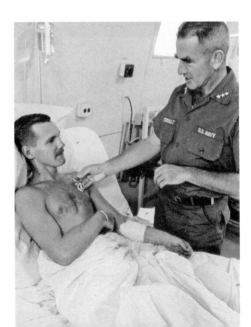

Admiral Tran Van Chon and Bud developed an extraordinary affinity and respect for each other. Each used the word "brother" to describe their lifelong relationship. *(Photograph VA015203, Admiral Elmo R. Zumwalt, Jr., Collection, The Vietnam Center and Archive, Texas Tech University.)*

Bud beams his pleasure to Chon with the product of a Vietnamese navy animal husbandry project at An Thoi naval base. Similar programs were under way at naval bases throughout South Vietnam as part of Operation Helping Hand. *(Naval Historical Foundation.)*

Bud described the relationship with Paul Nitze as "Plato and Socrates." The two men remained close personal friends and political allies throughout their lives. *(Photograph VA047216, Admiral Elmo R. Zumwalt, Jr., Collection, The Vietnam Center and Archive, Texas Tech University.)*

Bud receives his fourth star from Mouza and General Creighton Abrams at the change of command aboard the USS *Page County* (LST-1076) moored on the Saigon River, May 15, 1970. General Abrams also awarded Bud the Distinguished Service Medal. *(Photograph VA018656, Admiral Elmo R. Zumwalt, Jr., Collection, The Vietnam Center and Archive, Texas Tech University.)*

Secretary of the Navy John Chafee and Chairman of the Joint Chiefs of Staff Admiral Thomas Moorer welcome Zumwalt to Washington just prior to his assuming the position of chief of naval operations. *(Photograph VA014979, Admiral Elmo R. Zumwalt, Jr., Collection, The Vietnam Center and Archive, Texas Tech University.)*

With their hands on the same Bible that Admiral Faragut used aboard his flagship, the USS *Hartford*, during the Civil War, Secretary of Defense Melvin Laird and the navy's staff judge advocate administer the oath of office to the new CNO at the change-of-command ceremony in Annapolis. *(Photograph VA014965, Admiral Elmo R. Zumwalt, Jr., Collection, The Vietnam Center and Archive, Texas Tech University.)*

Titled "26 Years Later": Elmo's birth certificate bore his name, Elmo R. Zumwalt III; his father's, Elmo R. Zumwalt, Jr.; and the name of the doctor, Elmo R. Zumwalt, Sr., M.D. *(Photograph VA014775, Admiral Elmo R. Zumwalt, Jr., Collection, The Vietnam Center and Archive, Texas Tech University.)*

This family portrait was taken on Elmo and Kathy's wedding day in July 1970. Elmo had just returned from Vietnam and was beginning law school. *(Photograph VA047645, Admiral Elmo R. Zumwalt, Jr., Collection, The Vietnam Center and Archive, Texas Tech University.)*

It was another proud day for Bud when his son James G. Zumwalt II transferred from the U.S. Navy to the U.S. Marine Corps and was commissioned a second lieutenant. The ceremony took place on July 16, 1971, in the office of commandant of the Marine Corps, General Leonard F. Chapman, Jr. *(Photograph VA015156, Admiral Elmo R. Zumwalt, Jr., Collection, The Vietnam Center and Archive, Texas Tech University.)*

As CNO, Bud often returned to Vietnam. Here he talks with members of U.S. Navy SEAL Team One at Nam Can Naval Base. The photo also evidences Bud's relaxed regulations on facial hair and sideburns. *(Photograph VA014891, Admiral Elmo R. Zumwalt, Jr., Collection, The Vietnam Center and Archive, Texas Tech University.)*

Bud receives a briefing from his special assistant Lieutenant Commander William S. Norman before speaking to the National Newspaper Publishers Association Convention in Atlanta, Georgia. Bud credited Norman with helping him lay the foundation for a new navy family that recognized no barriers of race, color, sex, or religion. *(Photograph VA016623, Admiral Elmo R. Zumwalt, Jr., Collection, The Vietnam Center and Archive, Texas Tech University.)*

In Yokosuka, Japan, Bud speaks with the human relations council at the Fleet Activities. These meetings furthered the CNO's goal of achieving true equal opportunity by sensitizing navy personnel to the barriers that impaired effective race relations. *(Photograph VA014883, Admiral Elmo R. Zumwalt, Jr., Collection, The Vietnam Center and Archive, Texas Tech University.)*

With Secretary of the Navy John W. Warner looking on, Bud took a lot of ribbing for this photo of him congratulating Alene B. Duerk, the first woman to reach the rank of rear admiral. Bud liked to joke that one does not become CNO without having kissed a lot of admirals. *(Photograph VA016078, Admiral Elmo R. Zumwalt, Jr., Collection, The Vietnam Center and Archive, Texas Tech University.)*

Mouza played an integral role in Bud's career. When Bud became CNO, he saw to it that a ship was named after the brave navy SEAL John C. Brewton, whose death was the one Bud never forgot. Here Bud looks on with evident pride during the commissioning ceremony for the escort ship USS *Brewton* (DE-1086) in 1972. *(Photograph VA015858, Admiral Elmo R. Zumwalt, Jr., Collection, The Vietnam Center and Archive, Texas Tech University.)*

The most popular letter in the navy identifies its wearer, CNO Admiral Bud Zumwalt, talking with Marine Major Rusty Hansen, commanding officer of a helicopter attack squadron aboard the amphibious transport dock USS *Cleveland*. *(Naval Historical Foundation.)*

A view of the bulletin board where Admiral Zumwalt's Z-grams were posted. *(Naval Historical Foundation.)*

Bud and Admiral Hyman Rickover sparred throughout their careers. By his own admission Bud was fascinated by Rickover's style and the way he was able to achieve his goals. *(Photograph VA019541, Admiral Elmo R. Zumwalt, Jr., Collection, The Vietnam Center and Archive, Texas Tech University.)*

By 1973–74, there was no Nixon-Zumwalt relationship of which to speak. The president sent no thank-you note for Bud's service to the nation as CNO. Indeed, he instructed the secretary of defense not to present Zumwalt with the Distinguished Service Medal at the change-of-command ceremony. *(Photograph VA018495, Admiral Elmo R. Zumwalt, Jr., Collection, The Vietnam Center and Archive, Texas Tech University.)*

One of the cherished moments in Bud's life came when he and his seventy-nine-year-old father returned to Tulare for "Zumwalt Day." The city renamed a park in honor of both men, even unveiling a special Z-gram for the occasion. Four thousand citizens turned out to acknowledge the family's lifelong service to the community and nation. *(Photograph VA047594, Admiral Elmo R. Zumwalt, Jr., Collection, The Vietnam Center and Archive, Texas Tech University.)*

President Ford kept his promise to Bud by speaking at the change-of-command ceremony at the Naval Academy, Annapolis, Maryland. Bud was relieved by Admiral James L. Holloway III as chief of naval operations. *(Photograph VA014561, Admiral Elmo R. Zumwalt, Jr., Collection, The Vietnam Center and Archive, Texas Tech University.)*

Bud and Mouza depart the change of command, but this was hardly the start of retirement. Bud adhered to General Marshall's maxim that a commander of forces in wartime had two lifelong responsibilities: first, to care for those who fought under you, and second, to assist in the reconciliation process by healing the wounds of war with the enemy in order to build a stable peace. *(Photograph VA015302, Admiral Elmo R. Zumwalt, Jr., Collection, The Vietnam Center and Archive, Texas Tech University.)*

On the campaign trail for Senate in 1976. Looking on is a young Doug Feith, the future undersecretary of defense for policy from July 2001 until August 2005, then a volunteer on the campaign. *(Photograph VA018712, Admiral Elmo R. Zumwalt, Jr., Collection, The Vietnam Center and Archive, Texas Tech University.)*

Bud and Mouza at their fiftieth wedding anniversary celebration. *(Photograph VA017301, Admiral Elmo R. Zumwalt, Jr., Collection, The Vietnam Center and Archive, Texas Tech University.)*

In January 1998, Bud was one of fifteen Americans honored by President Clinton to receive the Presidential Medal of Freedom. Parts of the citation read, "In both wartime and peacetime, Elmo Zumwalt has exemplified the ideal of service to our nation. . . . For his dedication, valor, and compassion, we salute Bud Zumwalt." *(Courtesy of Zumwalt family.)*

Mouza greets President Clinton, who would preside over Bud's memorial service in the United States Naval Academy chapel on January 10, 2000. *(Photograph VA047601, Admiral Elmo R. Zumwalt, Jr., Collection, The Vietnam Center and Archive, Texas Tech University.)*

The president spoke of "a man who was our friend and whose love for his family, his Nation, and his Navy were as deep as the oceans he sailed." *(Photograph VA047283, Admiral Elmo R. Zumwalt, Jr., Collection, The Vietnam Center and Archive, Texas Tech University.)*

In 2003, Tran Van Chon, at the age of eighty-three, came to honor and say good-bye to his friend. Bud's son Jim is standing in uniform behind Mouza's wheelchair. "I loved him like a brother and I know that he felt the same way," said Chon. *(Getty Images.)*

Artist rendition of the DDG-1000 *Zumwalt*-class destroyer, America's next generation, multimission naval destroyer, serving as the vanguard of an entire new generation of advanced multimission surface-combat ships. *(Courtesy of General Dynamics Bath Iron Works.)*

Well before his successor had been announced, preparations were under way for Bud's change-of-command ceremony. Protocol dictated that the CNO inquire about the president's availability, but the last person Bud wanted at a ceremony honoring his career was Richard Nixon. Fortunately for Bud, these sentiments were reciprocal. In a March 16 telephone conversation with Alexander Haig, Bud asked, "Is there any possibility that the President would be interested in coming to my Change of Command ceremony on 29 June?" Haig said he would ask but knew that the president was scheduled to speak at the Naval Academy commencement in June and doubted he would want to return so shortly afterward. If Nixon could not speak, Bud planned to invite Secretary of Defense Schlesinger to be his major speaker.

Change-of-command ceremonies are steeped in navy tradition. As naval aide to the CNO, Dave Woodbury had been assigned responsibility for planning and organizing the change of command. Unbeknownst to Dave, Mouza had already contacted Lieutenant Commander Ned Muffley, director of the Navy Band, with instructions that the music should include "The Sound of Music," "Sunrise, Sunset," "The Impossible Dream," and "Moonlight and Roses." Dave thought it sounded like a reincarnation of *The Student Prince* and that "The Sound of Music" had no place at the ceremony. He told Muffley that none of the four songs fit the occasion, and they could not be played. Before speaking with Muffley, Dave ran the whole thing by his boss, who approved his decision.

A few days before the ceremony, Mouza and Muffley were discussing the musical selections when she learned what Dave had done. Mouza was still wife of the current CNO, so she sent for Dave, who explained Navy tradition and that the music just did not fit the occasion. The conversation ended with Mouza saying that she would discuss it further with Bud that evening. The next morning, Bud called Dave into his office to ask, "What's wrong with 'The Sound of Music'?" Dave asked him to hum a few bars, which Bud could not do. Instead, Bud said, "You know, I've been considered a maverick CNO throughout my tour. Why not go out that way?" Dave repeated the traditional rationale, emphasizing that "some of us weren't retiring and, given a choice, I'd just as soon not want to be remembered as the Naval Aide who played 'Sound of Music' at the CNO's change of command." Dave then informed his boss that when James Holloway heard the music plan, he said he would not show up under those circumstances. After thinking about it for a few seconds, Bud proposed a compromise. "The Impossible Dream" would be played after the "Navy Hymn," but only as the Zumwalts departed the dais. The other compromise was that, for this musical selection to be played, Bud would have to introduce it.[68]

It was already eleven a.m. on the morning of June 29, 1974, when Congressman Charlie Wilson and his wife, Jerry, arrived at Annapolis under rainy skies. Wilson recalled his own days as a plebe at Annapolis, but he had never seen so many admirals in one place. He overheard one officer saying, "When you have this kind of class guy, the ceremony is bound to be heavy." The hundreds of seats in front of Bancroft Hall were full when the sun began peeking through the clouds. "White was the color of the day," observed Jerry Wilson.[69]

As a congressman's spouse, Jerry had attended her share of official events, but she sensed something different today. "The speeches were unlike any at a formal ceremony that I can remember. The genuine and deep feelings of these men for Admiral Zumwalt set the tone that flavored the day. Respect. Admiration. A very real loss for the United States Navy." In his white suit with sword at his side, and with his well-known bushy eyebrows, Jerry thought that Bud looked like a movie admiral. Now under sunny skies at the steps of Bancroft Hall, the change-of-command ceremonies were about to begin.

Nixon's order that Schlesinger not attend the ceremony put only a minor wrinkle in the plans. Vice President Gerald Ford had been invited by Bud to be the main speaker. Throughout his tenure as CNO, Bud had always enjoyed working with Ford, who had been open, straightforward, and thoughtful. A few days before the event, Ford thanked Bud for his many contributions to the security of the United States and the cause of world peace. Citing the "many sacrifices and the honorable manner" in which he had served his country, the vice president observed, "You are a great American, a great sailor, a true servant to your country. I know I speak for all the people of the United States when I say,

'Well done.' Smooth sailing and best wishes for continued success."[70] Bud was unaware that Ford had received a call from Al Haig relaying a message from the president that he should not attend the change of command. "I told Bud that I would speak at his change of command and I will do so," replied the vice president.[71]

In his remarks, the vice president steered clear of any comments on people programs, choosing to state that "Admiral Zumwalt's greatest accomplishment was the realization that the United States was facing an ever-increasing block obsolescence as far as ships were concerned. . . . As I look at the four years of Admiral Zumwalt's outstanding career, I would say that his convincing of the civilian heads in the Department of Defense, his convincing of the several present, his convincing the Congress of the United States that something had to be done to get over block obsolescence, to initiate on a sizeable and substantial basis a new shipbuilding program, will be the hallmark of Admiral Zumwalt's career. This program that he sold to the Congress, to his civilian superiors in the Defense Department, will be a lasting accomplishment as our country faces the threats and the problems that we encounter in seeking to maintain peace and strength throughout the world."

With those words, the vice president turned to Bud Zumwalt and presented him, "on behalf of the President of the United States," the Distinguished Service Medal, the very one that Nixon had instructed Schlesinger not to present. Technically speaking, Schlesinger, whose name was not on the program but who was sitting on stage, did not present the award. They were all safe! Schlesinger even chose to say a few words at the end of the ceremony. "Ford gave a very laudatory speech which I considered an act of some courage," recalled Bud, "given his knowledge that the White House did not want me in any way to be praised."

In Bud's remarks, he reminisced about his career and teased the vice president about moving into Admiral's House, the home all CNOs for the past fifty years had occupied. He had, of course, testified against it and warned Ford that "even the roof leaks." (Ironically, Ford never moved in, as Nixon would soon resign and Ford's vice president, Nelson Rockefeller, refused to live there.) Turning to more serious matters,

Bud warned that "the consequences of conceding the world's oceans to others are simply too serious to be allowed to continue. Our countrymen have never before turned away from their responsibilities, nor do I expect them to do so now."[72] The nation needed to redirect its course "on a heading toward regaining the undisputed supremacy at sea [on] which our national survival demands and which we have surrendered temporarily."

He then read his discharge orders. "Admiral Holloway, the watch is yours. I pass on to you the absolute authority and responsibility of command." The band played "Ruffles and Flourishes," the cannons boomed a nineteen-gun salute, and Admiral Elmo Russell Zumwalt, Jr.'s personal flag was lowered for the last time. Breaking from tradition for a final time, he introduced his family, daughters Ann and Mouzetta and two sons, "the lawyer" and "the Marine." Rear Admiral R. W. Elliot thought that recognizing his family "was most appropriate since it was representative of that which had been the hallmark of your career as Chief of Naval Operations, thoughtfulness for the people with whom you worked and for whom you were responsible."[73]

As red roses were presented to Mouza, the U.S. Navy Band struck up "The Impossible Dream" in her honor. Bud joked with Vice President Ford, saying that the song was not intended for him. The Zumwalts headed off the dais to their waiting car, and the Holloways headed to the receiving line for the new CNO.

"The Change of Command at Annapolis was a thrilling affair," wrote John Chafee. "I thought your departure was done with the usual Zumwalt flair." Turning to Bud's legacy, the man who hired him observed that "there has been no CNO in modern times who has made as deep an impression on the navy as you have. Your actions are going to help the Navy, not only in the Seventies but also in the balance of the century. In personnel policies as well as the imaginative development of new ships and weapons, your mark will be felt for years to come."[74]

Bud Zumwalt was now a retired admiral. He had earned the navy's highest accolade, "Well done." In fact, that was the headline of the *Salt Lake Tribune* on June 29, 1974. Noting that the old brass would never understand the importance of what Zumwalt had tried to accomplish in his personnel policies, the opinion essay urged the continuation of those policies: "Failure would be detrimental to the service's future."[75]

In "A Farewell to Admiral and Mrs. Elmo R. Zumwalt, Jr.," his staff provided the following career summary tribute:

So Long. It's Been Good to Know You

You're from San Francisco, that gem by the bay.
You were marked for distinction at the U.S.N.A.
In the far Leyte Gulf you entered the fray,
Where Halsey and Kinkaid carried the day.

Now it's . . .
So Long. It's been good to know you.
So Long. It's been good to know you.
So Long. It's been good to know you.
It's a long time since you signed on,
And you're gonna be moving along.

In command of ATAKA you sailed up the Yangtze.
With our flag at the masthead for the whole world to see.
It was there you met Mouza, your bride soon to be.
You've an eye for grace, beauty, and charm, we agree.

[Refrain]
You've skippered the finest, ashore and at sea.
The Tills and the Isbell, and Dewey make three;
CRUDESFLOT Seven and COMNAVFOR V—
You've been all the things sailors set out to be.

[Refrain]
You became CNO in July Seventy,
With green stripes and grams that were signed with Z.
You pushed people programs and man's dignity
And sought to restore zest to life on the sea.

[Refrain]
You've pursued new initiatives with great energy.
PF, and LAMPS, and ASMD;
Hydrofoils, HARPOON, and STOLS with a V.
With a single objective, control of the sea.

[Refrain]
Now, Mouza and Bud, you've a great family.
There's Ann, there's Mouzetta, and Elmo the three,
And Lieutenant Jim in the USMC.
He's your contribution to projection from sea.

[Refrain]
It's been exciting with ne'er a dull day—
With you on the E Ring, we've come a long way.
Now whate'er the future, be what it may,
We wish you smooth sailing as once more we say:

[Refrain]
So Long. It's been good to know you.
So Long. It's been good to know you.

For certain there was no singing or joy in Moscow. Neither Nixon nor Kissinger sent Bud a congratulatory message.[76] Bud believed he was the first service chief in modern times to retire without a thank-you from the president. With impeachment proceedings impending, Bud had no doubt that the Soviets, keen observers of American politics, had hardened their line for negotiating with a president who was desperate to achieve a deal that he hoped might buttress his domestic support.

By BuPers order, "When relieved on 30 June detached duty as CNO, proceed to your home of selection. You will regard yourself relieved of all active duty effective 2400 on date of detachment." Bud and Mouza's destination was a rental home at 6509 Walters Woods Drive in Falls Church, Virginia. Bud planned to spend the next few months organizing his papers, giving some lectures, and exploring his new freedom.

After all the years of navigating charted waters, the prospect of setting a new course was an exciting one.

Bud Zumwalt left office as he entered it, fighting for a strong navy. He was only fifty-four years old, thirty-five of which had been spent in the navy. By his own admission, he was departing "with as much relief as regret." He needed to distance himself from a corrupt administration so that he would be free to criticize its policies, as well as its procedures that were "inimical to the security interests of the United States." The bugging and break-ins paled in comparison to the "deliberate, systematic and unfortunately, extremely successful efforts of the President, Henry Kissinger, and a few subordinate members of their inner circle to conceal, sometimes by simple silence, more often by articulate deceit, their real policies about the most critical matters of national security."

Appearing on *Meet the Press*, Bud was asked seven questions about SALT. He kept his word to Schlesinger by dodging each one. What ended up making headlines was that he had turned down the Veterans Administration position. When asked about the job, Bud explained that "the domestic political condition at the present time is such that important innovative programs have very little chance for success."[77]

Before leaving office, Secretary Schlesinger told Bud that, in his judgment, Nixon and Kissinger had already "set up a new plumbers operation, to start tracking where all the leaks were coming from. This was while they were dying, they set it up."[78] This was no surprise to Bud, who had become accustomed over the past two years to "a positive policy to destroy the adversary process. Advice was not only not wanted, it was manufactured to fit the mold fashioned by the president or his principal advisors. If it couldn't be molded, it was discouraged. If it couldn't be discouraged, there were threats. If the threats didn't work, they were sometimes carried out."[79]

Meanwhile, Henry Kissinger flew directly from his meeting in Moscow to China. In a recently declassified document, Ambassador Huang Chen mentioned to Kissinger that he had been reading about Paul Nitze and a former Joint Chiefs member's criticisms of Kissinger. Knowing immediately that the ambassador was referring to Zumwalt, Kissinger joked that "when I read them I get scared myself! We don't have the practice in our country of sending our military leaders off to the provinces. [*Laughter*] This is just nonsense."[80]

The United States might not send them to the provinces, but the Nixon administration had other ways of dealing with recalcitrant retired admirals. The day after the change of command, a navy captain arriving from the Pentagon at Bud's home with papers that needed to be signed reported that he had been followed. The next day, Bud and Mouza noticed a black limousine with a driver and "co-pilot" circling their home for almost forty-five minutes with the co-pilot taking notes. Bud was scheduled to leave home for a speech the next day. Knowing the White House sentiment toward her husband, Mouza placed a call to Michael Dingman, then CEO of Wheelabrator-Frye, asking whether or not she should be as concerned as she was feeling. When Bud returned home, he called Michael to apologize for Mouza's excessive concern. "Whereupon he reported that, after reflection on his conversation with Mouza, he had had second thoughts. He had driven from his headquarters in New Hampshire to see Governor Meldrim Thompson, Jr., about the matter." Thompson later told Michael, "Mouza may be right; he could be in danger. Tell them to move to New Hampshire and we will provide security by state security officials."[81]

Within six weeks, the Nixon administration would collapse. Henry Kissinger would emerge with even more influence.

In August 1974, Captain William Cockell reported that he had discovered being bugged and followed while going to meetings with Senator Scoop Jackson. On August 27, 1974, Bud called Scoop Jackson, who was on vacation. Jackson "had contacted the FBI at the highest levels to find out what was going on with regard to keeping track of people who were having contact with Senator Jackson. . . . Jackson is prepared, if necessary, to call Haig and Kissinger and put them under oath to find out if they initiated this tap and trail which he believes they have put on Captain Cockell. He also said that he plans to hold hearings beginning in another month or two on the whole question of where SALT is going."[82]

Bud Zumwalt was already focused on the target of his next war.

CHAPTER 14

The Watch Never Ends

You exemplify what is best and noble in our armed forces.
—PAUL NITZE[1]

Arleigh Burke warned Bud how difficult it would be adjusting to a completely different lifestyle.[2] After operating on forced draft for thirty-five years of navy life, and especially after carrying the burdens of senior leadership in the last six wartime years, the sudden absence of responsibility created a "vacuum and a sense of loss."[3] Two weeks into his retirement, Bud wrote to Representative John Hunt that he was "already experiencing the withdrawal symptoms of not working, but that will pass."[4]

Remaining on the sidelines was not an option. Bud recalled his 1950 meeting with General Marshall at Pinehurst, when he asked Marshall why he was not taking it easy, playing golf, and enjoying retirement after so many years of government service. "I never forgot his reply," recalled Bud. " 'When you have commanded forces in wartime, you have two lifelong responsibilities: first, to care for those who fought under you; and second, to assist in binding up the wounds with the enemy to ensure a stable peace.' "[5]

Committed to addressing the challenge and fulfillment of carrying out General Marshall's prescription, Bud knew he would remain involved in shaping the country's debate on national security policy. The only question was, *How* could he get into the game? The final collapse of the Nixon presidency occurred just six weeks after his retirement. "The uniqueness and the horror of the disregard of the constitution by senior officials of that Administration convinced me that all those who had been close to the events had an obligation to record their experiences," said Bud.[6] Thus began the eighteen-month process of writing his memoir, *On Watch*.[7]

Rumors swirled that Bud would return to California to seek elective office where he was being wooed by both political parties.[8] He had not run for election since 1942 when elected president of the Naval Academy literary society. Tulare was still Bud's legal residence, but for the past seventeen years he had lived in the Commonwealth of Virginia. After considerable thought, Bud decided to run for the United States Senate, believing that even if he lost, the media coverage would help him.[9]

Several major hurdles needed to be overcome, the first being that as a nonpolitical military officer, Bud did not belong to a political party. Both of his parents had been lifelong Republicans, and Doc Elmo had run unsuccessfully as a Republican for Congress. The Republican Party was clearly more aligned with Bud's thinking on foreign policy and military strength. Bud also felt comfortable with President Ford's approach on national security, especially when they had worked together during Ford's vice presidency. Moreover, his brother Jim had been a lifelong liberal, and throughout their adult life, Jim and Bud engaged in spirited debates on the war in Vietnam and the role of the military. Bud agreed with very little of Jim's views on foreign policy. Logic suggested that he join the Republican Party.

But there were two motivating factors for registering as a Democrat. The first was the overpowering need, bordering on obsession for the truth, to rebut Henry Kissinger's "Spenglerian" view that the United States was a civilization on the wane, facing a Soviet Union that was rising to supremacy, and that Kissinger's task was to preside over the smooth transition to the reversed power relationship between the United States and the Soviet Union.[10] Running as a Democrat and taking on Secretary of State Kissinger would provide Bud with a forum for stimulating public debate on the issues of military strength, expenditures, and direction of policy.

The other reason for joining the Democrats was that most retired military were Republicans. Bud hoped to "contribute within the Democratic party to moving the Party to take a stronger position on national security."[11] He saw himself aligned with his friend Henry Jackson and the Jackson element of the Democratic Party, which included Jeane

Kirkpatrick, Eugene Rostow, and Paul Nitze. "In the end I concluded that the best political leverage to highlight my issues could be achieved by campaigning as a Democrat."

The next decision was where to run. As early as December 22, 1974, columnists Rowland Evans and Robert Novak speculated that Bud would take a shot in 1978 at unseating the most conservative member of the Senate, right-wing Republican William Scott of Virginia. The combination of being viewed as liberal on social issues and a hawk on defense was seen as potentially lethal for Scott. Bud's dynamic personality and speaking skills also offered a sharp contrast with the incumbent. Evans and Novak downplayed the significance of Bud being a legal resident of California; he could scarcely be called a carpetbagger in Virginia, where he had lived for almost two decades.[12]

But 1978 was too far down the road for Bud. He could not wait that long, especially with a presidential election scheduled for 1976. He therefore decided to challenge incumbent Independent Harry Byrd, Jr., of Virginia, whose Senate seat would be up in 1976. In doing so, he would be challenging a dynasty in Virginia politics.[13] Bud was always realistic about his chances. "I believed that the book [*On Watch*], combined with the media coverage that I would secure during the Senate campaign, should I win, lose, or draw, would give me the best opportunity to highlight my concerns about our country's future in the world."[14]

Before politics, however, came the tug of Vietnam. Bud's thoughts were on Admiral Chon, whose country stood on the precipice. Bud's fear that the Paris Peace Accords negotiated by Henry Kissinger would be a suicide pact for South Vietnam had become reality.[15] Bud understood the depths of Kissinger and Nixon's deception ever since the November 30, 1972, JCS meeting where Bud listened to Kissinger and Nixon explain the draft treaty. The U.S. negotiating position for any settlement had always been premised on the concept of a return of American POWs, the creation of a political process allowing the Vietnamese to resolve their differences, and "mutual withdrawal" of troops from South Vietnam. In the secret negotiations, North Vietnam's Le Duc Tho insisted that the United States was the only foreign army in South Vietnam.

Kissinger soon capitulated by agreeing to a unilateral withdrawal in return for North Vietnam accepting that South Vietnam's president Nguyen Van Thieu could remain in power, pending elections. This was a condition that Tho had always opposed, but by late October 1972, the balance of forces in the South was decidedly in Hanoi's favor for an in-place cease-fire. The Politburo instructed Tho that he could concede on the point of President Thieu remaining in power, leading Kissinger aide John Negroponte to quip, "We bombed them into accepting our concessions." Bud shared Negroponte's view that Henry Kissinger seemed to have a "death wish for South Vietnam."[16]

On January 27, 1973, American involvement in the war ended with the signing of the Agreement on Ending the War and Restoring Peace in Vietnam. By the terms of the deal, over 150,000 North Vietnamese troops remained in the South, whereas the United States, over the course of Nixon's presidency, had unilaterally withdrawn over 500,000 of its own troops. President Thieu and his fellow countrymen understood that the diplomatic battle had been won by Le Duc Tho. President Thieu had agreed to nothing more than a protocol for American disengagement. Thieu only agreed to the deal because Nixon had guaranteed brutal retaliation if the North resumed any aggression. Nixon assured Thieu that if he went along with Kissinger's plan, he would be able to rally America's Silent Majority—"people with character of steel"—to support renewed bombing.

But could these guarantees be trusted? The fate of his country depended on them. Nixon was intent on maintaining South Vietnam until he turned the keys of the White House over to the next occupant in 1976. Nixon was even prepared to take on Congress in a possible war-powers battle involving the resumption of B-52 bombings. Kissinger and Nixon privately encouraged Thieu not to hold elections until the communist troops went home. Kissinger advised Thieu to use Vietcong political prisoners languishing in jails as hostages for getting the communist troops out of the South. There was no reason to risk a political solution until after a North Vietnamese withdrawal from the South.

Watergate derailed the plan. The resignation of Richard Nixon on

August 9, 1974, unraveled the private assurances of retaliation if the communists violated the paper truce. Moreover, Americans were ready to turn away from Vietnam, to bury the entire sordid history of the war. Political factors were already in play. The November 1974 congressional elections had given Democrats forty-three new seats in the House and an additional three in the Senate. In a final plea for assistance, President Thieu penned a personal letter to a man he had never met, President Gerald Ford. "Hanoi's intention to use the Paris Agreement for a military takeover of South Vietnam was well-known to us at the very time of negotiating the Paris Agreement. . . . Firm pledges were then given to us that the United States will retaliate swiftly and vigorously to any violation of the agreement . . . we consider those pledges the most important guarantees of the Paris Agreement; those pledges have now become the most crucial ones to our survival."[17]

President Ford had already accepted the political reality that Congress would not fund another supplemental budget request. While personally reviewing the first draft of his address to a joint session of Congress, the president read the proposed words "and after years of effort, we negotiated a settlement which made it possible for us to remove our forces with honor and bring home our prisoners." President Ford crossed out the words "with honor."[18]

By the final days of April 1975, a mass evacuation was under way from Saigon. Bud understood the danger Chon faced if he remained in Vietnam. He would be a prize catch for the invading North Vietnamese army. Bud arranged, through the Defense Attaché Office, seats on a flight for Admiral Chon and his entire family. Chon hurried to the coastal city of Vung Tau to fetch his parents for their flight to freedom. After explaining the situation to his eighty-eight-year-old father, Chon realized he faced a personal dilemma. His father could only sit in silence, with tears streaming from his eyes, distraught at the idea of leaving the place where his ancestors were buried. His father would not leave their family's homeland. Chon felt he had no choice other than to honor his father's wishes. Assembling the family, Chon advised them that they were remaining. "I hoped that the good deeds that my parents had done would be repaid to my family."[19]

But not all of Chon's family members were in Vung Tau. His eldest son, Lieutenant Chanh, an officer in the South Vietnamese navy, was commanding a gunship bringing CNO Admiral Cang and chief of staff Admiral Thuy to a destroyer that was part of a flotilla waiting to evacuate thirty thousand Vietnamese refugees. Another son, Lieutenant Junior Grade Tran Minh Truc, a recent graduate of Annapolis, was serving as an electronics officer on another Vietnamese ship joining the flotilla.

After completing his mission, Chanh sailed back into Saigon, still flying the colors of the flag of the Republic of Vietnam. Mooring his ship at Pier Five in Saigon Harbor, Chanh went directly to get his parents and family so they could all join the evacuation. "We are staying," Chon said. Chanh knew he could not abandon his parents, but he first returned to his ship where the crew was waiting to depart and join the flotilla. After deliberating with his crew, Chanh decided to sink the ship and dismiss the crew. "Chanh's loyalty and devotion makes me very proud of him and my Navy," recalled Chon. "What he and his PGM [patrol gunboat, missile] crew had done proves that the Republic of Viet Nam Navy had built up an extremely strong unyielding spirit among their naval officers, petty officers and enlisted men."

Bud was worried about having no word from Chon. On May 6 he received a letter from Captain Pham Manh Khue, fleet commander of the Vietnamese evacuation. Khue was already aboard HQ5, the flagship of the Vietnamese fleet. Admiral Cang was aboard HQ3. "As you know, our beloved Vietnam has fallen to the Communist's hand since 1st May 75," wrote Khue. "I feel my heart broken when I think that tomorrow will be the last day of the VNN Fleet, the day all of us will leave forever our dear warships on which we have fought for the last decade throughout the SVN territorial waters. . . . I think that all of you fellows in the U.S. Navy, ex-Naval Advisors and sailors who had fought hand by hand with us in Vietnam would have wanted to share our sorrow." Rightfully proud of the role played by the Vietnamese navy during the evacuation, Khue reported that all of the warships of the Vietnamese navy, the large percentage having been part of the ACTOV turnover, were just sixty miles from Subic Bay, where they planned to retire their fleet, disembark all crew and families, and "then the VNN Fleet will be dismissed and be turned [over] to the U.S. Navy."[20]

Khue beseeched "his Admiral" to use his influence to help in their resettlement and living in the United States. "I still remember many programs, such as 'Helping Hands,' the 'Buddy Base' and 'Navy Shelters' you had previously initiated and that were very fruitful with regard to the VNN during the period from 1969 to 1972." In closing, Khue said he had heard no news about the family of Admiral Chon, "but it seems to me that he was not able to leave Saigon."

The Chon family had taken refuge in Tay Ninh, where they eluded the new Communist regime until June. Then orders came for all retired officers to report for "reeducation." Chon was advised to bring rations for one month. His sons took him to report. They would not see one another for seventeen years.[21]

A few days after the fall of Saigon, the *Washington Post* broke a story about the secret correspondence between Nixon and Thieu promising to "respond with full force" and "take swift and severe retaliatory action" if North Vietnam violated the agreement. Bud's longtime friend and ally Senator Henry Jackson seized on the letters, charging that here was evidence that the Nixon administration had misled a foreign government and Congress about the nature of our commitment. Bud not only joined the chorus, but he provided Jackson with inside information on the betrayal of an ally. Bud had been keeping records of the deceit. "Kissinger and Nixon did not level with the Congress as to the commitments that were made," said Bud. "The Nixon administration must bear a large share of the blame for the fact that Congress failed to honor those commitments that had been made in the name of the country."[22]

On April 11, 1976, Bud filed the necessary papers to seek the Democratic nomination for Senate in Virginia. He easily won the Democratic primary against Perry Mitchell, a northern Virginia community college professor. Bud was given little chance against incumbent senator Harry Byrd, Jr. Wherever Bud went in the state, he was "driving down the

Harry Byrd highway, flying into the Richard E. Byrd airport, and . . . every city in Virginia south of the Rappahannock River had a 'Byrd Sanctuary.'"

During the primary period, Bud had been successful in raising major contributions from friends and businesses in outside states. Once he was the Democratic nominee, Bud faced two opponents, millionaire Republican Martin Perper and the incumbent, Independent Harry Byrd. Paul Edwards provided a contrast between the Zumwalt and Byrd campaigns. "On only four hours of sleep, the Democrat was shaking hands at a Norfolk shipyard at 6 a.m., and underwent a rigorous question and answer session at Old Dominion University at noon, appeared at a Democratic club in Norfolk at 6 and finished up with a . . . speech to a sparsely attended fundraiser in Newport News. In between appearances, he spent six hours on the telephone in Portsmouth's bare Democratic headquarters trying to raise enough money through personal appeals to keep the campaign going. On the same day, Byrd made a single speech at the exclusive Country Club of Virginia in Richmond and taped a television interview. He spent no time on the telephone raising money and, according to Byrd's staff, has never had to make a direct appeal for funds to anyone."[23]

Byrd had a lock on most of the big money.[24] "My personal funds were limited and the power of Senator Byrd, who served on the Senate Finance Committee, to influence the business community reduced significantly my ability to raise funds in the general election."[25] Bud thought he needed to raise $750,000 to wage an effective campaign against Byrd. In a May 10, 1976, solicitation letter to Paul Nitze, Bud wrote that "I am now the officially certified Democratic nominee and the next objective is victory in the general election on 2 Nov. 1976. If we can keep our momentum (dollars needed) we are confident we can win." Paul and Phyllis Nitze each immediately contributed $1,000.[26]

Bud managed to alienate the Democratic National Committee, however. In 1976 John Chafee was the Republican nominee for the Senate from Rhode Island. His Democratic opponent was attacking Chafee for being responsible, as secretary of the navy, for base closings in Rhode Island and neighboring Massachusetts. Chafee insisted that these recommendations had been made after he left office. The DNC contacted Bud to solicit a press release stating that as CNO he could confirm that Chafee had indeed been responsible for the base closings. "I sent back word that I could not do so because it wasn't true. The next day a rather threatening message from DNC suggested reprisal if I didn't come through."[27] Bud was so offended by the threat, he issued his own press release that as CNO, he had made the base closure recommendations to John Warner. Bud's punishment for his "political sin" was that the $5,000 the Senate Democratic Campaign Committee gave to each Democratic candidate was channeled to Independent Harry Byrd.

Bud's campaign manager, Tim Finchem, tried coaching Bud on how to work a room. This meant shaking a hand, turning away quickly, and moving on to the next handshake. "I found myself too interested in a subject to turn away or too concerned that to do so would hurt the feelings of a shy voter." When campaigning, he always introduced himself as Bud and seemed to never dodge a question or stop talking.[28] When asked about his position on abortion, he refused to give the safe answer. He felt strongly that "abortion is a question for a woman, her conscience, and her God."[29] He often joked that he was "to the left of Byrd on domestic policy and to his right on foreign policy."[30] A Democratic General Assembly member said, "The guy is everywhere, but he leaves no wake."

Instead of running so much against Byrd, Bud focused on Henry Kissinger. The main message of his campaign became the dangerous decline in the defense position of the United States relative to the Soviet Union, and the post-Watergate incapability to curb Soviet expansion. Bud called for Kissinger's dismissal and a redirection of foreign policy. The battle with Kissinger had become a very personal one. In December 1975, Bud decided to write President Ford on the crucial issue of SALT I and II and Kissinger's disingenuousness in critical parts of those negotiations. Noting that the continuing shift in the strategic and conventional force balance to the advantage of the Soviet Union was a cause for grave national concern, Bud urged Ford to ensure that any agreements being negotiated by Kissinger provide for essential equivalence in the strategic force capabilities. "We must, under all circumstances, avoid a repetition of the SALT I experience in which negotiations against a deadline produced technically imprecise agreements which had seriously detrimental effects on U.S. security."[31]

Taking dead aim at Kissinger, Bud warned the president that "any agreements reached must be completely drafted and clear on all significant particulars, leaving no room for evasion or circumvention. In addition they should be adequately verifiable by national technical means. . . . Past performance makes it clear that, to continue shifting the strategic balance to its advantage, the Soviet Union can be counted on to exploit every weakness of the structure, language and enforceability of such agreements." Bud told Ford that it was a matter of the "greatest national importance" that he go public and fully inform the Congress and American public of the deficiencies in SALT I. "I urge your adoption of a policy of frankness in discussing Soviet behavior."

Bud then released his letter to the media and soon received an anonymous phone call at home, "saying that Kissinger would soon hold a press conference and 'attack' Scoop Jackson and me for our SALT charges." The caller hung up.[32] A few days later, Kissinger held the press conference, to attack what he said were falsehoods in Zumwalt's charges. A few months later, Bud received another anonymous call. He was fairly certain it was the same voice as earlier. "You should know that on at least two occasions recently Kissinger has said to Dobrynin an accident should happen to Admiral Zumwalt." The caller then hung up.[33] Right about the same time, Bud's sixteen-year-old daughter, Mouzetta, drove home after spending the night sleeping at a friend's home. She reported being followed by an unmarked car. Bud saw the car driving off, ran to his car, and followed in pursuit. Bud was able to corner the car after the driver had turned into a dead-end street. "I blocked the driveway with my car, got out, and demanded to see the driver's identification. He showed me identification that indicated he was a security official from the Treasury department." When confronted, he said that Mouzetta had been driving too fast. Bud had no doubt Kissinger was behind the surveillance. He was also convinced that Kissinger was bugging his home and taping his calls. When going to sleep each evening, Bud would say aloud, "Goodnight, Henry."[34]

In Norfolk, Bud characterized the secretary of state as "an evil Chamberlain" who overlooked Soviet violations of SALT I in order to maintain détente. He accused Kissinger of denying key Pentagon officials, congressional leaders, and arms-control negotiators documentation of Soviet violations and of misleading the public.[35] "Henry Kissinger doesn't believe in his own country. He doesn't think the American people have the will. I've heard him say so himself," said Bud. Not to be outdone, Byrd was more than eager to attack Kissinger, who became a punching bag for both candidates. The *Washington Star* headline of January 11, 1976—THE KISSINGER FACTOR IN VIRGINIA POLITICS—questioned whether the Virginia senatorial race could be won on the issues of Kissinger's lies about détente and the Soviet threat.

In an attempt to broaden his national exposure, Bud was invited to deliver the defense plank at the 1976 Democratic National Convention in Madison Square Garden, which nominated Jimmy Carter.[36] His platform speech attacked Nixon and Kissinger, saying that their "secretiveness and single-handed policy-making are the ingredients of scandal, not security."[37] Bud was also able to negotiate some of his more hawkish views into the speech in return for endorsing a platform that called for less defense spending and an end to exotic arms systems.[38] The concessions to Bud in the defense plank involved a modernized fleet with "production of smaller, less costly warships." This ended up hurting him in Virginia, where many of Rickover's ships were built at Newport News Shipbuilding & Dry Dock Company, Virginia's largest single private employer.

Although Bud proved "indefatigable," he failed to energize voters.[39] In his first and only foray into the arena of electoral politics, Bud Zumwalt won just 38 percent of the vote against a man whose family has held the power in politics in Virginia for most of this century.[40] Bud was proud that the race was a model of respectability. About the worst Byrd could say was that his opponent was a "wandering Admiral who had wandered into Virginia to run against him." Bud's reply was that he was "a wondering Admiral, wondering why it was that when my Zumwalt ancestors were fighting in Virginia in the Revolutionary war, Colonel William Byrd was a Tory."

One of those joining the campaign was Douglas Feith, the future undersecretary of defense for policy from July 2001 until August 2005, who helped devise the U.S. government's strategy for the war on terrorism and contributed to policy making for the Afghanistan and Iraq campaigns. At the time, Feith was a law student at Georgetown University.

> I volunteered for Admiral Elmo Zumwalt's Senate campaign in Virginia. . . . Like his friend Scoop Jackson, Zumwalt was a liberal Democrat who was also hawkish on defense and skeptical of détente. It is rare for a military service chief to win the general public's attention, but Zumwalt had become a national celebrity—complete with a cover of *Time* and an interview in *Playboy*—for championing civil rights for blacks and eliminating what he called 'Mickey Mouse' regulation of sailors' beards, motorcycles, and civilian garb. What made him famous, though, also produced a boatload of enmity for him in conservative navy circles: Virginia was full of retired admirals, and a number of them actively campaigned for his defeat. Zumwalt lost the election resoundingly. It must have been an unpleasant experience for him, but for me the campaign itself was a treat and a boon. Zumwalt had personal traits I found uplifting to watch at close range: honesty, courage, learning, judgment, love of family, love of country, and kindness. I felt eager to write things for him because he was literate and exacting—and, when satisfied, generous with praise. I happily worked with him on speeches, letters, and policy papers, and we continued to collaborate on writing projects for years after the election.[41]

Following Zumwalt's defeat, Democratic senator Hubert Humphrey offered words of counsel and advice for the future. "You waged a good campaign. I hope you won't look upon it as a loss. You should see it as the beginning of a political career. It's not easy to run against an entrenched political machine, but it can be done. It will take perseverance, however, I know you have that. Be of good cheer. You carried the banner of the Democratic party with dignity and verve. I'm proud of you and honored to be included as one of your friends."[42]

Bud would weigh one more foray into the political arena. In 1977 Virginia senator William Scott announced his intention not to seek re-election in 1978. In a June 29, 1977, letter to Paul and Phyllis Nitze, Bud admitted, "I am not obsessed with the idea that I must run. My interest lies in serving the office, not holding it; in bringing about vigorous debate of vital issues facing Virginia and the United States; and in clarifying these issues so that we can all make better choices."[43] Bud thought his chances of winning in 1978 were much better than 1976. "Much has happened since then," wrote Bud. "I have learned political campaigning. My name is now known in the State, and I would not, in 1978, run against a family institution as I did against Harry Byrd, Jr."

Bud asked the Nitzes for financial support so that he could commission a private poll to determine the feasibility of moving ahead. By August Bud had the answer. In a "Dear Boss" letter, Bud thanked his mentor for another generous contribution. "Although the results do not warrant making the run, the investment in the poll means that an intelligent decision is to be made and that no other friends will be solicited for a forlorn cause. I look forward to exciting work with you and the Committee [Committee on the Present Danger] in the months ahead."[44] This work also involved moving the Democratic Party toward the center with regard to national security issues in the movement called the Coalition for a Democratic Majority.

Bud had tried convincing Elmo that he should remain in the navy as a career, but his son had his heart set on studying law and getting married.[45] Following their wedding at St. Anthony's Catholic Church in Falls Church, Virginia, Elmo entered law school at the University of North Carolina, and Kathy completed her undergraduate studies at Meredith College. Three years later, in an end-of-year wrap-up letter to the family clan, Bud wrote that "in many respects, 1973 was an exciting year for us for it saw Elmo graduate from law school and enter a fine practice in Fayetteville, North Carolina where he and Kathy are now happily settled in their new home."[46] Elmo and Kathy had taken almost a dozen Vietnamese into their home and supported them until they were able to provide for themselves.

Thirteen years after returning home from Vietnam, Elmo learned of the ticking time bomb inside his body.[47] By now, Elmo and Kathy had two children, Maya, age nine, and Russell, age seven. In January 1983, Elmo first started feeling ill, unable to shake a persistent cough. During a physical examination, he reported still jogging a mile a day with no diminished exercise tolerance. A series of tests followed, including a lymph node biopsy that revealed "Stage IV-A nodular poorly differentiated lymphocytic lymphoma."[48] Non-Hodgkin's lymphoma (NHL) had already spread throughout Elmo's lymphatic system to his spleen, bone marrow, and liver.

As soon as he heard the diagnosis, Elmo thought about his family, about his exposure to Agent Orange as a young naval officer, and about his seven-year-old son, Russell, born with a sensory dysfunction that caused developmental problems. Elmo located each one of the crewmen who had served with him in Vietnam, urging them to get a physical to see if they too had unknowingly been wounded.

Bud immediately grasped the terrible irony that by his own hand his son had been poisoned. "It was a haunting choice," said Bud. "It is the first thing I think of when I awake in the morning and the last thing I remember when I go to sleep at night."[49] Elmo's struggle for life spanned five and a half years and involved each member of the family in the search for a cure. In 1986, Bud and Elmo wrote *My Father, My Son*, recalling their efforts to win the biggest battle of their lives. "Because I have a disease that was probably caused by the military orders of my father, I feel that I have been singled out to tell my story," said Elmo in explaining his decision to go public.[50] They would soon have to deal with the advent of a second form of cancer—Hodgkin's disease, which was discovered a year after the onset of non-Hodgkin's lymphoma.

At the time of Elmo's diagnosis, there were more echoes from Vietnam. Admiral Chon's son, Truc, who had graduated from Annapolis and had been expecting to rendezvous with his family in Subic Bay, was now a

U.S. citizen residing in California. In a letter to Bud, Truc reported that his father was still languishing in prison and there was no word about his condition. The one bit of good news was that "my brother Chanh, his wife and daughter and two of my brothers and sister have escaped from Vietnam by boat. They are now in the Galang Island, Indonesia." Truc was trying to get the family into the United States, but the Refugee Resettlement Agency was insisting this would take well over a year. He wanted to know if there anything Bud could do to speed up the process.[51]

Bud did not hesitate, writing his friend John Holdridge, then ambassador to Indonesia, that the immediate family of "my esteemed counterpart in Vietnam, who himself remains in a Hanoi prison," had escaped Vietnam. Bud asked Holdridge to look into what could be done to expedite the process so they could join their brother in the United States.[52] Bud's intervention did expedite the process. By the end of the year, the family was permitted to enter Thailand and soon thereafter were in California. In a letter to their beloved "Uncle Bud," Chanh explained that his father had been able to sneak a letter out of prison, telling the family that they should make an effort to escape and be certain the first person they contacted was his friend Admiral Bud Zumwalt. Chon asked that Bud do whatever he could to secure his release from Hell.[53]

Chanh was overjoyed to have "placed my feet on American soil, the country of freedom."[54] Truc had told Chanh how Bud and the Zumwalt family had provided financial support to "help us get started." No words could express their gratitude for Bud's commitment to them, especially in helping them gain employment. Chanh reported that his mother had been allowed to visit her husband, who was "very thin, but well."[55] His mother had filed papers with the ODP (Orderly Departure Program) but would not leave Vietnam without her husband. Truc soon wrote Bud to say they had just heard about Elmo's illness and were very concerned.[56]

Elated with news that Chon was alive, Bud threw himself into the effort, enjoining others in influential positions to join him. Two months after hearing from Chanh, Bud was seeking an appeal for Chon's release on humanitarian grounds. In October, Paul Wolfowitz, then assistant secretary of state for East Asian and Pacific affairs, wrote Bud that ODP was aware of "his interest in Chon" but the prospects were not good. "I don't need to tell you about the people with whom we are dealing, or how difficult they can be in negotiation."[57]

On July 30, 1984, Elmo turned thirty-eight years old. "I find myself feeling very lucky that we have had 38 years to love and appreciate him," Bud wrote to Mouza. "Those years are made more precious by the fact that we might not have many more with him."[58] Elmo believed he could conquer his cancer with mental strength. He had sidestepped polio, heart surgery, and Vietnam, so there were few who would challenge the assumption. Elmo was one of the lucky "terminal" patients, having found a bone marrow transplant match in his family, sister Mouzetta. He had won the lottery, a 25 percent probability that a family member would be a match. As he was preparing for his procedure, Aunt Saralee wrote that "each day I say a prayer that you will beat your latest challenge. I truly admire your courage, Elmo. Anyone can be brave on a battlefield, but true heroes are those who face whatever vicissitudes life hands them with squared shoulders and a smile."[59]

Elmo was admitted to the Fred Hutchinson Cancer Center in Seattle on February 11, 1986. A two-day course of chemotherapy began the next day, followed by six days of radiation. Bud was elated to learn that the attending physician was Dr. Donnall Thomas, the originator of the bone marrow transplant process and the head of that branch at the Fred Hutchinson clinic. Bud started a "Dear friends of Elmo" letter to keep everyone updated on his son's progress. "Day zero" for Elmo was February 20, when Mouzetta entered the center for the procedure that would transfer her marrow as a replacement for her brother's dead marrow. The night before the procedure, Kathy and Maya spent the night with Elmo in his hospital room. A few hours before Elmo underwent his treatment, he gave his father a sealed letter, to be opened only upon his death.

> *Dear Dad,*
> *I saw the tears in your eyes as you read the letters I left the family.*
> *I know more than anything I wrote, it was the helpless desperation*

and agony as death approached someone you loved that caused those tears. . . . I also know you continued to help us with your spirit for our sake. You made a difference. You lightened our burden. Both in Vietnam and with my cancers, we fought battles and lost. Yet, we always knew even when the battle was clearly desperate, that our love could not be compromised, and that however bad the odds, we were incapable of ever giving up. After my death, your strong burning torch of love, which only death extinguished from my being, will light Maya, Russell, and Kathy. . . . How I loved you. How I would have loved to have continued to fight the battles by your side. You always made a difference. You made my last battle, the journey to death, more gentle, more humane.

I love you, Elmo.

Bud stayed with Mouzetta until three a.m. at the hospital. "All the Zumwalts salute a gallant Mouzetta who has never wavered from her strong desire to provide Elmo her life-giving sustenance," wrote Bud.[60] Elmo called her room to relay the message that "he was now receiving her beautiful, red, non-cancerous bone marrow and was deeply grateful." In his journal, Bud wrote that as day zero began, "the doctors predict that Elmo, whose blood counts are now close to zero, will feel progressively weaker each day for several days as those counts drop to zero."

For the next two or three weeks, Elmo would have no immune system whatsoever. The danger of infection would be high. Enclosed in a nylon eight-by-ten cubicle filled with high-pressure germ-free air, surrounded by family pictures and X-marked calendars, Elmo suffered horribly from the chemo and radiation damage to the mucous membranes of his mouth, throat, and esophagus. Morphine was regularly administered to subdue the horrible pain of the mucous passages. He was connected to a catheter in his abdomen because of the anomalous veins resulting from his cardiac birth defect.

By day six, doctors learned that Elmo's body was destroying platelets much faster than usual. It was normal for a patient to use up platelets at a high rate as the tissues damaged by radiation and chemotherapy were replaced. Mouzetta was sent to Puget Sound Blood Center for a procedure that removed platelets for Elmo from her bloodstream. During the session, Bud received a call from Elmo, who in a broken voice said, "Dad, maybe it's just the drugs or maybe it's psychological, but I am really feeling deeply distressed and hurt that you and Mouzetta went off to lunch and out of contact when I was suffering." Bud said, "Elmo, I'll be right over." At that point, Elmo broke into a laugh, having demonstrated once again the Zumwalt humor.

On day eight, Elmo lost all his hair within an hour. His mouth, throat, and esophagus were bleeding. "The pain is so excruciating that the nurse took him up to 12 milligrams of morphine per hour," Bud wrote in his journal. "This left him so groggy that he fell hitting his head on the wall and cutting himself with serious bleeding."

On day twelve, Mouzetta was asked to give more platelets. The next few days were critical for evidence of the graft going forward. "The next big test, after the graft taking, was to survive the graft vs. host disease that began as Elmo's body sought to resist the encroaching graft, while also trying to avoid infections that can result from having no immune system," Bud wrote in an update to Elmo's friends.

On day fifteen, Bud learned that he, too, would be needed to give platelets. The problem was that Mouzetta's provided a fully matched human leukocyte antigen (HLA) type; Bud's were a half-matched HLA type. Elmo called them both to his room and vetoed the idea of "changing his luck." On day nineteen, Elmo's brother Jim arrived in Seattle. Elmo confided in Jim the agony he was going through, "the skin around his groin coming off like sheets of tobacco a few days ago—blood bruises popping up in various parts of his body due to low platelets—the horrible sensation of thick, gooey slime sliding down his esophagus and coagulating in his stomach." He had been able to endure only because of his love for his wife and children. That same day, Elmo received the great news that his white counts and platelets had gone up. When Bud left that evening, Elmo asked him to drive by his hospital window. "He was standing in the window giving us a 'V for Victory' sign with both arms."

In a letter to Elmo's friends, Bud explained that the doctors thought Elmo was doing better than the norm. He was "always keeping his eye on the ball that nothing must interfere with his taking every recommended precaution in order to survive." Elmo was a fighter. He was in regular communication with most members of his boat crew and already planning a trip to San Diego. He never thought he would not beat the cancer. He could not allow that thought to ever enter his mind. "When I found out that I had a terminal illness, I realized that being told you are going to die is not the worst thing in the world, but living miserably, because you have that knowledge, could be . . . hope is a successful, pragmatic way of life in the real world."[61]

On March 24, day thirty-two, Elmo was allowed to leave his encasing bubble. Russell was given the honor of walking in first, followed by Maya, Kathy, and Bud. "The joy was emotional to behold," wrote Bud. Concerned that many of his fellow sailors in Vietnam might have also come home with time bombs inside them, Elmo agreed to allow ABC's 20/20 to film and interview him during his battle.

By day thirty-four post-transplant, Elmo was making "remarkable progress."[62] He'd had no fever since day eighteen; "his white counts, polys, platelets and hematocrit counts all rose dramatically." The greatest danger between days thirty and sixty was pneumonia. Elmo's original immune system had covered him for the first thirty days, but after that, his new immune system, not yet fully able to deal with the invasion, was not strong enough to fight off pneumonia. Still, Elmo was released as an outpatient, allowed to live in the couple's rented apartment in Seattle. Each evening Kathy administered intravenous feedings until Elmo's daily calorie intake reached 2,500. "We are all afraid to be as optimistic as we feel," wrote Bud.

On day ninety-six, four days earlier than anticipated, Elmo was discharged from the Hutchinson Center.[63] Bud arranged for a corporate jet from one of the boards he served on to take them all home from Seattle. Elmo had avoided being one of the 20 percent who fail to make it through bone marrow transplant and now had a 55 percent probability of total recovery. "They tell us if he makes it through the first year, his odds would be much higher and if he has not had a recurrence by three years from the date of the transplant, he would be considered cured," wrote Bud in his final update to friends. Bud ended with a touching comment for Elmo's wife, Kathy, "who guarded him, loved him, and shepherded him back to health like a protective tigress; to his sister Mouzetta, who so unselfishly gave of herself time and again to save him; to Elmo, for his superlative courage."

Elmo had been reluctant to write about himself because his story was so personal and painful. It was his friend Walter Anderson, a Vietnam veteran and at the time editor of *Parade* magazine, who convinced him that the story needed to be told. Bud and Elmo quickly became celebrities, known for drawing attention to what Vietnam veterans went through in Vietnam and what happened to them after they returned. Through a bestselling book, numerous television appearances, and a 1988 TV film based on the book, starring Karl Malden as Admiral Zumwalt and Keith Carradine as Lieutenant Elmo Zumwalt, the story became known to a wide audience in this country and worldwide.

When ABC's *20/20* segment, "Agent Orange and Cancer—the Aftermath of Vietnam," finally aired, Elmo said he was looking for a miracle. Shortly thereafter, letters poured in from around the globe. A person in Tulsa offered a food supplement called Traumune developed as an immune system booster, made from shark cartilage and only recently available in United States, writing, "We believe in miracles, too."[64] Dr. Bernard R. Leipelt, founder and president of the International Research Foundation for the Advancement of Preventive Medicine, offered to provide alternative medical experts in Germany. James Lee of Notre Dame, Indiana, offered biophysically efficient procedures to bring homeostasis, balance, back to Elmo's body while ridding it of cancer. "You should approach your cure from the perspective of the cell" by drinking only the purest of waters; eliminating meat; eating only raw fresh foods, fresh wheatgrass juice in particular; completely avoiding sodium fluoride in every form because Japanese findings indicate that sodium fluoride damages DNA and the immune system; considering beta-carotene, aloe vera juice, heavy dosages of ascorbic acid, B-17, and all B vitamins; and avoiding all radiation treatments.[65]

Bud and Elmo were also searching for their own miracle, joined by an extraordinary team of physicians and researchers in the Hematology/Oncology Division of the Naval Hospital and elsewhere. Bud's personal physician from Vietnam, Dr. Bill Narva, helped arrange for

Elmo to meet with the very best physicians and was constantly search-ing for new experimental programs for which Elmo might be a candi-date. With Elmo weak and bedridden, Bud redoubled his own effort to get Elmo into an experimental program. In desperation, Bud called Dr. David Duggan in Syracuse, who had been running trials of interleukin-2 and beta interferon.[66] In a sobering note to Elmo, Bud wrote that "with two separate types of lymphoma, it might be difficult to get you on a protocol since all the protocols desire pure data—i.e., how does it work on Hodgkin's or how does it work on NPDL [nodular poorly differen-tiated lymphocytic lymphoma]. The mixture would spoil their data." Elmo was not eligible for the experimental programs.

Sometime in 1987, while Elmo was still fighting his cancer, Bud was called by Dr. E. Donnall Thomas, who had cared for Elmo during his treatment. Thomas was a pioneer in marrow transplants who in 1990 received the Nobel Prize in Physiology or Medicine for this lifesaving work. He reminded Bud that ten thousand people were dying each year simply because they lost the family-match lottery. Thomas envisioned creating a data bank from a much larger population. The goal was to create a coordinated National Marrow Donor Program (NMDP) by tying together transplant centers, blood centers, and donor centers to create a national registry of volunteer donors. Bud agreed to serve as director of NMDP and was soon made chairman.

During the next decade, the Donor Registry grew from thirty thou-sand volunteers to four million; a million foreigners were also carried in overseas registries. The odds of finding a match multiplied greatly. Bud used his influence with Congress to secure an almost $60 million annual budget for a federally sponsored program. "One of the special sources of satisfaction for me is the significant number of Vietnam veterans and their children for whom we have identified matched donors and thus provided transplants through the NMDP."[67] In 1991 Bud established the Marrow Foundation, a tax-exempt foundation that raised money to add to the federal funds that support NMDP.

Since 1975 Admiral Tran Von Chon had been a prisoner in his home coun-try. Twelve years later, Bud was still working tirelessly for his release. In a letter to Richard Schifter, assistant secretary of state for human rights and humanitarian affairs, Bud tried getting Chon's name put "on the list of those who have been deprived of their human rights."[68] Bud wanted Schifter to understand that Chon "is a saintly man, a devout Buddhist, and was, throughout the years of US involvement in Vietnam, totally dedicated to the U.S./Vietnamese partnership and a firm believer in the democratic cause. He has, for these past twelve years, paid the price for that loyalty."[69] Was there anything Schifter could do to bring pressure or influence on the Vietnamese?

Two months later, Schifter reported that he had personally been in touch with the embassy in Bangkok, and they were willing to consider Chon and his family for admission into the United States whenever the Vietnamese issued exit permits. Everything still depended on the willingness of Vietnamese authorities.[70] One bit of good news soon arrived from Truc, who reported that his father had been released from prison but was under constant surveillance. Chon could not risk writing directly to Bud, but through a message relayed by his wife, Chon wanted Bud to know that he was "very sad" to hear about Elmo's illness.[71]

In the early-morning hours of August 13, 1988, Elmo's battle was coming to an end. Bud had devoted literally every minute of the past year to finding a miracle cure for his son. Now too weak to rise from bed, Elmo was down to less than a hundred pounds. Unable to retain nourishment, he began saying his good-byes, dictating a note to his uncle Jim and Aunt Gretli, "who between them, nurtured me at birth and in terminal circumstances, with love and appreciation for their lifelong support."[72] With his brother Jim at his bedside, Bud asleep on a couch nearby, and Kathy upstairs exhausted from the weeks of personal hospice care, Elmo passed away. "I looked over at Elmo again, only to see him take his last breath. His heart, strong in spirit but weakened by his exhaustive battle for life, had finally given out. There was one final sigh and then silence," said Jim, whose brother was finally released from the pain. Cradling his brother in his arms, Jim decided not to wake Bud and Kathy. They would need their rest.[73]

Elmo died at the age of forty-two. His death certificate recorded

August 13, 1988, 6:30 a.m., as the time and non-Hodgkin's lymphoma, Hodgkin's lymphoma as the causes of death. Five days later, his uncle Jim wrote a poem titled "A Magnificent Warrior," ending with the verse, "His life gave hope to thousands of suffering souls. / It remains a noble inspiration to us all." Shortly before he died, Elmo had sent each of his siblings a poem titled "Home from the Sea." The author of the poem was his father, who wrote it on the occasion of Elmo's bone marrow treatment five years earlier.

> *A sailor tossed by the seas of life,*
> *Tho torn and tired by the constant strife,*
> *Would cling to hope and his raft of dreams,*
> *Aware that life is a maze of streams,*
> *Each raging, churning and calling out:*
> *Come join my course—shoals are all about.*
> *Saw land and trees at the rim of sight,*
> *Knew day had followed a turgid night.*
> *Sensed rest and joy were a stroke beyond,*
> *Where life could be as a gentle pond.*
> *He turned his eye from the raging sea,*
> *And knew he's found all Eternity.*

Elmo always viewed his cancer as a matter of fate, a statistical incidence of something that happened to people in war. He understood the probabilities were that he would not survive, but he never gave up hope. He also never blamed his father for what happened to him, believing that the defoliation was necessary to save lives. Whenever asked if he had any regrets, Elmo's answer was the same: "The saving of American lives was always his first priority. Certainly thousands, perhaps even myself, are alive today because of his decision to use Agent Orange."[74]

"I absolutely believe, there's no doubt in my mind that Elmo's cancer had to be the result of exposure to AO," Bud readily acknowledged.[75] He never expressed any sense of guilt about being the instrument of such suffering, insisting that, given the same circumstances, he would do so again because it saved American lives. "I do believe

Agent Orange induced the two kinds of lymphomas from which my son died," Bud wrote in a letter to a veteran. "I regret that we fought the war. I regret we lost the war. I do not regret the many lives we saved using Agent Orange, even though some of those saved are dying later from exposure."[76]

Many have found this position difficult to fathom, coming from a man who had experienced the trauma of losing a son and being the instrument of that terrible weapon for so many others. "I think we did the right thing. In an identical situation, even knowing it was carcinogenic, I would use it again," said Bud. "We took 58,000 dead. My hunch is it would have been double that if we did not."[77] Bud's reasoning went as follows: "If I were faced with the same type of situation in which the situation required that I defoliate and if I knew that Agent Orange was carcinogenic, I would again use it for the reason that out of every 4000 sailors exposed to Agent Orange, one would have had lymphoma if not exposed, eight would have cancer after exposure, out of that same 4000 sailors, 2800 would have been killed or wounded had we not used it. Thousands and thousands are alive because of the use of it."

Bud always saw his son as a casualty of the war. He requested that the following inscription be placed on a chair donated by the family to the Navy Memorial: IN MEMORY OF LIEUTENANT (J.G.) ELMO R. ZUMWALT III, JULY 30, 1946–AUGUST 13, 1988. A CASUALTY OF THE VIETNAM WAR.[78] In Elmo's memory, the navy later authorized the creation of a bronze frieze at the U.S. Naval Memorial in Washington, depicting his Swift Boat, PCF-35, engaging the enemy on a river in the Mekong Delta. Part of the inscription reads: "At the age of 39, the younger Zumwalt died of cancer believed to have been caused by Agent Orange, a defoliant used by the U.S. Armed Forces in Vietnam."

Elmo's illness caused serious problems within the family. "I recall the horror of the 5 long years watching Elmo die and the tragic problem among us that that battle caused," Bud wrote to Mouza on February 12, 1995. Each member of the family dealt with Elmo's struggle differently. Kathy was losing a husband; their two young children a father; Jim, Ann, and Mouzetta a brother; Mouza and Bud their firstborn son. On July 30, 1984, Bud wrote his "dearest Mouzenka" on the occasion of

Elmo's thirty-eigth birthday. "As I think about what a remarkable boy and man Elmo was and is, I have to be grateful to you for giving him to me." Elmo was born at a time when Bud could not be with Mouza; Elmo came down with polio at a time Bud was at sea; Mouza had to manage young Elmo's health problems while dealing with three other children, again when Bud was not there. "You instilled in him the strength that has made it possible for him to deal with his tragedy," wrote Bud.[79]

Evidenced by Elmo's many letters to Mouza from Vietnam, the bond between mother and son was especially close. As Elmo's illness progressed, Mouza sought to protect her son, but that was now Kathy's job. The more Mouza inserted herself into the dynamics of Elmo's care and by default, his marriage, the more difficult the situation became for Elmo, who ultimately asked his mother to understand he was in Kathy's care. There was little Mouza could now do other than pray for her son.

Bud always felt that as a commander of men in battle he had made the right decision to use Agent Orange, but that did not make it any easier watching his beloved Elmo slip away. "In private, Bud expressed enormous guilt feelings," said Dr. Bill Narva.[80] In "A Last Letter to a Valiant Son," Bud wrote, "You were diagnosed in 1983 as having one form of lymphoma, and then a second in 1985. Our research, as well as that of many respected experts, revealed a high probability of a causal connection between your cancers and your exposure to Agent Orange in Vietnam—a chemical used on your father's orders . . . a tool your father opted to use only after he received representations from the chemical manufacturers and the government that its use posed no threat to human life."[81]

Bud and Mouza never stopped grieving the loss of Elmo. "After Elmo's death, my father championed a new cause—recognition by the U.S. government of Agent Orange–induced health problems and financial compensation for those Vietnam veterans affected," explained Bud's surviving son, Jim.[82] He felt a tremendous responsibility to the veterans who served under him, to see that the government did not fail in its responsibilities. "I decided that I would focus more precisely on the problems of the less fortunate among those who had served under me in war and among our former allies. At the same time, I would work on binding up the wounds with the regime in Vietnam." He came to see his work as a way of memorializing his son.[83]

Once again Bud Zumwalt needed to find a way to get into the game so that he could make a difference.

In October 1989, a year after Elmo's death, Edward Derwinski, a former congressman who was serving as the first secretary of veterans affairs in the Bush cabinet, requested that Bud come and speak with him. Bud's relationship with Derwinski dated back to the CNO years. The congressman had always been a strong supporter of the navy and had endorsed the changes that improved the lives of those in the service.[84] Derwinski laid out the increasing scientific evidence emanating from thousands of pages of transcripts of the Committee on Environmental Health Hazards (CEHH), which had been created by Congress to advise him on scientific judgments bearing on the causes of diseases in Vietnam veterans. It was his statutory responsibility to make determinations as to whether specific diseases in Vietnam veterans were "as likely as not" the result of exposure to Agent Orange. The law only required that there be as much evidence for as against such an association in order for compensation to be provided to the affected veteran or surviving spouse and minor children. In the fourteen years since the war ended, no such finding had been made.

Derwinski lacked the staff resources and time to go through the evidence, so he asked Bud to serve as an unpaid special assistant on the Agent Orange issue. Bud never hesitated, having found a way to enter the most important phase of his life. On October 6, 1989, at the age of sixty-eight, Bud was appointed special assistant on Agent Orange matters to the secretary of veterans affairs. Mary Stout, president of Vietnam Veterans of America (VVA), wrote Bud, saying, "This appointment certainly shows a new direction on the part of the Department of Veterans Affairs in addressing this very important issue for the Vietnam veterans."[85] The same man who had refused to accept Haig's offer to serve as administrator of the Veterans Administration in 1974 was now working as an unpaid assistant in a position to help veterans.

Bud was infused with energy and commitment to find answers for those still living. He worked twenty-hour days with limited staff support, writing members of Congress, asking Freedom of Information officers at the Environmental Protection Agency to provide their files pertaining to the latest available material on the toxic effects of 2,4-D, 2,4,5-T, and dioxin on humans and animals.[86] By his own admission, Bud hoped not to find a link, sparing himself the agony of knowing he was responsible: "When I began the formidable task assigned to me by Secretary Derwinski, I hoped against hope that I would not find a discernible association between illnesses experienced by Vietnam veterans and exposure to Agent Orange."[87] The most obvious reason was that he had approved of and ordered the extensive spraying as a proven means of reducing combat casualties and second, while he and Elmo always suspected the link, "Both he and I believed, as did many others, that there was insufficient scientific evidence to support a linkage between his illnesses and Agent Orange exposure. That was, of course, the conventional propaganda at the time."[88] It became a seven-month battle, first for the truth, then for justice, and ultimately for personal redemption.

Bud quickly enlisted the support of Dr. Robert Gray, executive director of the Fred Hutchinson Cancer Research Center in Seattle, along with two associates, Dr. Ken Kopecky and Dr. Scott Davis. He also received support from army lieutenant colonel Richard Christian of the American Legion national staff, whose last active-duty position had been head of the Environmental Support Group in the Pentagon—the group responsible for reconstructing where military units had been located on a daily basis and the correlation with herbicide spraying. "He had shown great courage in defying the efforts of certain of his superiors to have him give misleading testimony to Congress on the Agent Orange issue," said Bud. "He had become an expert on the government's machinations to manipulate studies on this issue. His help became essential to me in treading through the chicanery of the government agencies involved and in critiquing my report to Derwinski."[89]

Bud personally wrote his former subordinate commanders in Vietnam, Art Price and Roy Hoffmann. "My recollection is that there were occasions when we got access to Agent Orange and used it by spraying from some of our boats and helicopters."[90] Price wrote back that in Operation Giant Slingshot along the Vam Co Dong and Vam Ca Tay rivers, where the banks were heavily camouflaged with brush, they defoliated from air force planes, but missed spots were sprayed from some of "my TF-117 boats on loan for the operation."[91]

The more Bud looked at the facts, the more disconcerted he became about the corruption of science by politics. He learned that several members of the Committee on Environmental Health Hazards had been receiving compensation from chemical manufacturers and corporations producing Agent Orange. These "company docs" always found a way to offer an "inconclusive" resolution and "negative" correlation. They were manipulating the process and, in a world of uncertainty in data and findings, chose on the side of the manufacturers and not the veterans. This manipulation led the committee to never recommend to the secretary of veterans affairs that a specific disease be correlated to exposure to Agent Orange. Bud also saw that Monsanto, a producer of Agent Orange, was relying on grossly unscientific reference studies from the 1980s that were frequently used by CEHH "company docs" as the basis for what they declared were scientifically sound studies that found no positive linkage between exposure to dioxin and diseases. Meanwhile, Bud discovered several Swedish studies that had posited the correlation between exposure and diseases, but these had been dismissed by CEHH doctors on the payroll of Monsanto.

One of the greatest obstacles in establishing chemical company liability for the effects of Agent Orange was the "contractor defense," under which any private entity contracted by the federal government to produce military weaponry generally is not responsible for the effects of that weaponry's use. The one caveat is that contractors must be producing items exactly to the specifications of the government. Agent Orange manufacturers, such as Dow, Monsanto, and Diamond Shamrock, held from the onset of allegations of torts that the government knew dioxins were present in Agent Orange and authorized production and inclusion in the final product. Zoltan Merszei, former chairman and chief executive of Dow Chemical who became Bud's close personal friend, insisted that Dow produced what the government asked for.[92] Unlike civilian applications, Agent Orange in Vietnam was sprayed undiluted, at six to twenty-five times the manufacturers' suggested concentration. Manufacturers maintained that the government knew that the defoliants it was ordering contained dioxin and that workers exposed to dioxin during the production of 2,4,5-T—one of the components of Agent Orange—had experienced certain health effects, mainly a skin condition known as chloracne, and that dioxin was hazardous in pure form. The Department of Defense did not consider this sufficient to merit a warning or even mention of a danger to humans and did not advise against exposure.

The more research he conducted, the more astounding the discoveries. Bud learned that in the 1960s, chemical companies and military scientists knew that Agent Orange was harmful. "When we (military scientists) initiated the herbicide program in the 1960's, we were aware of the potential for damage due to dioxin contamination in the herbicide," testified Dr. James Clary, an air force scientist who served in Vietnam and worked with Operation Ranch Hand. "We were even aware that the 'military' formulation had a higher dioxin concentration than the 'civilian' version, due to the lower cost and speed of manufacture. However, because the material was to be used on the 'enemy,' none of us were overly concerned. We never considered a scenario in which our own personnel would become contaminated with the herbicide. And, if we had, we would have expected our own government to give assistance to veterans so contaminated."[93]

As early as 1966, Bionetics Research Laboratories, under government contract, informed the National Cancer Institute that female lab mice injected with even small doses of 2,4,5-T gave birth in "very high ratios" to offspring with birth defects. In massive doses, 100 percent of female rats produced either stillborn or mutated young. These findings went to the surgeon general, the National Institutes of Health (NIH), and Dow Chemical. "Everyone agreed to sit on the report."[94] In 1969, scientist William Haseltine was privy to information suggesting that government officials were suppressing information about the teratogenicity (birth-defect causation) of 2,4,5-T—the compound that comprised half of Agent Orange. The information was based on a secret study, "2,4,5-T:

Teratogenetic in Mice," conducted by the Food and Drug Administration's Genetic Toxicology Branch.

Perhaps most disturbing for Bud was the discovery that there had been several high-level memorandums issued during the Reagan administration to government agencies working on the dioxin issue. From the discovery process, Bud learned that the policy of the U.S. government during the Reagan years had been to instruct government agencies involved in studies of Agent Orange that it would be most unfortunate if a correlation between Agent Orange and health effects was found. This was because the Reagan administration had adopted the legal strategy of refusing liability in military and civilian cases of contamination involving toxic chemicals and nuclear radiation. As a result, the government sought to suppress or minimize findings of ill health effects among Vietnam veterans that could be linked to Agent Orange exposure, because this could set a precedent for government compensation to civilian victims of toxic contaminant exposure at such places as Love Canal and Times Beach. This would have "enormous fiscal implications, potentially in the hundreds of billions of dollars," wrote Office of Management and Budget attorneys in their secret communications.[95]

One smoking-gun memo to OMB director David Stockman stated the strategy quite clearly. "Dioxin—is a major issue in this area (Love Canal and Times Beach are largely Dioxin exposure cases); we will be in the tenuous position of denying dioxin exposure compensation to private citizens while providing benefits to veterans for in many instances lower levels of exposure."[96] Bud considered these activities "disgraceful" for putting saving money and the protection of corporations from liabilities ahead of scientific accuracy.

The fact that this had occurred under Reagan's stewardship disturbed Bud no end. Dating back to their first meeting in the early 1970s arranged by Bill Thompson, Bud and Reagan had enjoyed a solid friendship. In 1980 Bud broke with Jimmy Carter and led Democrats for Reagan. Bud believed that the final version of SALT II signed by Carter gave advantage to the USSR, giving it twice the area of destruction capability, twice the throw weight, three times the megatonnage, and five times the hard-target kill capacity. "I felt that had Mr. Carter been reelected, my children would not have lived out their lives in freedom."[97] The battle for and against ratification culminated in the Senate Armed Services Committee hearings of October 1979. Bud and several members of the Committee on the Present Danger (COPD) testified against ratification. When the Senate failed to ratify the treaty, key members of the COPD—Paul Nitze, Gene Rostow, George Shultz, Richard Perle, and Ronald Reagan himself—helped produce the Strategic Arms Reduction Treaty (START) negotiations. Bud believed that Reagan would close the "window of vulnerability" so that the United States would have a true deterrent posture. In 1983 Bud correctly predicted that if Reagan served two terms, "we will have achieved the capabilities of maintaining general world stability, and that the days of the Soviet empire will be numbered."[98]

Reagan had taken a special interest in Elmo's health. On the occasion of Elmo's fortieth birthday, celebrated in Coronado with the family, President Reagan wrote, "You proved yourself a hero in Vietnam, carrying out dangerous assignments with great courage and devotion to duty. I know how proud your father is of you. So am I. Now you're in another battle, fighting against cancer. That also, I know, takes quite unflinching courage. Keep up the fight, Elmo, trust in God, and know that your family and friends are in your corner."[99] Elmo did not know that the government was not in his corner. Following Elmo's death, Reagan wrote directly to Kathy in Fayetteville. "I'm certain that your husband, despite his valour and the decorations earned, would never have considered himself unique among those who have worn this country's uniform. In him, however, and the story he told with his father of his experiences and his lack of regret, our nation does have an exceptional account of the human spirit in its triumph over pain. It is often in the crucible of suffering that a man's character is revealed to himself and to others."[100]

Bud submitted his report to Derwinski in May 1990, identifying twenty-eight diseases that met the test of the statute that they were "as likely as not" the result of exposure to Agent Orange. Bud also recommended eliminating the CEHH.[101] After reviewing the scientific literature and the available data, he wrote, "I conclude that there is adequate evidence for the Secretary to reasonably conclude that it is at least as likely as not that there is a relationship between exposure to Agent Orange and the following health problems: non-Hodgkin's lymphoma, chloracne and other skin disorders, lip cancer, bone cancer, soft tissue sarcoma, birth defects, skin cancer, porphyria cutanea tarda and other liver disorders, Hodgkin's disease, hematopoietic diseases, multiple myeloma, neurological defects, auto-immune diseases and melanoma, pancreatic cancer, stomach cancer, colon cancer, nasal/pharyngeal/ esophageal cancers, prostate cancer, testicular cancer, liver cancer, brain cancer, psychosocial effects and gastrointestinal diseases."

Bud reached these conclusions in consultation with his informal scientific advisory board and by contesting corporate and government studies that had, in his judgment, been manipulated to avoid a positive correlation of health effects related to exposure to Agent Orange. Bud's yearlong study was the first systematic and comprehensive effort by the government to determine the magnitude of exposure. "The sad truth which emerges from my work is not only that there is credible evidence linking certain cancers and other illnesses to Agent Orange, but that the government and industry officials credited with examining such linkage intentionally manipulated or withheld compelling information of the adverse health effects associated with exposure to the toxic contaminants contained in Agent Orange."

Bud offered a blistering indictment, charging that the White House, primarily through the OMB, had adopted the legal strategy of refusing liability in military and civilian cases of contamination involving toxic chemicals and nuclear radiation. The report also concluded that by suppressing and minimizing findings of ill health effects among Vietnam veterans that could be linked to Agent Orange exposure, the government had caved in to corporate lobbying, destroying government objectivity in critical research studies.

A few months following his report to Secretary Derwinski, the House Committee on Government Operations, chaired by Congressman Ted Weiss, issued *The Agent Orange Cover-up: A Case of Flawed Science and Political Manipulation*.[102] The subcommittee report noted, "The White House compromised the independence of the CDC [Centers for

Disease Control] and undermined crucial decisions guiding the course of research at the same time it had secretly taken a legal position to resist demands to compensate victims of Agent Orange exposure and industrial accidents." Newspaper accounts focused on how the House committee concluded "that White House officials during the Reagan administration 'controlled and obstructed' a federal study of Agent Orange exposure among Vietnam War veterans. The congressional panel said a secret White House strategy to deny federal liability in toxic exposure cases led to the cancellation of the Centers for Disease Control study in 1987. The report by the House Government Operations Committee bolsters arguments of two veterans groups, the American Legion and the Vietnam Veterans of America, who filed a lawsuit last week seeking to have the CDC resume its study of the health effects of Agent Orange exposure during the Vietnam War."[103]

Secretary Derwinski ended up recommending that all of the problems listed be covered. President Bush approved compensation for three diseases: non-Hodgkin's lymphoma, soft tissue sarcoma, and chloracne. Congress also eliminated the CEHH and assigned statutory responsibility to the National Academy of Sciences (NAS) for scientific advice on health effects of Agent Orange. The NAS contracted with the Institute of Medicine (IOM) to review all scientific evidence and studies and provide advice every two years. Bud met with Dr. Kenneth Shine, president of the Institute of Medicine, and they agreed that the scientists being appointed to the IOM panels should not include "company docs" or those whose work already showed that dioxin was a human carcinogen. The IOM issued its first report in 1993, and as a result, seven more diseases were approved for compensation—Hodgkin's disease, multiple myeloma, porphyria cutanea tarda, lung cancer, bronchial cancer, laryngeal cancer, and tracheal cancer. The report stated, "The result is that Vietnam veterans have been denied for over 20 years the benefits which the law would have provided had scientific truth prevailed over pseudo-scientific manipulation." In 1996 three more diseases were made compensable—prostate cancer, peripheral neuropathy, and the first of the birth defects affecting children of Vietnam vets, spina bifida.

Once the CEHH was dismantled, Bud formed the Agent Orange Coordinating Council, which he chaired. Membership consisted of representatives from most of the national veterans organizations and Vietnam veterans' widows. Bud's persistence never waned. "I have been carrying on a running battle with the government concerning the need to provide compensation for the diseases resulting from the exposure to Agent Orange," Bud wrote to Ron Kirkwood, who served on rivers near Saigon, where he was certain he had been exposed to Agent Orange. "Thanks to President Bush and many others we have been able to win compensation for four such diseases. If I live we will win more such rulings."[104] He would live long enough to help many others, writing longtime friend Jerry Wages, who was fighting his own battle with prostate cancer, that twenty-eight health effects resulting from exposure to Agent Orange had been identified. "One by one we have flagged nine of them into the system over the bodies of the dead bureaucracy."[105]

Bud's handwritten notes from a talk he gave at a December 1992 Renaissance Weekend offer a glimpse into his public battle and private agony. Titling his remarks, "Whoops! Mistakes and Their Consequences,"[106] Bud explained, "As many of you know . . . I was the Commander of the U.S. Naval Forces in Vietnam who decided in 1968 to use Agent Orange to defoliate the vegetation along the banks of the narrow canals and rivers of the Vietnam Delta. At the time it seemed to be an intelligent decision. The U.S. Army had been using Agent Orange for three years. They believed their experiences had confirmed what the military had been assured by the chemical companies—that the only known human ill effects was the development of chloracne on the skin of some exposed individuals. . . . As is well known, twenty years later, in August 1988, Mouza and I lost our first-born son, Elmo III, from both Hodgkin's disease and non-Hodgkin lymphoma. . . . I have been deeply saddened by the additional insights I have gained: Chemical companies producing dioxin by-products have known for many years that these substances were harmful, have exchanged data amongst themselves about such harmful effects, and have delayed, in some cases, many years, in making reports to government concerning these harmful effects."[107]

Bud had asked the right questions but received dishonest answers. He was intent on not allowing that to happen again. At the time of the Persian Gulf War, Bud's thoughts were on the fact that it took the government fifteen years to deal with Agent Orange. Adding to his and Mouza's anxiety was the fact that their surviving son, Jim, was deployed in the Gulf. As soon as President Bush began deployments, Bud called Secretary of Defense Caspar Weinberger to urge that the daily positions of each tactical unit be recorded so that future studies of health effects could take place. Despite assurances that this would be done, it was not. Bud met several times with Assistant Secretary of the Navy Bernard Rostker, the overall Pentagon coordinator for studies on Desert Storm syndrome, as well as Coast Guard Rear Admiral Paul Busick, who was on the National Security Council staff to oversee government agency work on Desert Storm syndrome.

There was still more work to be done, including a much heralded return to Vietnam. At one of the Renaissance Weekends at Hilton Head, Bud had mentioned to President Clinton that the time had come to lift sanctions against Vietnam and begin the process of binding the wounds of war. The president was favorably disposed to the idea and asked Bud to get General Westmoreland and others to support the decision.

In his capacity as chairman of the Agent Orange Coordinating Council and chairman of the Advisory Committee of the Vietnam Assistance for the Handicapped (VNAH) foundation, Bud led a much publicized four-member team to Vietnam. He was accompanied by Dr. Arnold Schecter, who had been conducting pilot studies on Agent Orange; businessman Ca Van Tran, president of VNAH; and his own son, Lieutenant Colonel James G. Zumwalt, as a representative of his late brother. "It was against this backdrop then—the pain and anguish of having lost a brother to the war in Vietnam still embedded in my mind—that we returned to the country where the seeds of Elmo's death were sown," recalled Jim.[108]

The goal of the trip was to achieve funding for research by U.S. medical organizations on populations in Vietnam exposed to Agent Orange, in order to provide more definitive scientific answers on exposure. This would entail the establishment of cancer and birth-defect registries throughout Vietnam to be used as a starting point for many avenues of research: studies designed to show whether dioxin was an initiator or promoter (or both) of cancers in humans and how it does so; studies of the effects of dioxin on reproduction and development and on the endocrine system—on the sex hormones, on the thyroid, and as a causal factor in diabetes; studies of immune deficiency and neurological damage from dioxin, especially on cognitive and behavioral deficits in children; studies to establish biomarkers of exposure and genetic sensitivity; and studies of dioxin's effect specific to women—endometriosis and cervical, breast, and ovarian cancers.[109]

The agenda also provided Bud with a chance to come face to face with General Vo Nguyen Giap, the North Vietnamese army commander in chief during the war, and with Tran Van Tra, the wartime commander of the Vietcong in the South. Bud would also confer with the president of the Socialist Republic of Vietnam, Le Duc Anh, and with the cabinet minister responsible for Agent Orange research, Tran Dinh Hoan. There were additional meetings with Dr. Hoang Dinh Cau and Dr. Le Cao Dai, the two senior officials on the 10-80 Committee (Committee for Investigation of Consequences of the Chemicals Used during the Vietnam War on Human Health), the government body in charge of Agent Orange/dioxin research in Vietnam. The Vietnamese were especially curious about the trip because in Vietnam the name Zumwalt was associated with responsibility for the use of this destructive chemical weapon.

In 1993 a Vietnamese American by the name of Ca Van Tran came to visit Bud with a remarkable story. Ca had served as an interpreter for U.S. Marine units during the war. Arriving at a refugee camp penniless and without any personal effects, Ca had only his U.S. Marine Corps identification card. Processors told him he would not need a sponsor and offered Ca and his wife airline tickets to any U.S. city of their choice. Ca selected Rochester, New York, because he remembered a marine who lived there. When they arrived at the home of the marine, it was packed with refugees. The next day, they decided to hitchhike south. A truck driver picked them up and took them all the way to Springfield,

Virginia, depositing them at the Springfield Mall. Still without any money, they slept in the mall while looking for work during the day. The mall manager soon hired Ca as a janitor, where his diligence impressed the manager of a Mexican fast-food franchise, who hired Ca as an assistant chef. He soon became head chef and then manager. Ca and his wife saved their money, and before long they owned their own Mexican fast-food restaurant. By the time he came to see Bud, Ca owned five thriving restaurants and had become very prosperous.

Ca had a dream and needed Bud's help. He wanted to take a portion of his wealth and help the Vietnamese people. He planned on selling three of his restaurants and using the proceeds to hatch a plan. He needed Bud to help him get funds from the Agency for International Development (AID) in order to establish a prosthetics facility in Vietnam to provide limbs for war victims and asked Bud to serve as chairman of the advisory board. "I told him that I would be glad to try to open doors for him at AID and in Congress, but that I thought the odds against success were very high." Bud underestimated Ca's tenacity. Within months he had generated support from Congress and the executive branch for seed money.

They named the new organization Vietnamese Assistance for the Handicapped (VNAH). Initially, wheelchairs and artificial arms and legs were purchased or solicited as charitable contributions and shipped to Vietnam. Later, VNAH supported the production of prosthetic devices at prosthetic facilities, first in the Mekong Delta town of Can Tho and later in Saigon. During their September 1994 trip, Bud, Jim, and Ca visited the facility where Bud had the thrill of fitting the limb on the twelve-millionth victim. "It was a moving experience to see my father, then 73, lift a double leg amputee out of a chair and place him in the first wheelchair the veteran had ever owned." It was then that Jim understood that his father "had exorcized himself of any demons related to putting the tragedies of war behind him."[110]

The meeting with General Giap allowed two old warriors to speak of their losses and the ravages of war. "It is time for you and me," said Bud, "who have presided over so much destruction, to work together for reconstruction." Giap took special note of Elmo's death: "So you, like the people of Vietnam, suffer from that cruel, inhumane weapon." Bud felt he had "a very special responsibility to help deal with the wounds of war." He presented Giap with a signed copy of *On Watch*, with the inscription, "To Gen. Vo Nguyen Giap. With respect to a former adversary and new friend." Bud joked with Tran Van Tra that he was surprised how easy it was to find him today. "I spent two years trying to find you the last time I was here and could not." Giap laughed, telling Bud that the effectiveness of his programs during the war led him to approve the assassination attempt. Bud asked if he was still on their hit list. "No, Admiral; I give you my personal guarantee; now that you have returned to Vietnam as a tourist, not a military invader, you are most welcome." On September 21, 1994, Bud wrote Giap, thanking him for the warm reception. "I was deeply moved by being able to meet you in peace and respect. . . . As a military man, I have always spoken my mind. But I regret that I dealt with the past since my desire is to have a friendly future. I intend to continue my efforts to bind the wounds of our two peoples and I am delighted that that is your wish too."[111]

The year 1994 was a busy one for Bud, who worked ardently for the reelection of Senator Chuck Robb, running against Oliver North in a hotly contested race. Robb had been one of Bud's strongest supporters during the Agent Orange hearings. In a widely circulated letter to veterans, Bud urged veterans to support Robb, not only because of his position on veterans' causes, but also because, "He is running against a man, Oliver North, who has disgraced the military uniform by his violation of his constitutional oath and who continues to dissemble in his military life."[112] Virginia's other senator, John Warner, had also opposed North, by endorsing an independent candidate, Marshall Coleman. In a razor-close finish, Robb defeated North, but conservative Republicans never forgave Warner for not supporting North, because Coleman had drawn support from North, not Robb.

Giving proof to the adage, "The enemy of my enemy is my friend," Bud came to the support of his old nemesis John Warner. Following Robb's 1994 election, Bud wrote Warner that he was signing up to be a "Democrat for Warner" in 1996. "I want to give you my personal appreciation as a fellow Virginian for the key role you played in the defeat of Oliver North. It was courageous and necessary. Please sign me up as a Democrat for Warner in 1996."[113] Warner soon replied, "Our friendship strengthens as the years roll by."[114] After winning the Republican primary in 1996, Warner wrote Bud that "we sailors stick together. You have been a stalwart as I have faced the 'heavy seas' of politics."[115]

This was an extraordinary turnaround from the time in 1982 when Warner stopped Bud's appointment to an arms control advisory board, the ACDA's General Advisory Committee. Bud's nomination had been approved by the Senate but returned at the request of Warner, based on Bud's role as a syndicated newspaper columnist. The nomination had cleared the Senate, but Byrd and Warner got the White House to call it back and return it to the Senate as a gesture of courtesy. This was an unusual request. By December the nomination was dead in lame-duck session because both Virginia senators wanted Bud questioned on a potential conflict of interest as a journalist since Bud was contributing weekly columns to the "Zumwalt/Bagley Report." Warner warned the Senate that Bud's nomination constituted a "very important precedent" and wanted him to appear before the Senate Foreign Relations Committee before proceeding to vote on the appointment to the General Advisory Committee of the Arms Control and Disarmament Agency. Warner insisted it was not personal; rather, he did not want a journalist given access to confidential materials or sensitive national security documents.[116]

President Bill Clinton, with Bud Zumwalt as his counselor, became the strongest advocate veterans had ever had. On May 28, 1996, Bud, serving as chair of the Agent Orange Coordinating Council, joined President Clinton, Vice President Al Gore, and Secretary of Veterans Affairs Jess Brown in the White House. The press room was filled with veterans, many of whom had experienced firsthand the ferocious power of Agent Orange. They were there to hear President Clinton expand medical benefits for Vietnam War veterans by adding prostate cancer and peripheral neuropathy to the list of diseases covered. Clinton had come to announce that for the first time in history, the offspring of combatants would get benefits for birth defects they suffered. Clinton had sided with the veterans. His goal was "to ease the suffering our nation unintentionally caused its own sons and daughters."

Before the president spoke, Vice President Gore introduced the "tireless advocate" who "has helped guide us down the road to truth." Then, after thanking President Clinton for his support, the vice president spoke about the fact that for many years the government did not listen to arguments that veterans exposed to Agent Orange were entitled to disability payments as a result of their exposure. "I want to thank my longtime friend Admiral Bud Zumwalt," said Clinton. "Vietnam veterans have had no greater champion." The president joked about the "proselytizing of Admiral Zumwalt" over the past ten years. "No one has done more to demand that all of us do better for our veterans. No one has done more or shown more to take personal responsibility for our actions. Every single American with a heart and soul to love this country is in your debt today and we thank you." The president then saluted his friend. The audience broke into spontaneous applause.

The relationship between Bud Zumwalt and President Clinton had evolved into a remarkably close one. In 1996 Bud served as an honorary member of the Veterans for Clinton/Gore National Steering Committee. Bud's respect for Clinton went well beyond the Agent Orange issue.[117] Bud never hesitated to offer ideas on national security issues that could contribute to the crafting of defense and security strategy for the next millennium. He was also unfailingly frank, as expressed in a November 1994 letter urging the president to recognize that with Republican control of the Congress, he would have far more freedom in foreign policy. It was imperative that he "devote increasing concentration to those issues and to enhancing your foreign policy team. In my lifetime I have found the brightest chapters often follow the darkest. I know you will find the same as you apply your great talents in the two years ahead."[118]

The president consulted with Bud on a range of topics, starting with the lifting of sanctions against Vietnam but moving quickly to suggesting a replacement for Secretary of Defense Les Aspin, dealing with

North Korea, enhancing the military budget, replacing Warren Christopher as secretary of state, and dealing with issues related to Bosnia, Somalia, Haiti, and Agent Orange. Bud's primary focus, aside from Agent Orange, was on the need to develop and articulate an overarching national security vision that demonstrated the inseparable linkages between domestic interests and international stability. He saw the task as analogous to Paul Nitze's authorship of NSC 68, which provided such a vision prior to the Korean War. This vision could then serve as the foundation for all national security decisions, particularly for defense planning.

In November 1996, Bud took a first stab at highlighting these enduring strategic realities for the president.[119] Bud suggested that the new vision should address the challenges and opportunities of the twenty-first century and their implications for the shaping of our military capabilities. "The increasing likelihood that terrorist actions involving biological weapons can be used against our population can be prepared [for] only by your personal involvement because such readiness requires government-wide coordination." Bud advocated a vision that "should highlight America's enduring strategic realities, such as: America is an island continent on an ocean planet. Seventy-five percent of the planet's population lives within a few hundred miles of a seacoast. Most economies, especially our own, rely upon the sea. To shape our global environment, and to protect our citizens and interests, we must project America's power and influence overseas. These foregoing factors have led to the free world alliances, which have linked NATO nations and the Pacific rimland nations to the U.S. with many interests in common. To nourish these common interests, while access to overseas bases is declining and international stability continues, presages a growing need for the forward presence of U.S. forces. The prospect of terrorism grows greater and the capability of rogue nations to support terrorists with nuclear, chemical and—most easily—biological threats increases."

Bud wanted the president to understand that "your second term will present an historical opportunity to shape anew a national security vision for the 21st Century and to shape the armed forces of the next

century. Both efforts will require your personal involvement to succeed, but both can provide America with a lasting legacy of your Presidency. Indeed, you have a once-in-a-generation opportunity to make
significant changes in defense/security policy, and I urge you to take
advantage of it. A unique opportunity is at hand, and I am eager to assist
you in any way I can."[120] Bud's loyal marine aide, Colonel Mike Spiro,
believed that "if you peel layers away from Bud and Clinton, they are
very much alike. Both men had the courage to tackle big issues in the
public arena and a willingness to fight their enemies."[121]

By early October 1988, Bud had learned from Dick Schifter that Chon
was finally at the top of the list for emigration to the United States,
but the Vietnamese kept dragging their heels until 1990, when Chon received authorization to leave.[122] "Finally, the Communists have allowed
him to leave Vietnam," an ebullient Truc wrote to Bud. "He, my mother
and my sister have received their exit visas."[123] Yet again, the release was
delayed. A year later Dick Schifter wrote to say that Chon's name was
not on the most recent list.[124]

On December 9, 1992, a large crowd was assembled in San Francisco
International Airport, awaiting the arrival of a flight carrying some four
hundred refugees, including Admiral Tran Van Chon and his family.
At a nearby hotel Bud waited for Chon to work his way through immigration processing, wondering what effects the brutal imprisonment
had had on the man Bud considered a brother, who had taught him to
love Vietnamese culture and the Vietnamese people. One hour later,
upon completion of the immigration paperwork, the admiral appeared
with a happy smile, looking at everyone carefully, trying to identify
each person whom he had not seen for over seventeen years. People
were dumbfounded when they found that he looked unchanged, except
for his shoulder-length white hair. He was emotionally embracing his
children and grandchildren and shaking the hands of his friends. A few
hours later Chon was with Bud, his "personal political hero."[125] Tears
trickled down their faces. Chon's journey from hell was over. Bud had
made a difference. When once asked how he would like to be remembered, Bud answered, "someone who cared."[126]

In January 1998, Bud was one of fifteen Americans honored by President Clinton to receive the Presidential Medal of Freedom. With his wife, three children, and five of his grandchildren in the audience, Bud watched as the other recipients were escorted by a military aide to the stage, where the president and Mrs. Clinton stood. Another military aide read the citation and the president placed the Medal of Freedom on its blue and white lanyard around the recipient's neck. For some reason, when Bud's turn came, the military aide had not come as promptly as he had for the other recipients. The president seemed to be approaching the end of his remarks, and Bud decided not to wait any longer. Unlike the solemn procession of the other recipients, Bud's walk to the stage was rapid and unescorted. The president was so amused that he joked, "What's the matter, Admiral; did you think I might change my mind?"

Having introduced some levity into the program, Admiral Elmo R. Zumwalt, Jr., proudly accepted the award and the citation: "In both wartime and peacetime, Elmo Zumwalt has exemplified the ideal of service to our nation. A distinguished veteran of World War II and Korea, he served as Commander of U.S. Naval Forces in Vietnam and rose to become the Navy's youngest chief of Naval Operations in 1970. As CNO, he worked vigorously to improve our sailor's quality of life and devoted himself to eliminating discrimination in the Navy. In a life touched by tragedy, he became a great champion of veterans afflicted by ailments related to service in Vietnam. For his dedication, valor, and compassion, we salute Bud Zumwalt."

Following the ceremony, Bud wrote a personal letter to the Clintons, thanking both for "the grace and dignity of your presentation and by the warmth of your personal greeting to all of us." Bud was always going to cherish the moment and memory. "Notwithstanding the separation which the presidency must require, my family and I consider the two of you to be dear friends."[127] The president had much earlier confided similar sentiments. "Hillary and I have greatly enjoyed your and Mouza's friendship, and I have benefited from your wise counsel."[128]

In 1998 President Clinton appointed Bud to serve as a member of the Special Oversight Board for Department of Defense Investigations of Gulf War Chemical and Biological Incidents, chaired by Warren Rudman. Bud passed away three months before the report was completed. On April 4, 2000, at a board meeting held at the White House Conference Center, the board passed a motion dedicating their final report to "the memory of Admiral Elmo 'Bud' Zumwalt, Jr. He was a patriot and a gentleman."[129]

Six months after Bud Zumwalt's death, on July 4, 2000, aboard the carrier USS *John F. Kennedy* anchored in the Hudson River, President Clinton announced that the navy would honor Bud Zumwalt by naming the twenty-first-century land-attack destroyer (DD21, now DDG-1000) after his friend. The USS *Zumwalt* "will be a platform that values its crew more than any other ship on which sailors have ever lived, fought, and worked," said Secretary of the Navy Richard Danzig. "It is a fitting tribute to the legacy of Admiral Bud Zumwalt."

Six months after President Clinton's announcement, Mouza returned to Hilton Head, South Carolina, during the December 2000 Renaissance gathering. She came to thank her friends for their thoughts, notes, and prayers during Bud's illness. "I am not here to give a speech," said Mouza. "I am standing here in front of you, as a widow of a remarkable man with a great mind and sensitivities, Bud Zumwalt. He did not have any limitation of love for all peoples of every race and creed. The valedictory speech, which he gave at his high school graduation as a seventeen-year-old young man, gives us some insights into his heart and mind: "We stand at the threshold of a new world. With the light hearts of youth, with the joy of righteous struggle, we shall plunge into the intangible wilds, resolving that courage, eagerness and intelligence—the heritage from a pioneer past—shall continue the progressive civilization of our America."

Just two years earlier, Bud had written the president on the topic of operational flexibility and the global reach of sea power that characterized his foreign policy and national security vision, perhaps more than that of any other peacetime president. Like President Teddy

Roosevelt's Great White Fleet, which signaled America's entry onto the world stage, Clinton's reliance upon naval forces ensured that the United States was ready to enhance security, bolster prosperity, and promote democracy in the new millennium. But the country and the navy were at a critical stage. The president could not afford to be distracted by anything. "I strongly urge that you find time to work with Secretaries Cohen and Dalton and Admiral Jay Johnson to add the resources ($5 billion per year above currently programmed levels in my judgment) which will lay the foundation for the 21st Century Navy—the ships, aircraft, weapons systems, and especially the people—that will be the sine qua non for America's continued greatness in the decades to come."[130]

The tribute to Bud Zumwalt came close to being scuttled when an effort was undertaken by a group of retired admirals and other opponents to prevent naming the new class after "that man." There had been some opposition when the DD-21 (21st Century Destroyer) was first announced, but in 2003 as part of broad budget reductions, the DD-21 was going to be pared down and resignated DDG-1000. Opponents used the redesignation phase to argue that the name should be changed. Rear Admiral Charles Hamilton, who was Program Executive Officer for Ships (PEO Ships), recognized the historical significance of the moment. Hamilton had been in the pews at Bud's memorial service and aboard the *Kennedy* when President Clinton made the announcement. He had been a strong proponent of naming the DD-21 after Zumwalt because having a name attached always helped it survive the budget process, but in this case Hamilton had always admired Bud for "lancing the boil that forced us to re-examine race." A meeting was called of all the principals, including those representing "voices" of the past. Hamilton told the assembled group that changing the name was dishonorable. Taking his ID card from his pocket and placing it onto the conference room table, he said that if the name was changed, he could no longer serve and "they needed to find a new guy." If Hamilton was going to be rolled, he would do so with his dignity intact. The response from the group was that Hamilton should remain to oversee construction of the DDG-1000 *Zumwalt* class.

Several years later on November 18, 2011, Rear Admiral Hamilton was understandably jubilant on the occasion of the keel-laying ceremony of the USS *Zumwalt*,[131] Former president Clinton sent a letter that was read by Mouzetta to those hundreds of people assembled at Bath Iron Works in Bath, Maine. "I couldn't then, and can't now, think of anyone better than the late Admiral Elmo 'Bud' Zumwalt to be its namesake. Its technological advancement reflects Admiral Zumwalt's commitment to modernizing the U.S. Navy; the unprecedented level of integration in its power system reminds us of his own integration efforts in making the Navy one of the most color-blind institutions in America; and the high quality of life it will offer is a fitting way to honor a man who called himself not the former Chief of Naval Operations, but a former sailor. He believed deeply in a strong Navy worthy of our great nation, and that anyone who chose to serve in it was deserving of respect and dignity. Even after retirement, Bud and his wonderful wife, Mouza, advocated vigorously for men and women in uniform. I was proud to present him with the Presidential Medal of Freedom in recognition of his lifetime of service, and I will always be grateful for the blessing of his friendship. . . . May the future sailors of this magnificent ship find inspiration in Admiral Zumwalt's legacy of honorable service, physical bravery and moral courage, and character and conscience."[132]

On a warm July day in 2003, the eighty-three-year-old Vietnamese admiral walked slowly through the cemetery at Annapolis, his long white hair flowing over his shoulders. Chon had come to the hilltop cemetery to honor his friend.[133] "I loved him like a brother and I know that he felt the same way." Chon had been unable to attend Bud's memorial service because of his wife's health. On this day, he had come to thank his friend who was not forgotten.

At his age, Chon was aware this was his final opportunity to thank Bud for all he had done, but his thoughts were also focused on a dream he had had the night before Bud had passed away. He and Bud were navigating together aboard a double-hull ship. The ship landed in a remote area, and Bud opened the door and went out for a few seconds. He returned with a large strange fish that hung from his chest to his feet. Its head looked like a dragon. "He waved his other hand as a sign of goodbye and disappeared from view." Chon interpreted the dream as symbolizing that Bud was king of the sea and he now had returned to his realm for command of the oceans.[134]

Chon saluted his friend with the words, "Fair winds and following seas and long may your big jib draw!"

ACKNOWLEDGMENTS

It is a pleasure to acknowledge those who helped me along the way. In this case, my debts are many and significant. At the start is my agent, John Wright, who believed in the book, and Bruce Nichols, who signed us to Harper. My former editor, Elisabeth Dyssegaard, provided encouragement and advice in the formative stages. I was living under the right star when Bill Strachen inherited the book. Bill's support, guidance, and common sense enabled me to reach the finish line. Associate editor Kathryn Whitenight provided indispensable support. Copy editor Gary Stimeling did a masterful job helping me to clarify points throughout the narrative.

I am especially pleased to single out a few friends and colleagues for their help. Jim Reckner was a constant source of advice on all things navy and read multiple drafts of the manuscript. Paul Stillwell read the complete draft, catching many errors and proffering valuable suggestions. Larry Serra, a former NILO, also read the entire manuscript, providing cogent insights and commentary.

I heard from many of "Big Z's" sailors. With sincere apologies to anyone I may have inadvertently omitted, I acknowledge Tom Glickman, Ralph Christopher, John Woody, Jerry Wages, Joe Muharsky, Jimmy Bryant, Bob Powers, Jim Davie, Kirk Ferguson, Bob Matthews, James "Oke" Shannon, Richard Della Valle, Joe Ponder, Ken Delfino, Jim Thompson, Bob Matthews, Jim Morrison, Joe Calamia, Gary Holmes, Bob Tipton, Mike Wiley, Max (Gab) Gabriel, Bill Ferguson,

Wade Sanders, Curtiss Johnson, Don Mathews, Jim Morgan, Mike Worthington, Al DeRoco, Bob Monzingo, and Richard Anderson.

Contributors to the research for the NILO section included Larry Serra, Phil Babb, Paul Baker, Don Brady, Nick Carbone, Tim Corcoran, Pete Decker, Ed Dietz, Bob Doyle, Jess Foster, Jim Glavin, Allan Grace, Rich Gragg, Charlie Kirchoff, Lou Lesesne, Merek Lipson, Ed McDaniel, Bob Melka, Billy Roberts, John Vinson, Robert Laney, and Phil Ziegler.

I conducted numerous interviews and conversations during the course of my research. I especially thank Walter Anderson, Stephan Minikes, Charles DiBona, William Narva, Bill Thompson, Roberta Hazard, Zoltan Merszei, Robbie Robertson, Chick Rauch, Charlie Hamilton, Chuck Wardell, Joe Roedel, David Halperin, Howard Kerr, Tran Van Chon, Philip Lader, Qui Nguyen, Richard Schifter, Robert Sam Anson, Rex Rectanus, Donna Weakley, Arnold Schecter, Mike Spiro, Nguyen Trong Nhan, Phu Nguyen, Weymouth Symmes, Emmett Tidd, Bob Powers, Burt Shepherd, Homer Murray, S. Scott Balderson, Leslie Cullen, Nicholas Thompson, Mark Feldstein, Matt Dallek, Seymour Hersh, John Sherwood, John Prados, Ron Spector, Steve Sherman, William McQuilkin, Mark Clodfelter, Bill Burr, Susan Hammond, Richard Eger, Joe Kinsella, Joe Sandell, Roger Thompson, David Winkler, Robert Ancell, David Zierler, Edward Miller, Edward Marolda, and Maggie Tate.

The Zumwalt family was unfailingly helpful to me, especially when they became aware of my problems with NHHC. Bud and Mouza's three surviving children, Mouzetta Zumwalt-Weathers, James G. Zumwalt, and Ann Zumwalt Coppola, provided me with boxes of papers, documents, and supporting materials. Bud's brother James G. Zumwalt, a retired schoolteacher and self-professed amateur historian, invited me to San Diego and provided dozens of his unpublished family vignettes. Jim answered every one of my e-mail requests for additional information. Dr. Michael Coppola helped me understand issues related to Bud's final days and executing a health-care proxy. I benefited from conversations with Lauren and Camille Zumwalt Coppola, Barbara Zumwalt, Gretli Zumwalt, Fran Zumwalt, and James P. Zumwalt. I owe a special debt to Saralee's son, Richard Elmo Crowe, who came to visit me in Davis (he is an alumnus of UCD) and shared with me the results of his family research on Anna Frank and later his extensive research files on Mouza's lineage. Lavinia Mohr, who with Richard provided the conclusive DNA testing results, provided much helpful insight on Frances. Saralee's other son, Fred Crowe, was a source of realism and wit.

The Vietnam Archive at Texas Tech University is the depository Admiral Zumwalt chose for his personal papers and memorabilia. He did so because he wanted historians to have access to his materials. Director Steve Maxner did everything possible to facilitate my research in a proactive and professional manner. Vietnam Archive staff, Khanh Cong Le, Ty Lovelady, Victoria Lovelady, Justin Saffell, Mary Saffell, Kelly Crager, Kevin Sailsbury, Sheon Montgomery, Katrina Jackson, and Amy Mondt, were always helpful and proactive.

My problems with the command structure at NHHC were mitigated by the professionalism of archivists who actually wanted to help me and expressed frustration with how others were thwarting my research. I thank Dan Jones, Gregory Ellis, Tim Pettit, John Greco, and John Hodges. Photo curator Robert Hanschew and Marissa Knack, an undergraduate intern in the photo archive, assembled photo materials for my review. Frank Arre of the Navy Historical Foundation helped me secure images for the book. At the United States Naval Institute (USNI), Janis Jorgensen, manager of the Heritage Collection, expedited my purchase of oral histories. Benjamin Peisch at the *Washington Post* writer's desk was instrumental in securing the photo of Tran Van Chon visiting Admiral Zumwalt's grave.

Along the way I have also benefited from the support of friends Andy Burtis, Arthur and Cathy Delorimier, Ron Smith, Ed Costantini, Stephen Routh, Bruce Murphy, Ron Milum, Ron Reisner, Charlie Benn, and my mentor, Fred Greenstein. I also thank John Shelton and his chariot for making certain that I never missed an interview and was always on time at the Navy Yard.

I started *Zumwalt* while a faculty member at the University of California, Davis, and finished it as dean of the Honors College at Georgia State University. I have incurred debts on both sides of the continental divide.

At Davis, my former chair John Scott allowed me to retain my office for six months after retirement so that I could write each day before starting at GSU in January 2012. Davis colleagues Alan Taylor and Miko Nincic offered advice and support whenever I asked. My undergraduate research assistants, Veronica Cummings, Brendan Ripicky, Leslie Tsan, James Baker, and Liron Feldman, all made significant contributions. Baker's work on Agent Orange and Tsan's in-depth research on multiple assignments merit special distinction. Finally, I want to thank Michelle Hicks, Cindy Simmons, and the entire staff at UCD for their daily assistance and good cheer.

In the Honors College at GSU, special thanks to administrative assistant Lannetta Somerville for preparing the final manuscript and photos for production. I could not have met my deadline without Lannetta's diligence. My undergraduate research assistant Shelby Lohr did a terrific job tracking fugitive materials and proofreading. Whenever I asked, student Jerel Marshall of the Honors College was ready to assist. I also want to thank my entire Honors College team for understanding that their founding dean was working a new day job as well as finishing his book early morning, late evening, and all weekend. The support of Karen Simmons, Jeffrey Young, April Lawhorn, Greg Chisholm, and Annahita Jimmerson was indispensable to surving the transition.

My adult children, Scott and Lindsay, as well as son-in-law Juan, were always sources of encouragement. This book is dedicated to my two grandchildren, Isabel and Ian. Both could find no better role models than Bud and Mouza Zumwalt, for making a difference in the world.

Most indispensable is Nicole, who understands how important this book is to me. For the past year she gracefully accepted my refrain, "I am almost finished." But I wasn't. Now that *Zumwalt* is done, it's time to find out what weekends are all about.

Larry Berman
Atlanta, Georgia

Chronology of the Career of
Admiral Elmo R. Zumwalt, Jr., USN

1939	Entered United States Naval Academy, Annapolis
1942	Graduated from the Academy, seventh in his class
1942–43	Ensign, USS *Phelps*
1943–44	Operational Training Command, Pacific
1944–45	Lieutenant (junior grade), USS *Robinson*
1945–46	Executive Officer, USS *Saufley*
1946–48	Executive Officer, USS *Zellars*
1948–50	Assistant Professor, naval science ROTC at Chapel Hill, NC
1950–51	Commanding Officer, USS *Tills*
1951–52	Navigating Officer, USS *Wisconsin*
1952–53	Student at Naval War College
1953–55	Bureau of Naval Personnel, Washington, D.C.
1955–57	Commanding Officer, USS *Arnold J. Isbell*
1957–59	Assistant for Naval Personnel, Washington, D.C.
1959–61	Commanding Officer, *USS Dewey*
1961–62	Student at National War College
1962–63	Desk Officer, Arms Control unit, Washington, D.C.
1963–65	Executive Assistant to Secretary of the Navy
1965–66	Commanding Officer, Cruiser-Destroyer Flotilla Seven
1966–68	Director, Systems Analysis Division
1968–70	Commander, U.S. Naval Forces Vietnam
1970–74	Chief of Naval Operations
1976	Democratic candidate for Senate, state of Virginia

DATES OF RANK

Midshipman, June 7, 1939
Ensign, June 19, 1942
Lieutenant (junior grade), May 1, 1943
Lieutenant, July 1, 1944
Lieutenant Commander, April 1, 1950
Commander, February 1, 1955
Captain, July 1, 1961
Rear Admiral, July 1, 1965
Vice Admiral, October 1, 1968
Admiral, July 1, 1970

DUTY ASSIGNMENTS

Aug 1942–Nov 1943	USS *Phelps* (DD-360), watch officer
Nov 1943–Dec 1943	Operational Training Command Pacific, San Francisco, student
Jan 1944–Oct 1945	USS *Robinson* (DD-562), watch officer
Oct 1945–Mar 1946	USS *Saufley* (DD-465), executive officer
Mar 1946–Jan 1948	USS *Zellars* (DD-777), executive officer
Jan 1948–Jun 1950	NROTC Unit, University of North Carolina at Chapel Hill, assistant professor of naval sciences
Jun 1950–Mar 1951	USS *Tills* (DE-748), commanding officer
Mar 1951–Jun 1952	USS *Wisconsin* (BB-64), navigator
Jun 1952–Jun 1953	Naval War College, Newport, Rhode Island, student
Jun 1953–Jul 1955	Bureau of Naval Personnel, Washington, D.C.
Jul 1955–Jul 1957	USS *Arnold J. Isbell* (DD-869), commanding officer
Jul 1957–Dec 1957	Bureau of Naval Personnel, lieutenant detailer
Dec 1957–Aug 1959	Office of the Assistant Secretary of the Navy for Personnel and Reserve Forces, special assistant, executive assistant

Aug 1959–Jun 1961	USS *Dewey* (DLG-14), prospective commanding officer, commanding officer
Aug 1961–Jun 1962	National War College, Washington, D.C., student
Jun 1962–Dec 1963	Office of the Assistant Secretary of Defense (International Security Affairs), desk officer
Dec 1963–Jun 1965	Office of the Secretary of the Navy, executive assistant
Jul 1965–Jul 1966	Commander, Cruiser-Destroyer Flotilla Seven
Aug 1966–Aug 1968	Director, Systems Analysis Division, Office of the CNO
Sep 1968–May 1970	Commander, U.S. Naval Forces Vietnam and Chief, Naval Advisory Group Vietnam
Jul 1970–Jun 1974	Chief of Naval Operations

AUTHOR'S RESEARCH NOTE

Before the Freedom of Information Act, I used to say at meet-
ings, "The illegal we do immediately; the unconstitutional
takes a little longer." [laughter] But since the Freedom of In-
formation Act, I'm afraid to say things like that.

—HENRY KISSINGER[1]

I first met Bud Zumwalt in the mid-1990s at a triennial conference at Texas Tech University's Vietnam Center. I was writing my book *No Peace, No Honor: Kissinger, Nixon and Betrayal in Vietnam*. We shared a similar perspective concerning the endgame in Vietnam and the duplic-ity encapsulated in Richard Nixon's three words, "peace with honor." The admiral served as a valuable source for my book, and we partici-pated in a C-SPAN Book TV series focusing on the process of writing *No Peace, No Honor*.[2] We later worked together for the Vietnam Center at Texas Tech University, where Zumwalt chaired the National Advisory Board on which I still serve.

No Peace, No Honor was published in 2001, and I next hoped to write this biography. Unfortunately, the Zumwalt papers were not yet declas-sified. Seven years later I received news from the acting director of the Naval Historical Center (NHC) that the Zumwalt materials were open. I was given authorization to begin working on what he and other histo-rians at NHC saw as an important contribution. I could never have an-ticipated my journey into the navy's equivalent of Alice in Wonderland.

I began research in the Operational Archive of NHC on January 25, 2008. I was especially excited about access to the files used by Zumwalt when writing his 1976 memoir, *On Watch*. By my second day of research, I had made hundreds of copies of declassified documents and written several pages of notes. My euphoria was shattered when the head archi-vist approached me to say that I had inadvertently been given access to classified materials. All copies were taken from me and the next day I learned that fifty-six pages were being withheld pending further review by other agencies. They have never been returned.

It was puzzling, since some of the documents withheld had been used and quoted extensively in *On Watch*, published thirty-two years earlier and available in libraries throughout the world; the same documents were also available either online or in the research room at the Vietnam Archive at Texas Tech University, the archive chosen specifically by Zumwalt over both the National Archives and the U.S. Naval Academy as home for his personal papers.

My puzzlement quickly turned to bewilderment when I was contacted by a special agent of the Naval Criminal Investigative Service (NCIS). In our meeting, the agent directed me to turn over all copies I made from my first day of research so they could also undergo "declassification review." The agent soon presented me with a "Classified Information Non-Disclosure Agreement," essentially a legal agreement between an individual who has been given access to classified materials and the United States government. It is a "house rules" agreement by which the signatory acknowledges that "I understand and accept that by being granted access to classified information, special confidence and trust shall be placed in me by the United States Government." The agreement spells out serious consequences for unauthorized disclosure of classified materials and assigns "to the United States Government all royalties, remunerations, and emoluments that have resulted, will result or may result" from the release of classified materials.

After consulting with my legal counsel, I told the agent that I would not sign this agreement. I was not a member of the government; nor had I received security training. Indeed, I had been given access to these materials by people in NHC who presumably had received training in handling classified materials. The agent could not force me to sign the agreement, but that spurred another request from her: I would have to turn over any handwritten notes I had taken during my research in the archive. NHC then shut off my access to unclassified materials on the grounds that these materials needed to undergo a page-by-page inspection as part of what is known as a Kyl-Lott review. It was nothing personal, just business.

Let me explain the Kyl-Lott reference. On April 17, 1995, President Bill Clinton signed Executive Order 12958 titled "Classified National Security Information." The intent of the executive order was to foster transparency, openness, and access to government records by prescribing a uniform system for classifying, safeguarding, and declassifying national security information. A central provision was the requirement that U.S. government agencies declassify all of their historical records that were twenty-five years old or older by the end of 1999, "except for those documents that fell within certain specified exempt categories of records, such as documents relating to intelligence sources and methods, cryptology, or war plans still in effect."[3] It was this executive order that allowed me to see the Zumwalt materials in the first place.

By 1998, a controversy involving Chinese nuclear espionage and a spy scandal led the U.S. Congress to tighten reclassification efforts associated with Clinton's executive order. Congress took aim at the Department of Energy's sensitive nuclear secrets and nuclear weapons design–related information.[4] All of these materials were removed from public access, pending review. A year later, Congress adopted an amendment to the 1999 Defense Authorization Act that terminated all automatic declassification activity while a "plan to prevent the inadvertent release of records containing Restricted Data" is developed. This amendment was further amended, named after its sponsors, Senators John Kyl and Trent Lott, requiring review of all previously declassified documents, meaning that any documents released into the public domain since 1995 would need to be re-reviewed on a page-by-page basis for "restricted data or formerly restricted data unless such records have been determined to be highly unlikely to contain restricted data or formerly restricted data."[5] The Zumwalt materials fell under this broad umbrella.

My options were now narrowing.

In my next move, I requested an expedited review of the materials because they had been generated from the years 1970 to 1974. Zumwalt had obviously dealt with "nuclear issues," so I focused on the "highly unlikely" part of the amendment's language by requesting access to the CNO files on race relations, racial incidents at sea, retention study groups, and people programs. I was willing to wait a few years for the review of defense-related files and never expected to see nuclear-related materials. My request was denied on the grounds that there might be a misfiled nuclear secret in a retention study report. The risk to our nation's national security was deemed too great. I was in a state of disbelief and wondered if this book would ever be written.

My best option was to generate a Freedom of Information Act (FOIA) request, using the federal law that gives the public the right to make requests for federal agency records. Throughout my career, I have used FOIA to gain access, most notably in my case seeking Presidential Daily Briefs (PDBs) from 1965 to 1968, the so-called crown jewels of intelligence reports. I lost that case, *Berman v. Central Intelligence Agency*, on appeal in the California Ninth Circuit.[6] But I learned much from the experience.

Before filing this FOIA request, I needed to see the Folder List or Finding Guide for Admiral Zumwalt's Office Papers. I had already used this folder list on my first day of research, but now needed it simply to identify files for my FOIA. The head of Operational Archives tried to dissuade me from filing the FOIA because of the drain it would place on her small and beleaguered staff. She also told me that filing a FOIA would slow down the Kyl-Lott review. I knew that denying a researcher a finding aid or folder list because of fear of a FOIA request is a violation of Department of Defense regulations. When I raised this issue, the entire Folder List was declared Classified. Since I lacked the necessary security clearances, I could not see the folder list and therefore would be unable to identify documents for a FOIA. If this folder list was indeed classified, it had been mishandled for many years within NHC, constituting a serious security violation.

I offered another compromise: What if NHC allowed me to see unclassified and declassified materials in the folder list and blacked out the classified materials? That request was also denied. What NHC did not know was that I had by now obtained a copy of Admiral Zumwalt's official list of documents stored in the navy's "secure safe" upon transfer of his papers to NHC. These were available to any researcher at the Vietnam Archive as part of the papers that Zumwalt had given them. I still

needed the navy's current folder list, however, because during processing from the secure safe into the navy archive, a new numerical system
for identifying the materials had been created.

I understand why the navy must comply with the legislative requirements delineated in Kyl-Lott, but observance of law can and should be
consistent with meeting the needs of historical researchers. I went up
the chain of command to request that Operational Archives conduct
"review on demand," just as the National Archives had done in its Kyl-
Lott review. That is, why not prioritize reclassification reviews so that
historians needing materials for projects could get priority over files
that no one was likely to ever use?

The navy did not support my request. Vice Admiral J. C. Harvey,
Jr., director of navy staff in the Pentagon, wrote that "previously established government declassification priorities" related to Kyl-Lott prevented review on demand. The deputy director of the Naval History
and Heritage Command (the renamed Navy Historical Center) wrote
me that NHHC "does not have the authority to provide the expedited
declassification review requested in your letter." I was informed that
only the navy declassification program manager could make decisions on the review of records subject to the provisions of the Kyl-Lott
Amendment. My appeals to this individual went unanswered.

When in a hole, stop digging.

I filed a blanket FOIA on April 3, 2009. The FOIA included a request
for the complete folder list for the 14 boxes of the Zumwalt Personal
Papers used for writing *On Watch*, as well as all documents in that list.
I also requested the folder lists for Zumwalt's CNO office oo files (the
central files of the CNO, 1970–1974, which contained over 750 boxes).
I requested access to all documents and materials in approximately
300 containers identified as Admiral Zumwalt's Personal Papers. I also
requested access to all documents withheld from me on my two research days in the archive. Finally, and perhaps most vital to this story,
I requested a public fee waiver, meaning that the documents would be
furnished without charge if the navy believed that disclosure of the information was in the public interest and likely to contribute to public
understanding of the operations or activities of the government.

The FOIA was referred to the director of the Naval History and Heritage Command. On July 28, 2009, I was informed that my request for a fee waiver had been denied. The estimated cost for conducting a FOIA review of approximately 295,000 pages would be $44,235. I was somewhat bemused to read, "Please notify my office if you agree to pay these estimated costs and we will immediately continue processing your request."

It was clear to me that the navy, intent on denying me access to the Zumwalt materials, had stooped to the lowest level of subterfuge by concluding that a book about Admiral Zumwalt would not be in the public interest; however, if I wanted to pay $44,235, they would facilitate the project by processing the unclassified papers. I had one option left: a final appeal to the Office of the Judge Advocate General (JAG), the navy's office for resolving legal issues involving military operations, organization, and personnel.

My appeal focused on the Naval History and Heritage Command's denial of a fee waiver as a deterrent to my right of access under FOIA. I described the continued pattern of manipulating the Code of Federal Regulations (CFR) to deny me timely access to government information. This was not within the spirit of the DOD's published policy that aims to restrict procedural obstacles and promote openness, transparency, and the public's right to access U.S. government information. I went on to document a two-year pattern of procedural deceit and noncompliance intended to thwart my research, focusing on the denial of folder lists and finding aids. The Naval History and Heritage Command had not been acting in good faith per Title 32 of the CFR by creating numerous procedural obstacles that had impeded my right to prompt and timely access to DOD records. I closed with the following plea: "I beseech the JAG appeal officers to recognize the broad public interest that will result in disclosure of the requested information."

On October 9, 2009, a letter arrived from Captain M. G. Laverdiere, JAGC, U.S. Navy, deputy assistant judge advocate general (general litigation), the secretary of the navy's designee for determining my rights under the FOIA. My eyes focused immediately on four words in the second paragraph, "I grant your appeal." A JAG lawyer acknowledged the obvious: "disclosure of the information was clearly in the public interest and likely to contribute significantly to public understanding of the operations or activities of government and is not primarily in the commercial interests of the publisher." The Naval History and Heritage Command was informed that I had met the criteria for a waiver of fees, and they were instructed to process my FOIA request immediately.

The JAG's ruling had ripple effects throughout the chain of command. The FOIA coordinator and archivist in NHHC's Histories and Archives Division began copying whatever responsive records were available and expedited the review of CNO materials, meaning that those unclassified and not exempt from release would be made available to me. I received folder lists and made several research trips to the archive. Certain materials still needed to undergo the Kyl-Lott review. I was given no date for when that would begin.

In May 2010, I received the Naval History and Heritage Command's Vice Admiral Edwin B. Hooper Research Grant to assist in my research and writing for this biography about the life and times of Admiral Elmo Russell Zumwalt, Jr.

Bud Zumwalt understood that the disposition of his papers depended on a system that not even Kafka could have dreamed up. He anticipated that the bureaucracy might do everything possible to prevent these materials from reaching public disclosure. He did not think that Henry Kissinger should get to write the history of the Cold War. He sought to ensure that his extensive set of records, including private tapes as well as personal and official papers, would be made available to historians in a timely manner. "He was meticulous about dictating what I guess would amount to journal notes every night," observed Admiral Harry Train. "He had extensive records that he himself kept."[7]

Bud Zumwalt made multiple copies of many documents, including remarkably candid materials from his own intelligence network of loyal spies reporting on the inner secrets and operations of the Nixon White House. He also made dozens of tapes while his memory was fresh from the events of the day. These tape transcripts were later used as raw data for the ghostwriter of *On Watch*. The transcripts have been invaluable to me in reconstructing virtually all aspects of Zumwalt's life, from childhood memories of growing up in Tulare and the influences in his life, up through the CNO years.

I have been able to use the declassified materials from my FOIA request, the personal papers and materials in the Zumwalt Collection of the Vietnam Archive at Texas Tech University, dozens of personal interviews, access to personal family materials and records, the Agent Orange and Agent Orange Coordinating Committee files at the Vietnam Archive, the Papers of Paul Nitze in the Library of Congress, Naval Institute Oral Histories, and files within the National Archives as well as NHHC, and also the valuable interviews conducted by naval historian and author Paul Stillwell. I have triaged the primary source materials, that is, drawing my portrait of a man's remarkable journey through life from the materials available to me. I envy the next generation of historians, who should have greater access to the records that are rightfully ours.

There is one final sad exchange in my story of struggle for access to the materials. In November 2011, I was notified by Captain Jeffrey L. Gaffney, deputy director of the Naval History and Heritage Command, that I would no longer have access to materials. "I had wanted to touch base to explain why we were less able to help you on your research. I understand that you were quite disappointed in us, which of course was never our intention. We are proud to help with research such as yours, particularly your research because Admiral Z is such an important, complex and controversial figure. As you must have noticed during visits here to the Archives, we are in sad shape physically and cannot maintain the collection in the environmental conditions that it requires to prevent mold and other deterioration. This was pointed out to us very clearly by a recent Inspector General's visit which cataloged our problems and brought them to the attention of the highest Navy leadership. . . . The bad inspection results, our duty of stewardship of the collection, and a plus-up in our budget (thanks to Admiral Greenert) together pushed us to make hard decisions about the preservation of the collection. It needed to be moved to proper storage and it had to be done immediately, so we struck while the fire was hot and Navy leadership was ready to support us. Unfortunately, preparing this huge collection

(4.7 miles of shelving) for movement necessitates inventory and other preparation which takes up the time of our archivists. In order to preserve the collection for future naval historians, we have to curtail the availability temporarily. The collection will be moved next year to a temporary facility where we will still be able to access it, although not with the ease we are used to."

I am not surprised that the navy's Inspector General produced such a scathing report on the Naval History and Heritage Command's management. In light of my experience, it is essential that the new leadership at History and Heritage Command be more sensitive to the needs of researchers and find ways to establish the right balance between openness and security requirements. I hope that researchers are treated better than I was.

The battle never ends, and I just keep leaning forward.

NOTES

CHAPTER 1: CONSCIENCE OF THE NAVY

1. Zumwalt had made hundreds of tapes while his memory was fresh from the events of the day. Many of the recollections and assessments on the tapes never made their way into *On Watch*. I have made frequent use of these candid and unedited recollections from the Zumwalt Personal Papers (ZPP) at the Vietnam Archive, Texas Tech University. Many of the tape transcripts have no identifying markers, so they are cited as Zumwalt tape transcript (ZTT).
2. United States Naval Academy Chaplain's Center, www.usna.edu/Chaplains.
3. *Time* cover story, "The Military Goes Mod," Dec. 21, 1970.
4. William J. Clinton, "Remarks on Presenting the Presidential Medal of Freedom," Jan. 15, 1998. The ceremony can be viewed at www.c-spanvideo.org/program/98416-1.
5. Author interview with Ambassador Philip Lader.
6. These remarks were made at Hilton Head, South Carolina, Jan. 1, 1999. A significant portion of primary source materials used in this book was provided by the Zumwalt family, who understood my frustration in dealing with the Navy History and Heritage Command (NHHC). I cite these materials the first time as Zumwalt Family Collection and afterward as ZFC. By agreement with the family, the bulk of these materials will be given to the Vietnam Archive as part of the Larry Berman ZUMWALT collection, so that scholars may make use of these materials.
7. July 25, 1968, personal letter to "My Dear Bud," signed "Much love, Jim," ZFC.
8. June 3, 1970, personal letter, ZFC.
9. I am especially grateful to Bud's brother James G. Zumwalt for providing me with his extensive collection of unpublished family vignettes, from which this recollection comes.
10. Nov. 5, 1995, letter from Saralee to Bud and Mouza, ZFC.
11. James G. Zumwalt funeral vignette plus personal interview with author.
12. Letter to Roger Himmel, May 24, 1994, ZFC.
13. These are the words on Mouza's tombstone at the Naval Academy, where she is buried next to her husband.
14. Remarks of Admiral Mike Mullen at Mouza Zumwalt Memorial Ceremony, United States Naval Academy Chapel, September 3, 2005.
15. Author interview with Philip Lader, former ambassador to the United Kingdom.
16. Radarman second class (RD2) Joe Muharsky, U.S. Navy, Vietnam, USS *Brister*, Destroyer Escort Radar (DER) 327, 1967; Patrol Craft Fast (PCF) 78, Danang, 1968; PCF 94, An Thoi, 1969.

17. Jan. 4, 2000, personal letter to Mouza, ZFC. The letter can be found at www.mwweb .com/ndc/zumwalt/zumwalt.htm.
18. This reference is found on an undated Zumwalt tape transcript titled "Paul Nitze."
19. Letter to Mrs. Adjemovitch, Hebrew Immigration Aid Society, Fort Indiantown Gap, June 10, 1975, ZFC.
20. Qui Nguyen, personal correspondence with author.
21. Phu Nguyen, personal interview with author in Saigon, Dec. 2011.
22. Elmo Russell Zumwalt, Jr, "After *On Watch*," unpublished manuscript. In July 1997, the U.S. Navy's senior enlisted man, the master chief petty officer of the navy (MCPON), John Hagen, requested that Bud write an update to *On Watch* as a report to sailors on his activities since retirement on July 1, 1974. The unpublished manuscript, dated "as of September 9, 1999," provides an extraordinary window into Bud Zumwalt's twenty-five years of retirement. I have drawn extensively from this manuscript. He hoped to revise the chapter in 2010 at the age of ninety to cover future activities, ZFC.
23. The Vietnam Center and Archive collects and preserves the documentary records of the Vietnam War and supports and encourages research and education regarding all aspects of the American Vietnam experience: www.vietnam.ttu.edu. My debt to Founding Director Jim Reckner and his successor, Steve Maxner, is expressed elsewhere.
24. See Elmo Zumwalt, Jr., and Elmo Zumwalt III, *My Father, My Son* (New York: Dell, 1987).
25. Award of the Silver Rose to Lt. (JG) Elmo Russell Zumwalt III, Nov. 11, 1997, ZFC.
26. Bob Hartzman letter to Mouza following Bud's death, undated, ZFC.
27. Letter from Togo West, Secretary of Veterans Affairs, Jan. 6, 2000, ZFC.
28. Zoltan Merszei, letter to Citizen's Stamp Advisory Committee, Sept. 1, 2009, ZFC. Also author's interview with Merszei.
29. Zumwalt letter to President Clinton, Oct. 19, 1994, Zumwalt Personal Papers, Vietnam Archive, hereafter ZPP.
30. When Zumwalt became CNO, he asked Wages to serve as a personal aide and special assistant. Wages letter to President Clinton, May 31, 1996, ZPP. Wages was buried with full military honors on Aug. 20, 2008, at Arlington National Cemetery.
31. Wages thanked the president for adding prostate cancer to the list of conditions linked to Agent Orange.
32. Elmo Zumwalt, Jr., *On Watch* (New York: Quadrangle/New York Times Book Co., 1976), 183.
33. "Regrettably, I was right. He did backtrack, and the good guys he promoted to Admirals, ahead of their senior peers, backtracked (read "betrayed") him. He was a good man, but as I described them, the interbred, intermarried and intellectually sterile establishment destroyed him." http://forums.military.com/eve/forums/a/ tpc/f/7010097960001/m/9440045361001/p/2.
34. Churchill's assistant, Anthony Montague-Browne, said that although Churchill had not uttered these words, he wished he had: www.winstonchurchill.org/learn/myths/ myths/quotes-falsely-attributed-to-him.
35. Letter to Ruth Kaplan, editor of *BNS News*, 000 box 86, folder 5, Navy History and Heritage Command, Operational Archive, hereafter NHHC.
36. In Z-gram 117, Zumwalt said, "Admiral Ernest King, in speaking to my graduation class, stated that true military discipline is the 'intelligent obedience of each for the effectiveness of all.' As I have said before, it is through enlightened leadership that we obtain that true military discipline about which Admiral King spoke some 30 years ago."
37. I am grateful to Howard Kerr for first making this observation in Paul Stillwell's *Reminiscences by Staff Officers of Admiral Elmo R. Zumwalt, Jr., U.S. Navy* (Annapolis, MD: U.S. Naval Institute Press, 1989), and then elaborating during our interviews.
38. Ibid.

39. Bill Norman interview, ZTT.
40. Ibid.
41. Dov Zakheim, Remarks to the Executive Program for General Officers of the Russian Federation and the United States, JFK School of Government, Jan. 10, 2000, ZFC.
42. David Woodbury, "Musings of an Ancient Mariner, on Being a Naval Aide," unpublished manuscript supplied to author.
43. Prior to the surgery, Bud was told he needed to gain weight in order to increase his stamina. His longtime physician, Dr. William Narva, ordered him to eat bowls of ice cream and begin a walking regimen to build up strength. Rear Admiral Horace B. Robertson, who served thirty-one years on active duty in the navy, first as a general line officer and then twenty-one years as a judge advocate before joining the faculty at Duke Law School, came to the hospital every morning to take Bud on walks around the hospital grounds, trying to prepare his friend for the next battle. Robertson served as Bud's legal counsel during the CNO years when racial disturbances on the USS *Constellation* threatened to end Bud's career. "He is the best example I have ever followed in life," said Robertson. "I patterned my life after the standards he set."
44. From Scott Davis, Oct. 7, 1999, ZFC.
45. From Rosemary Mariner, Oct. 12, 1999, ZFC. See also Rosemary Mariner, "Adm. Zumwalt Changed My Life," *Washington Post*, Jan. 9, 2000.
46. From Joe Ponder, Sept. 29, 1999, ZFC.
47. Joe Ponder, e-mail to Mouza, Jan. 13, 2000, ZFC; also personal correspondence with author.
48. Bud to Joe Ponder, undated, ZPP.
49. Harry Train, undated, and Arthur Price, Oct. 8, 1999, ZFC.
50. Nov. 18, 1999, ZFC.
51. Undated letters, ZFC.
52. Author interview with S. Scott Balderson.
53. An undetected anomaly in the configuration of his heart resulted in a surgical error being made that drastically reduced Bud's chances for survival.
54. Letter from Paul Nitze, Nov. 29, 1999, ZFC.
55. Note from Bill Clinton, Dec. 14, 1999. Another note came from Gerald Ford, who had spoken at Bud's retirement change of command in 1974. "With great recollections of our longtime friendship . . . the Navy was in good hands under your leadership. As a four year veteran naval officer in WW II, I was proud of the Navy with your stewardship." ZFC.
56. The letter is in the Leslie Cullen Collection at the Vietnam Archive.
57. Author interview with Ambassador Philip Lader.
58. This personal note was inserted into the coffin; copy courtesy of Ann Zumwalt Coppola.
59. Mike Spiro, personal interview with author, and letter from Mouzetta Zumwalt to Mike Spiro, ZFC.
60. Letter from Rick Hind, Jan. 4, 2000, ZFC.
61. Letter from Bill Clinton to Mouza, Jan. 4, 2000, ZFC.
62. Author personal interview with Burton Shepherd.
63. *Oral History of Admiral William Crowe*, Naval Historical Foundation, p. 612.
64. Philippians 4:8–9, New International Version.
65. "From the U.S. point of view, Israel was a stationary aircraft carrier in the Middle East." Author personal interview with Richard Schifter. Former prime minister Ehud Barak said, "Admiral Zumwalt was a man of honor and a great friend to Israel." Relayed by Ehud Barak to Jim Zumwalt at Grand Hyatt, Feb. 19, 2000.
66. William J. Clinton, "Remarks at Funeral Services for Elmo R. Zumwalt, Jr., in Annapolis, Maryland," Jan. 10, 2000, online ed. Gerhard Peters and John T. Woolley, American Presidency Project, www.presidency.ucsb.edu/ws/?pid=58232.

67. Bill Pawley, who served in Vietnam from 1969 to 1970, wrote to the Zumwalt family, "If I were to put an epitaph on his tombstone it would read something like this: 'Returned to Sender, with Great Pride and Love.'" ZFC.

CHAPTER 2: THE ROAD FROM TULARE TO ANNAPOLIS

1. I again express debt to James G. Zumwalt for sharing dozens of his unpublished written family vignettes from which I draw extensively.
2. The account was provided by Bud's father in a letter dated Nov. 29, 1952, as he remembered the moment: "I gazed upon a rounded bit of humanity that was ruddy, chubby and in spite of a formal introduction, totally ignored me as he continued to sleep peacefully thru 'ohs and ahs' of a doting parental gaze." ZFC.
3. Letter to Captain E. R. Zumwalt, Jr., Nov. 16, 1960, aboard USS *Dewey*. "I doubt that little in the world know your real name," his father wrote. ZFC.
4. Letter to Admiral Rucker, May 19, 1973, responding to handwritten letter of May 8, 1973, emphasis added, NHHC.
5. "Augusta Jane Evans (Wilson) (1835–1909)," http://public.wsu.edu/~campbelld/amlit/evans.htm.
6. See Paul Stillwell, *The Reminiscences of Admiral Elmo R. Zumwalt, Jr., U.S. Navy* (Annapolis: U.S. Naval Institute Press, 2003). I make extensive use of Stillwell's oral history interviews throughout the book. Fellow historians owe him an immense debt. Indeed, this book could not have been written without benefit of his materials. Excerpts and printed volumes of USNI oral histories are available at www.usni.org/heritage/oral-history-catalog.
7. Letter to Miss Suzy Clegg, Sept. 13, 1972, ZPC.
8. James G. Zumwalt vignette, "A letter to Andreas zum Wald," June 9, 1992.
9. James G. Zumwalt vignette, "James Brown Zumwalt." James Brown Zumwalt became president of the Sunset Oil Company, with wells in the Lost Hills of California that he sold to Shell Oil.
10. "Kings Canyon National Park: Zumwalt Meadow," www.shannontech.com/ParkVision/KingsCanyon/KCZumwalt.html.
11. Brooks Gist, *Echoes of Yesterday* (Tulare, CA: Gist, 1979), 21.
12. The home became the landmark Renaud Ranch.
13. Letter, "Dear Gang," on occasion of Elmo's fifty-eighth birthday, with reflections on his life, ZFC.
14. "Brilliant Military Wedding for Young Officer and Bride," *Richmond Daily Independent*, Feb. 8, 1918.
15. Letter, "Dear Gang," ZFC.
16. *Daily Tulare Register*, Feb. 1, 1919.
17. "Through the Eyes of a Son," *Tulare Advance-Register*, Nov. 25, 1961, 6.
18. Editorial, "'Doc,' the Complete Citizen," *Tulare Advance-Register*, Sept. 6, 1973, 10.
19. "Through the Eyes of a Son."
20. Frances's and Bruce Craig's ashes had been scattered previously.
21. Geraldine (Gerry) Eyer Soults, Mar. 21, 2009; Tulare's "Women of Destiny" featured thirty-six women in various categories. One was medicine, in which were exhibited numerous items and photos of Frances Pearl Frank Zumwalt.
22. I am indebted to the extensive family research conducted by Bud's nephew Richard Crowe, son of Saralee. This entire section draws from Richard's work, which will be available in the Tulare Historical Museum, www.tularehistoricalmuseum.org/index.html. "All my mother would say during my growing-up years was that she was an orphan—that's all she would say," recalled Bud.
23. Frances died in 1939, so it appears Sarah was unaware of this or chose not to change her will.

24. Related to author by Richard Crowe.

25. Anna Rich to Admiral Elmo, Jan. 10, 1971, sent to Doc Elmo at China Lake Naval Air Weapons Station, ZFC.

26. Dov Zakheim, Remarks to the Executive Program for General Officers of the Russian Federation and the United States, JFK School of Government, Jan. 10, 2000. Boorda raised his children as Protestants but is buried with a tombstone marked with the Star of David.

27. James G. Zumwalt vignette, "H Street Memories."

28. Bud recalled that during the Depression his parents never asked for money from farmers going through tough times, and his mother almost always packed two sandwiches for lunch so that he could pretend at school to be too full and give one to a friend who did not have very much. Stillwell, *Reminiscences of Admiral Zumwalt, Jr.*, 8–10.

29. *Tulare Advance-Register*, Nov. 5, 1935. "Former President Herbert Hoover stopped off at the Tulare hotel accompanied by Earl Warren, district attorney for Alameda county. While in Tulare, they were dinner guests at the home of Dr. and Mrs. Elmo Zumwalt."

30. "The President," vignette by James G. Zumwalt.

31. Letter to friend, Sept. 28, 1933, ZFC.

32. Ibid.

33. Author interview with James G. Zumwalt.

34. Stillwell, *Reminiscences of Admiral Zumwalt, Jr.*, 4.

35. Ibid.

36. Ibid.

37. Letter, "Dear Gang," on occasion of Elmo's fifty-eighth birthday, ZFC.

38. Stillwell, *Reminiscences of Admiral Zumwalt, Jr.*, 55.

39. Letter to Bud, Dec. 8, 1967, Naval Weapons Center. Congressman A. J. Elliott committed to giving Bud one of his appointments.

40. Stillwell, *Reminiscences of Admiral Zumwalt, Jr.*, 29.

41. Interview, Admiral Elmo Zumwalt, *Playboy*, June 1974, 73–90.

42. Stillwell, *Reminiscences of Admiral Zumwalt, Jr.*, 30.

43. Letter to Bud, Dec. 8, 1967, Naval Weapons Center, ZFC.

44. Stillwell, *Reminiscences of Admiral Zumwalt, Jr.*

CHAPTER 3: EDUCATION OF A NAVAL OFFICER

1. *Lucky Bag* yearbook, 1943. The rest of the quote reads: "Bud insisted youngster year was fruit and starred to prove it. Not being content with being great academically, he was twice winner of the Quarterdeck Society's public-speaking contest. Like all men of genius, Bud leaned a bit toward the absent-minded side. Few of us will ever forget his solo 'column right' in the middle of a company mass. But we who know Bud are satisfied that great success will follow him wherever he goes."

2. Most letters and correspondence in this chapter were provided by the Zumwalt family (ZFC). Some letters are also available at the Vietnam Archive in the Zumwalt Personal Papers (ZPP). Letters, Mar. 17, 1940, and Apr. 28, 1940. "The Academy is getting beautiful again . . . the sky blue, flowers are blooming, grass is green, trees are bearing leaves. The Severn has regained its healthy blue. The low cliffs that rise from the opposite shore are beginning to doll up in their springtime greenery." ZFC.

3. W. D. Puleston, *Annapolis: Gangway to the Quarterdeck* (New York: Appleton-Century, 1942).

4. In a letter written thirty years later, Bud described his feelings to a young man about

to embark on a naval career. "When I was your age, I wanted to be a doctor, because I very much wanted a profession which offered more than just a job, which offered me the chance to do something different, something worthwhile, something to help my fellow man. I found that profession in our country's naval service." Letter to Stephen Taylor, June 8, 1972, ZFC.

5. As recalled in a Dec. 8, 1967, letter from father on the occasion of Bud's forth-seventh birthday, ZFC.

6. Zumwalt correspondence file, Dec. 3, 1970, NHHC.

7. Stillwell, *Reminiscences by Staff Officers*, 36.

8. Robert J. Schneller, Jr., *Breaking the Color Barrier* (New York: New York University Press, 2005), 91.

9. Stillwell, *Reminiscences of Admiral Zumwalt, Jr.*, 42–44.

10. June 9, 1939, ZFC.

11. Tape 6230607037, Vietnam Archive, ZPP.

12. Between 1935 and 1939, Jim had prayed for Frances's recovery. Jim became agnostic after his mother died in 1939. Bud added that "it had a significant impact on me. Not only those events transpiring as early as they did, but also my father's feeling obviously had some impact on me. I saw him leading what I considered to be an extremely model life without it, and I'm sure it had some impact on me, both with regard to trying to do good work for other people and at the same time not getting myself carried away with religious fervor." Stillwell, *Reminiscences of Admiral Zumwalt, Jr.*, 45.

13. Tape 6, undated, ZTT.

14. James G. Zumwalt vignette.

15. Stillwell, *Reminiscences of Admiral Zumwalt, Jr.*, 35.

16. Letter, Nov. 10, 1972, en route from Washington to China Lake, 30,000 feet, ZFC.

17. On July 23, 1985, in a "dear children" note, Bud attached copies of these letters (parts of which are also in his personal papers). "I attach copies of the letters I wrote to your granddad from the naval Academy when I was younger and foolisher than you. I hope you enjoy them, Love."

18. Letter to Elmo Zumwalt, Apr. 17, 1940, ZFC.

19. Letter to Elmo Zumwalt, Feb. 18, 1940, ZFC.

20. "I must confess that in regard to my weight, temper, etc., you were right—I've slacked up on my physical training since X-mas," wrote a contrite Bud in reply.

21. Jan. 1940, ZFC.

22. Jan. 28, 1940, ZFC.

23. Jan. 29, 1940, ZFC.

24. Feb. 3, 1940, ZFC.

25. Midshipmen calendars referred to exams as *rivers*, and there was a countdown on exams. Plebes were required to sing, "We are almost out of the wilderness, / Out of the wilderness, / Out of the wilderness, / We are almost out of the wilderness, / Just *x* more rivers to cross." The graduating class put on a big show called *There Are No More Rivers*.

26. Bud discusses how poorly Dave Bagley, who would become a lifelong friend and member of Bud's inner team or kitchen cabinet, is doing. "He not only failed to raise his marks, he sank lower . . . I feel terribly sorry for him—he will live to suffer more genuine regret than any other plebe who bilges. Every contact with his family, every mention of a relatives' name will call to mind the beloved profession that he hadn't the guts to learn."

27. Feb. 18, 1940, ZFC.

28. Undated, ZFC.

29. "Finns Battle On as the War Goes into 4th Month," *Chicago Daily Tribune*, Mar. 3, 1940.

30. "Empires Offer to Give Finland Vast Aid in War," *Chicago Daily Tribune*, Mar. 12, 1940.

31. "Finland War Ended by Pact," *Los Angeles Times*, Mar. 13, 1940.

32. "Weygand Stresses Near East Defense." See also, "Near East Is Nervous: War Weather Coming in the Near East," *New York Times*, Mar. 17, 1940.

33. Letter, Mar. 23, 1940, ZFC.

34. "Gamelin Commands All Allied Forces," *New York Times*, May 13, 1940.

35. "Next week is exam week and I am worried about my particularly poor daily grades in Mech Drawing." He had only a 2.7 GPA and feared being unsat (unsatisfactory). "I can't study the stuff, so all I can do is pray. Be sure to write me often, because next week is a depressing one."

36. Bud received his marks: Bull 14, Math 87, Dago 123, Skinny 42, Steam 611.

37. "Weygand, Disciple of Foch, Named to Command Allies," *Chicago Daily Tribune*, May 20, 1940.

38. Letter, May 14, 1940, ZFC.

39. "Well the Allies seem to have gotten a half way break in this latest phase of the war. Germany has always jumped first. This time she found a Holland and Belgium ready, armed to the teeth, calm and organized. The 1,900,000 men of these low countries is more men than Germany can break thru in 8–10 days at least. Thus the traditional slowness of the British is given a time-lag factor of lee-way." Letter, May 17, 1940, ZFC.

40. Ibid.

41. Written by Ray as a vignette for Bud and Mouza's fiftieth wedding anniversary, Oct. 2, 1995, ZFC.

42. Letter to Elmo Zumwalt, circa May 1940, ZFC.

43. He also attributed the two ideas or themes to his father's notebook. "You will notice that the theme of this talk is stolen from the notebook you sent me. You used the phrase 'brave, homogeneous people,' which I borrowed, and 'it can't happen here.' Which I also used—in short, the whole idea grew up around your story."

44. May 25, 1940. "Something seems to have snapped within me. I have gotten back a good share of my old drive." Grades were good—3.88 in the Math final, which placed him twenty-eighth for the past two months. He was now sixty-third in his class. . . . In Skinny he got 3.8 on the final, and his class standing was twentieth.

45. Letter, June 1, 1940, ZFC.

46. See K. Jack Bauer and Stephen S. Roberts, *Register of Ships of the U.S. Navy, 1775–1990* (Westport, CT: Greenwood Press, 1991), 145: "Formerly the Spanish unprotected cruiser Reina Mercedes, she was scuttled in the entrance to Santiago Harbor on 4 Jul 1898. Wreck captured when the port was captured, 17 Jul 1898. Initial plans to convert her to a sea-going training ship proved impractical. In 1912 she became a stationary receiving ship at the Naval Academy, and remained in that duty until 1957, when she was scrapped."

47. Letter, June 10, 1940, ZFC.

48. Letter, June 16, 1940, ZFC.

49. The four years were reduced to three—freshman, sophomore, and senior.

50. Letter, June 29, 1940, ZFC.

51. His dad did send money so that during thirty-six hours' leave in New York, Bud and his friends went to the Hotel Astor, near Times Square, rented a suite for $10, and just hung out. They skipped the World's Fair. Instead, each took a hot bath and spent the day lounging in the suite's soft chairs, listening to the radio. "It was the most luxurious afternoon of the summer." In Boston on June 20, ten of them slept in the Hotel Statler.

52. "I want to thank you all again for a perfect Sept. leave. I can't imagine having a more perfect time. The fishing trip was the highlight of the month—a chance to get away from the routine—close contact with a worthless 'od ma' and a 'little squirt' of a brother. . . . And the last week—the thrill of falling for a wonderful girl—it was all just like a story book."

53. Arrived by train on Sept. 12 and went to the campus on the 13th.

54. Letter, Sept. 30, 1940, ZFC.

55. Ibid.

56. "Dear Dad" letter from Annapolis, Oct. 5, 1940, ZFC.

57. "My Dad is still carrying on—the same dauntless oak—unbowed by the lightening [*sic*] of Fate that has struck so often in the same place—his place in the community is still a Holy Grail to inspire me in my quest."

58. Elmo wrote to "My dear Bud" a few evenings later on his office stationery in Tulare, Oct. 10, 1940, ZFC.

59. Admiral Harold Stark was being badgered by Congressman Fish on the *Forum* radio show.

60. Oct. 10, 1940, ZFC.

61. Letter, Oct. 30, 1940, "I hope you take Jim's advice and don't sell the home. . . . Tulare is booming and that house is a good investment to hang onto." ZFC.

62. There are four generations of Zumwalt family Bronze Star recipients.

63. Bud refers to FDR as "the personality expert" and laments the "damn shame that the thinking classes don't swing a majority." Oct. 30, 1940. ZFC.

64. Letter, Nov. 7, 1940.

65. Doodle to Saralee, Nov. 1940, ZFC.

66. Undated letter, circa Nov./Dec. 1940, ZFC.

67. Ibid.

68. See letter to dad, February 23, 1941, "If I hadn't been through all this once before, I'd be pretty discouraged. As it is, I console myself somewhat in the loss of past experience. It doesn't help much, but some." ZFC.

69. The *Log*, Jan. 10, 1941; also see Stillwell, *Reminiscences of Admiral Zumwalt, Jr.*, 55–56.

70. Stillwell, *Reminiscences of Admiral Zumwalt, Jr.*, 56.

71. Letter, undated reply, ZFC.

72. Letter, Oct. 7, 1941, ZFC. Indeed, Jim did not like Doris from the start, "out of loyalty to my mother I suppose." Jim first met Doris when he went to Fort Lewis to be with his dad, who had arranged for Jim to have a civilian job hauling supplies for the army.

73. Letter, undated reply, ZFC.

74. Letter, Apr. 24, 1941, ZFC.

75. Letter, June 24, 1941, ZFC.

76. Stillwell, *Reminiscences of Admiral Zumwalt, Jr.*, 56.

77. Ibid.

78. Quoted in the *Washington Post*, Oct. 10, 1972.

79. Signed "the son who can never equal his father in anything except finance." ZFC.

80. Bud gave the second watch to his father. Western Union telegram to Capt. Zumwalt at the Post Hospital at Fort Lewis: "Congratulations on the victory of your eldest in fierce finals. Now there's a watch for both of us."

81. Professor of thermodynamics, L. S. Kintberger, letter on behalf of Bud for Rhodes Scholarship, ZFC. Bud "stood No. 34 in a class of 615 members."

82. Four years later, Bud served as executive officer under Kintberger's command on the USS *Zellars*. "His orations dealt on both occasions with international relations. I was impressed by the fine balance of idealism and logic in his treatment and by his poised delivery. . . . One of the important reasons was the candidate's leadership. He had a capacity for organizing and stimulating popular interest along literary lines. The field of international relations was used as a springboard for debate, oratory, open-forum discussion and composition." ZPP.

83. Letter to Doctor Zumwalt, June 8, 1942, ZFC.

84. The *Washington Post* reported on June 19, 1942, that "New Ensigns Answer Love's Call Before Sailing War-Swept Seas." The weddings went from dawn to dusk.

85. *Tulare Advance-Register*, June 27, 1942. "My graduation day was, therefore, particularly memorable in that we not only had all the usual ceremonies surrounding that, but then the subsequent glamour of a military wedding in the chapel with classmates holding the swords and so forth." Stillwell, *Reminiscences of Admiral Zumwalt, Jr.*, 58–61.

86. Undated letter, circa June 1944, ZFC.

Chapter 4: War Years

1. Recommendation by Commander Grantham that Bud receive the Bronze Star. "As evaluator in the Combat Information Center he furnished information indispensable to the success of the attack. This involved recommending courses and speeds and time to fire and time to retreat from combat. His outstanding skill and judgment as well as his exemplary conduct under fire were an inspiration to the officers and men."

2. "We were together then until I deployed for the Aleutian campaign, and our marriage broke up at that time . . . when I deployed for the Aleutians campaign, she went to Dos Palos, California, where my sister and brother-in-law lived and remained there for a number of months, working as an employee at Eagle Field, an Air Force training airfield. She then went back to her parents in Yeadon, Pennsylvania, for the balance of the war—or until our divorce." Stillwell, *Reminiscences of Admiral Zumwalt, Jr.*, 61.

3. Bud recalled this initial trip as a pleasant one with plenty of room and wonderful food, with time for sunbathing and sea stories. Ibid., 63.

4. See James Grace, *The Naval Battle of Guadalcanal* (Annapolis, MD: U.S. Naval Institute Press, 1999). Also, Thomas J. Cutler, *The Battle of Leyte Gulf* (Annapolis, MD: U.S. Naval Institute Press, 1994), 66–67, www.usni.org/selected-writings-lieutenant-commander-thomas-j-cutler-us-navy-retired.

5. See Colin G. Jameson, *The Battle of Guadalcanal, 11–15 November 1942* (Washington, DC: Naval Historical Center, 1994), and Richard Frank, *Guadalcanal: The Definitive Account of the Landmark Battle* (New York: Random House, 1990).

6. Ibid. The Guadalcanal supply operation was assigned to Task Force TARE under Rear Admiral Richmond K. Turner. The combined task force carried six thousand men to be put ashore on Guadalcanal and was protected by twenty combat ships.

7. Denis Ashton Warner, Peggy Warner, and Sadao Senoo, *Disaster in the Pacific: New Light on the Battle of Savo Island* (Annapolis, MD: U.S. Naval Institute Press, 1992).

8. James D. Hornfischer, *Neptune's Inferno: The U.S. Navy at Guadalcanal* (New York: Bantam, 2011).

9. Mikawa's dragon formation—the way the ships swooped in with a dense fan of torpedoes ahead of them, then added quick, accurate nighttime gunfire and quick escape—was a brilliant naval tactic. It also underscored the disorganization of the pre–World War II U.S. Navy, which was not prepared for night fighting, battle damage, or emergency ship maneuvers.

10. Warner et al., *Disaster in the Pacific*, 3.

11. "Cut of My Jib," tape, ZPP.

12. The log for the *Zeilin* on that night reads, "Heavy gunfire, three ships in flames." Twenty-four enemy dive bombers got through, the *Enterprise* taking three bombs, which he personally observed. Stillwell, *Reminiscences of Admiral Zumwalt, Jr.*, 64.

13. Stillwell, *Reminiscences of Admiral Zumwalt, Jr.*, 67.

14. Richard Crowe, "The Occasion of Midshipman Bud Zumwalt Meeting Lord Louis Mountbatten," ZFC.

15. Stillwell, *Reminiscences of Admiral Zumwalt, Jr.*, 70.

16. The transfer took place on August 25. Ibid., 75.

17. Ibid., 76.

18. Ibid.

19. Sturgeon's account is in the *Washington Post*, Dec. 10, 1972.

20. Stillwell, *Reminiscences of Admiral Zumwalt, Jr.*, 76.

21. Bud's first fitness report said something to the effect that "Ensign Zumwalt may be a very good naval officer, but it has not been possible to ascertain the facts, because he has essentially been in his bunk seasick since reporting aboard." It went on to recommend that he ought to be transferred to a bigger ship. Bud attached a letter to the report saying that he was gradually becoming acclimated. Stillwell, *Reminiscences of Admiral Zumwalt, Jr.*, 76.

22. In preparation for joining Task Unit 1.1.2, conducting antisubmarine patrol and screen duties for battleships.

23. Stillwell, *Reminiscences of Admiral Zumwalt, Jr.* Jane arrived on Oct. 8, 1942. Letter from Mary Crowe in San Francisco to Portland, ZFC.

24. Leave started Nov. 2, 1942.

25. Letter from Elmo to Jim, July 1, 1945, ZFC.

26. When they returned to Long Beach, Bud persuaded Jane to return to Dos Palos, where she could resume a job. Bud would have preferred that she return to her parents in the East, but Jane refused to do that. Bud initiated an interlocutory decree of divorce, with his sister serving as a witness. Stillwell, *Reminiscences of Admiral Zumwalt, Jr.*

27. The *Phelps*'s log shows that Zumwalt was promoted to lieutenant on May 1, 1943, and that on that day initiated a next-of-kin change, so that would establish the day that the interlocutory-decree divorce became effective. Stillwell, *Reminiscences of Admiral Zumwalt, Jr.*

28. Nov. 19, 1943, ZFC.

29. Undated, ZFC.

30. July 1, 1944, ZFC.

31. Oct. 19, 1944, ZFC.

32. Stillwell, *Reminiscences of Admiral Zumwalt, Jr.*, 62–65.

33. Ibid.

34. The log of the *Phelps* on Feb. 3, 1943, reads, "Ensign Elmo R. Zumwalt, Jr., U.S. Navy, returned from the nearest naval hospital, with entry in medical record." Zumwalt told Paul Stillwell, "I had spent the night in the Hilton Hotel in Long Beach with my wife. I got up at 4:00 in the morning to return to the ship, which was scheduled to get under way early for a several-day exercise. The next thing I knew, I awakened in the Long Beach Naval Hospital, to which I had been taken in an ambulance, having been found in the alley that one goes through to get from the Hilton Hotel to what was then either a bus station or a taxi station—I can't recall which. The next thing I recall is that a message came in from the squadron commander on board the USS *Phelps*, addressed to the shore patrol saying, 'Ensign E. R. Zumwalt, Jr., absent without leave, investigate.' Then, eventually, from the shore patrol reporting that I was in the hospital. A subsequent investigation concluded that I had apparently been struck over the head by someone, probably having robbery in mind, although I was discovered by an individual and an ambulance called with my pocketbook still on me. So I would assume that whoever did it was disturbed. There was no evidence of damage other than the temporary concussion that was discovered, and I went back to the ship most fully and honorably exonerated." Since there was no robbery, the author's theory is that Bud was beaten by one of Jane's suitors.

35. James G. Zumwalt, interview with author.

36. *The Aleutian Campaign, June 1942–August 1943* (Washington, DC: Naval Historical Center, 1993).

37. Stillwell, *Reminiscences of Admiral Zumwalt, Jr.*

38. Ibid.

39. *The Aleutian Campaign*, 88.

40. Stillwell, *Reminiscences of Admiral Zumwalt, Jr.*
41. Ibid.
42. Vignette, James G. Zumwalt.
43. "Col. E. R. Zumwalt Eye-Witness to Nazis' Atrocities: Takes Pictures of Scenes," *Tulare Union*, undated, circa Apr.–May 1945.
44. Letter to Bud, Apr. 26, 1945, ZFC.
45. See "History of the USS *Robinson* (DD 562)," Division of Naval History, Ships' History Section, Navy Department.
46. After shakedown training in San Diego, the *Robinson* was under way for Pearl Harbor on April 16.
47. Cutler, *The Battle of Leyte Gulf*, preface.
48. James C. Heinecke, "The Exploits of the U.S.S. *Robinson* (DD 562), 1944–1945," kept in violation of ship rules. Sonarman Third Class Heinecke died in 1994. ZPP.
49. The bombing was extremely effective. "The Japs have been pushed from the town and are crowded on the tip of Saipan. . . . Rather than be taken prisoner, they stripped themselves and started swimming for this reef . . . the beach is dotted with wounded, which we can't get at. What a hell of a way to die. One of the men said that on the island the smell of dead is unbearable," wrote Heinecke. ZPP.
50. The *Grant* sighted about twenty Japanese swimming in the water and used depth charges to take care of them. "Depth charge, very effective," recalled Bud. "I don't know how well it's known that we did some of that in the war. We also did it in Surigao Straits. Not a very pretty performance." Stillwell, *Reminiscences of Admiral Zumwalt, Jr.*, 151.
51. "History of the USS *Robinson* (DD 562)," 2.
52. "Our force consisted of five battleships, the *California, West Virginia, Maryland, Tennessee*, and the *Mississippi*, seven cruisers, one was Australian, and 12 destroyers including the *Robinson's* 112th division." Paul Stillwell, *The Reminiscences of Admiral Zumwalt, Jr.*
53. "History of the USS *Robinson* (DD 562)," 2.
54. This account draws from Jason Hammer, "The Night the Island Blinked." I also draw from a story in the files written by H. H. Buck Bedford, Ship Laundry, 3rd class, who was also on board that evening. In a letter to the Zumwalt family on Jan. 18, 2000, Hammer wanted the family to know "how this remarkable man touched my life." He concluded, "Thank you, Bud. I am certain a loving God has greeted you with a well-deserved: 'Well done, Admiral.'" ZFC.
55. "History of the USS *Robinson* (DD 562)," 3.
56. Jan. 18, 2000, ZFC.
57. At ten thousand yards, the Japanese detected the *Robinson*, and shelling began. The Japanese battleships opened up on the *Robinson*. "I felt the butterflies in my belly when I saw three salvos, three shells in each, come sailing toward us, and you know there isn't a thing you can do except watch with your fingers crossed. They sailed over us, hitting one of our division astern on our port quarter, the USS *Grant*."
58. See letter to William Stanhope, Jan. 10, 1990. Of all the battle situations he had been in, the one that stood out most vividly was the Battle of Surigao Strait. "I was serving as battle evaluator on the USS ROBINSON (DD562), leading a section of three destroyers down the east side of the Straits to attack the Japanese force with torpedoes," recalled Zumwalt in a letter. "The Japanese battleships opened up on us. Their first salvo was short. The second salvo was over. We were all holding our breath waiting for the third salvo which would have done it when suddenly our battleships opened fire and the Japanese trained their guns to return that fire. There was an immense feeling of relief followed by the guilty feeling that someone else was being shot at." Stillwell, *Reminiscences of Admiral Zumwalt, Jr.*
59. Letter to Paul Rifkin, Nov. 21, 1980, ZPP.

60. According to Hammer, "No waiting around to see the results, if any. We turned sharply and, at flank speed, rushed for the relative safety only distance could provide. . . . Enemy star shells hung in the sky directly overhead pointing a bright accusing finger of light at our naked vulnerability as the crew heard the ominous sounds of shells chasing our wake and often landing dangerously close to our thin-skinned hull. As the seconds ticked by, our lease on life seemed to lengthen in proportion to the distance we rapidly put between us and our desperate foe in the throes of its own funeral pyre."

61. Ibid. See Samuel Eliot Morison, *Leyte* (Edison, NJ: Castle Books, 2001).

62. ZTT, undated.

63. Ibid.

64. Two future CNOs were serving on destroyers during the engagement, Bud aboard the *Robinson* and James L. Holloway III, a gunnery officer on the USS *Bennion*. Holloway would succeed Bud as CNO in 1974.

65. Admiral Grantham, as related by Jerry Wages.

66. Letter postmarked Sept. 3, 1945, ZFC.

67. Letter, June 1, 1945, ZFC.

68. Letter, July 19, 1945, ZFC.

69. Letter postmarked June 5, 1945, ZFC.

70. Letter postmarked Sept. 3, 1945, ZFC.

71. Ibid.

72. Sept. 27, 1945, ZFC.

73. "Jap Gunboat Crew Sifted at Shanghai," *Washington Post*, Sept. 15, 1945, p. 2.

74. "190 Suspects Are Found Aboard Japanese Vessel," *New York Times*, Sept. 15, 1945, 5.

75. Bud's comments in fiftieth wedding anniversary video, ZFC.

76. Zumwalt, *On Watch*, 3.

77. Sept. 15, 1945, report from Lt. Zumwalt to Commander Task Group 73.2, "Narrative of Prize Crew of H.I.J.M.S. ATAKA Covering Trip to SHANGHAI, CHINA, and Return." ZPP.

78. Milton E. Miles's posthumous memoir was *A Different Kind of War: The Little-Known Story of the Combined Guerrilla Forces Created in China by the U.S. Navy and the Chinese During World War II*, ed. Hawthorne Daniel from the original manuscript (Garden City, NY: Doubleday, 1967).

79. Bud told Paul Stillwell, "Rear Admiral Miles's orders were to keep this Task Unit under heavy guard of fifty men supplied by the land based Naval forces, to continue the policy of denying the captured crew all contact with their Headquarters."

80. Letter to Mel Knickerbocker, Aug. 3, 1970, ZFC. The sword is now with Bud's son Jim Zumwalt.

81. Dr. Mel White, "The Admiral's Wife: A True Story of Mouza Zumwalt's Fifty Year Marriage to Admiral Elmo Zumwalt, Jr.," unpublished book proposal manuscript. I am grateful to Ann Zumwalt Coppola for sharing tape transcripts of her mother's recollections.

82. Ibid. Mouza's aunt had seen her husband and brother killed by the Russians in the revolution. She had remarried a man who had been in the Zabaikalsky Cossacks, whose first wife had been killed and their baby taken by the Soviets. He escaped to Shanghai and then married the aunt.

83. Interview notes of Marcia Risaino, Mar. 8, 1996, Next Book file, ZFC.

84. Undated letter provided to the author by Richard Crowe and Zumwalt family.

85. I am grateful for having the opportunity to read a draft of Richard Crowe's Zumwalt Family History (in preparation) with comments by Emile and Stephanie Ninaud.

86. *Washington Post*, Dec. 1, 1972. Decades later Bud quipped, "Wrong gesture, but the right effect." Dinner was a true feast: lots of vodka and wine, multiple courses "so delicious that the mind refused to acknowledge the stomach's defeat."

87. Bud's letter is reprinted in *On Watch*, Nov. 10, 1945.

88. "The interest in studying Russian began in the last six or eight months of the war, when it was clear that was coming to an end, and I could see that our next major adversary would be the Soviets. My own view was always that we were going to have very great trouble after the war. I felt that I had done a lot more reading than the average naval officer with whom I associated." Bud later requested assignment as assistant naval attaché to Moscow. This request was disapproved. Stillwell, *Reminiscences of Admiral Zumwalt, Jr.*

89. Tape transcript provided by Ann Zumwalt Coppola.

90. Renaissance Weekend, Mouza speaking about "the toughest decision of my life." Author interview with Phillip Lader.

91. Oct. 10, 1985, letter, Mel Knickerbocker and Jim Good's account, "As We Remember," sent to Bud following *Robinson* fortieth reunion. ZFC and ZPP.

92. Ibid.

93. From the handwritten diary of Augie Carrillo, Oct. 22, 1945, Shanghai, ZPP and ZFC.

94. Tape transcript of Mouza's recollections, provided to author by Ann Zumwalt Coppola.

95. Fitness report, Oct. 24, 1945, ZPP.

CHAPTER 5: CROSSROADS

1. Admiral Robert Hanks, fiftieth wedding anniversary reminiscence, ZFC.

2. Paul Stillwell, *Reminiscences by Staff Officers*, 159.

3. Letter to Mouza, Feb. 13, 1990, reflecting on forty-four years of marriage, ZFC.

4. Letter, Nov. 6, 1945, 6 a.m., ZFC.

5. Letter, Nov. 6, 1945, 2 p.m., ZFC.

6. Letter, undated, ZFC.

7. Letter, Jan. 26, 1945, ZFC.

8. Letter, Jan. 9, 1946, ZFC.

9. Letter, Feb. 3, 1946, ZFC.

10. Stillwell, *Reminiscences of Admiral Zumwalt, Jr.*, 167–69.

11. Letter from Everett F. Drumright to Lietenant Zumwalt, Feb. 19, 1956, ZFC and ZPP. See also "USS *General H. L. Scott* (AP-136)," www.navsource.org/archives/09/22/22136 .htm.

12. Mouza taped interview transcript provided by Ann Zumwalt Coppola.

13. Letter, Feb. 8, 1946, on USS *Saufley*, ZFC.

14. Mar. 11, 1946, ZFC.

15. Mar. 16, 1945, Tulare, CA, in care of Irene Fluckinger, ZFC.

16. Mar. 15, 1946, ZFC.

17. She had a particularly hard time with money and making change, because the difference between nickels and dimes made no sense to her—that the larger coin was worth less than the smaller one. Irene devised a plan for taping the coins together and writing their value on a piece of paper.

18. "I would say that she felt quite dependent on me . . . there was great apprehension on her part about being alone at those times, and she still didn't understand well our check system and so forth. I remember writing out all the checks that she would have to sign for utilities and that sort of thing, and showing her how to put in the date and amount." Stillwell, *Reminiscences of Admiral Zumwalt, Jr.*, 170.

19. Letter to Pricilla Barry with recollections for fiftieth anniversary, Sept. 25, 1995, ZFC.

20. Letter from the *Zellars*, July 14, 1946, ZFC.

21. Letter, July 15, 1946, ZFC.

22. Letter, "My Dear Bud," July 18, 1946, ZFC.

23. Letter from the *Zellars*, July 21, 1946, ZFC.

24. Letter recounting the conversation, July 22, 1946, ZFC.

25. Aug. 5, 1946, ZFC.

26. Ibid.

27. Stillwell, *Reminiscences of Admiral Zumwalt, Jr.*, 239–40.

28. "Mouza and I leave for Dos Palos tonight," telegram from Jim to Bud, Aug. 29, 1946, ZFC.

29. James G. Zumwalt, interview with author and vignette.

30. Below signature: "dictated to my dad on August 10, 1987," ZFC.

31. Stillwell, *Reminiscences of Admiral Zumwalt, Jr.*, 180.

32. Jan. 10, 1947, ZFC.

33. Aug. 9, 1947, ZFC.

34. Stillwell, *Reminiscences of Admiral Zumwalt, Jr.*, 178.

35. Letter, J. M. Reid, Jr., Mar. 18, 1947. "It sounds like fatherly advice," wrote Reid, who added, "was so sorry to hear about Mussa's [sic] father—after waiting all these years, to have such an ending." Reid described the tough economic situation in the States, with more and more people looking for work, prices rising, and the economy "not in a healthy condition." ZFC.

36. Stillwell, *Reminiscences of Admiral Zumwalt, Jr.*, 172; there is also an extensive discussion on Zumwalt tape 1, ZTT.

37. From the earliest draft of "The Cut of My Jib," ZTT.

38. R. R. Conner to Lieutenant E. R Zumwalt, USS *Zellars*, Aug. 22, 1946. "Rhodes Scholarship—Permission to Compete." ZPP.

39. Nominated by vice admiral of the navy and superintendent of the academy Aubrey W. Fitch "to represent the U.S. Naval Academy as an applicant for a Rhodes Scholarship from the state of Maryland."

40. He wanted to acquire knowledge "to aid me in dealing with the problems of the present and the future . . . I will be able to study the people of the one major nation I have not yet visited." Application copy. ZPP.

41. Sept. 9, 1947, ZFC.

42. Oct. 27, 1947, ZFC.

43. Dec. 1, 1947, ZFC.

44. Dec. 11, 1947, ZFC.

45. Letter to Mouza, Dec. 31, 1947. The *Zellars* was at Boston Naval Shipyard; Bud told Mouza he was staying aboard to save money. ZFC.

46. Stillwell, *Reminiscences of Admiral Zumwalt, Jr.*, 183.

47. "A Remembrance of Elmo Russell Zumwalt III, July 30, 1946–August 13, 1988." Eulogy, ZFC.

48. Stillwell, *Reminiscences of Admiral Zumwalt, Jr.*, 183.

49. Bill Edsel, "The 'It' Factor: Why Marshall and His Wife Chose Pinehurst," http://archives.thepilot.com/May2004/05-19-04/051904MyTurn.html.

50. "Why a Military Career," Aunt Tina tape transcript 2, p. 2, A-1, ZTT.

51. Ibid.

52. "In 1946–47-48, I applied to both Law and Medical School and was accepted. Each year when the time came to make the decision to leave the service, I found myself deferring out of concern for the trend of events in the world around me." Tape transcript, "The Cut of My Jib," ZTT.

53. ZTT with reference to the hearings.

54. "Why a Military Career," ZTT.

55. The ship's history reads, "On November 21, 1950 she was placed in full commission at Charleston, South Carolina, with Lieutenant Commander Zumwalt commanding officer."

56. Evaluation, United States Pacific Fleet, Commander Seventh Fleet, ZFC.

57. Stillwell, *Reminiscences of Admiral Zumwalt, Jr.*, 229.

58. Letter, "Dear Scattered Clan," Apr. 27, 1953.

59. "What Is Patent Ductus Arteriosus?" National Heart Lung and Blood Institute, www .nhlbi.nih.gov/health/health-topics/topics/pda.

60. Stillwell, *Reminiscences of Admiral Zumwalt, Jr.*, 229.

61. Ibid., 245.

62. Ibid., 246.

63. Ibid., 247.

64. Ibid.

65. Ibid., 247–48.

66. Ibid., 248.

67. Holloway letter, Executive Correspondence file, Plans and Policy Division of the Bureau, Zumwalt Papers, NHHC.

68. Ibid.

69. Stillwell, *Reminiscences of Admiral Zumwalt, Jr.*, 253.

70. Letter, Oct. 2, 1995; Rear Admiral Robert Hanks, Oct. 2, 1995, ZFC.

71. Hanks came to see Zumwalt as a superlative commanding officer, from whom he learned seamanship, leadership, and compassion for those under him. He also felt that Mouza was for the ship's wives what Bud was for the men aboard ship—always there for them. Fiftieth wedding anniversary reminiscence, ZFC.

72. Ibid.

73. Stillwell, *Reminiscences of Admiral Zumwalt, Jr.*, 275–78.

74. See Zumwalt, *On Watch*, 186–89. An E, for excellence, is generally awarded to a ship or component of a ship as a result of top performance in competition with other ships during a given time period.

75. Letter to Admiral Bruton, Jan. 30, 1956: "We had, as I recall, four of the six reserve officers on the *Isbell* extend or go career while I was on board, and that took a lot of doing, both on my part and on my wife's part, who worked with their wives to help them understand some of the great fun that you can have in the Navy and the togetherness that goes with being part of a Navy team. And that was at a time when not very many ships were having that kind of success." Stillwell, *Reminiscences of Admiral Zumwalt, Jr.*, 278.

76. Stillwell, *Reminiscences of Admiral Zumwalt, Jr.*, 278–79.

77. Draft, "The Cut of My Jib," ZTT.

78. Personal memo from Vice Admiral Holloway to Richard Jackson, assistant secretary of the navy, at the Pentagon, Oct. 2, 1957, ZPP.

79. Ibid.

80. Stillwell, *Reminiscences of Admiral Zumwalt, Jr.*

81. First draft "The Cut of My Jib," ZTT.

82. Zumwalt, *On Watch*, 86.

83. From 1955 to 1957, as CO of DD, he had qualified three different chief engineers and felt that he had been his own chief engineer for those two years.

84. Norman Polmar and Thomas B. Allen, *Rickover: Controversy and Genius* (New York: Simon & Schuster, 1982), 215. I draw extensively from this pathbreaking biography.

85. Bud later overtook Peet because Rickover required a five-year commitment from Peet.

86. Bud and Jim Calvert were first in class to make Rear Admiral in 1965, a year ahead of Peet.

87. Letter, Charles Dewey to Bud, Nov. 19, 1970. "How prophetic can one be?" "Keep the Z-grams coming!!" ZFC.

88. Stillwell, *Reminiscences of Admiral Zumwalt, Jr.*, 308–9, 316.

89. E-mail to author from Joe Roedel, Oct. 23, 2011. "He loved his men, his ship, his

duties. . . . Many times Capt. Zumwalt would go up to the flying bridge and get a couple of seamen, including me, and we would toss the medicine ball back and forth during our lunch break. He was never afraid to mix with the ranks of the seamen and always seemed to enjoy it. He had a great re-pore [*sic*] with the men and they in turn had the same for him."

90. Letter, Dec. 9, 1959, from S. M. Alexander, supervisor of shipbuilding, regarding Bud's performance at Bath and *Dewey*, ZPP.
91. Letter, "Dear Bud," W. L. Read, lieutenant commander, Flag Secretary Destroyer Force, Atlantic Fleet, ZFC.
92. Letter to Saralee, Sept. 8, 1960, ZFC.

CHAPTER 6: PLATO AND SOCRATES

1. "Paul Nitze," ZTT.
2. In the class were Naval Academy classmates James Holloway, Jim Calvert, and Richard Armitage.
3. Paul Stillwell, *Reminiscences by Staff Officers*, 338–40.
4. In a January 23, 1962, letter to Charles Bohlen, then special assistant to the secretary of state, Bud explained his research and sought authorization to use materials from the interview. The interview occurred on January 2, 1962. ZPP.
5. "Paul H. Nitze," Academy of Achievement, www.achievement.org/autodoc/page /nitobio-1.
6. Paul H. Nitze, *From Hiroshima to Glasnost: At the Center of Decision—a Memoir* (New York: Weidenfeld, 1989), 181.
7. Ibid., 182; "Paul Nitze," ZTT.
8. Paraphrased from Nitze evaluation of Captain Zumwalt, ZPP, and Paul Nitze, remarks at Bud and Mouza's fiftieth wedding anniversary celebration.
9. Ibid., fiftieth wedding anniversary celebration.
10. People at the War College "thought that it was dullsville, that it was committee solutions, constant wrangling for this phrase and that comma in the inter-service rivalries." Stillwell, *Reminiscences of Admiral Zumwalt, Jr.*, 350.
11. "And, you know, who's to say that they weren't right, had it not been for the accident of Paul Nitze becoming Secretary of the Navy? I think that by the time I came up for rear admiral, that the thinking had changed enough, under McNamara's imprint, that it was recognized that the policy really was set in OSD, not on the Joint Staff. I believe that even had Paul Nitze not become Secretary of the Navy, that a selection board would have placed the ISA duty above the previously selected for that rank." Ibid.
12. Ibid.; "Paul Nitze," ZTT.
13. Nitze evaluation of Captain Zumwalt, ZFC.
14. "Paul Nitze then decided to move me over and put me in as director of arms control, which took me from an o-6 slot to an o-8 slot." The pay grade for a navy captain is o-6; o-8 is for a rear admiral. Stillwell, *Reminiscences of Admiral Zumwalt, Jr.*, 353.
15. "Paul Nitze," ZTT.
16. Nitze regularly consulted on general foreign policy problems, seeing Bud as an expert on the Soviet Union. Nitze then took Bud with him to be his executive officer when Nitze became deputy secretary of defense and later when he became secretary of the navy.
17. "Paul Nitze," ZTT.
18. This became most apparent when serving as Nitze's Cuba contingency planner.
19. Memorandum of conversation with Daniel Ellsberg, May 13, 1974, ZPP.
20. Nitze, *From Hiroshima to Glasnost*, 256.

21. "Paul Nitze," ZTT.
22. Stillwell, *Reminiscences of Admiral Zumwalt, Jr.*, 354.
23. Nitze, *From Hiroshima to Glasnost*, 218.
24. "Paul Nitze," ZTT.
25. "Bud told me later that he memorized an acrostic so that he could recreate the instructions in their correct sequence since he was without pencil and paper." Nitze, *From Hiroshima to Glasnost*, 228.
26. Nitze offers a similar account. "Bud, I think you got it. I want you to know that McNamara told me to fire you if you missed a single instruction." Ibid.
27. See Deborah Shapley, *Promise and Power: The Life and Times of Robert McNamara* (Boston: Little, Brown, 1993), 241. "He stood at attention while the secretary barked at him and indirectly bawled out the entire navy. Zumwalt refused to bend over to pick up the pad and pencil to make notes, to avoid debasing the Navy."
28. Stillwell, *Reminiscences of Admiral Zumwalt, Jr.*, 371. Bud always thought this minor change was crucial to the outcome.
29. Nitze, *From Hiroshima to Glasnost*, 228.
30. Ibid., 230.
31. Ibid., 230–31.
32. "Paul Nitze and Cuba," ZTT; Stillwell, *Reminiscences of Admiral Zumwalt, Jr.*, 371–73.
33. Letter, Jan. 16, 1997, ZFC.
34. Cyrus Vance, letter of commendation, Nov. 7, 1963.
35. "Robert McNamara," ZTT.
36. See Andy Kerr, *A Journey Amongst the Good and the Great* (Annapolis, MD: U.S. Naval Institute Press, 1987). On August 1, 1963, Admiral David L. McDonald succeeded Admiral Anderson as chief of naval operations. Anderson became U.S. ambassador to Portugal.
37. January 15, 1987, materials provided by Bud for Nitze's memoirs; prepared by Robbie Robertson. ZPP.
38. Ibid.
39. The special assistant for legal counsel was Commander Horace B. Robertson. Robbie was the judge advocate general (JAG) officer. The administrative aide was Commander Dick Nicholson, a surface-warfare officer.
40. Nicholas Thompson, *The Hawk and the Dove: Paul Nitze, George Kennan, and the History of the Cold War* (New York: Henry Holt, 2009), 195.
41. Robert Kaufman, *Henry M. Jackson: A Life in Politics* (Seattle: University of Washington Press, 2000), 57.
42. Ibid.
43. Zumwalt draft chapter on Paul Nitze, ZPP.
44. Nitze, *From Hiroshima to Glasnost*.
45. These details are in the Zumwalt draft chapter as well as Stillwell, *Reminiscences of Admiral Zumwalt, Jr.*, 375–80.
46. "I remember being in Paul Nitze's office on an arms control issue, however, when the phone rang. Margaret, his secretary, said something to him. I saw Paul's face go white, which I'd never seen, and he kept saying, 'No, no. Oh, my God, no, no, no.' It went on for about three minutes, and he put down the phone and said, 'President Kennedy's just been shot' and dismissed everybody. We turned on the radio, and we just sat there totally almost unable to function for the next several hours while we listened to the news reports." Stillwell, *Reminiscences of Admiral Zumwalt, Jr.*, 379.
47. "There was absolutely no doubt about where Paul Nitze stood on defense, and the Committee gave him its endorsement but it was an education to me to see the way in which dissent is registered by interested groups and how carefully their concerns are reflected in the Congressional questioning." ZTT, "Paul Nitze."

48. Tape 33, side A, ZTT.

49. Tape 33, ZTT.

50. Kerr, *A Journey Amongst the Good and the Great*, 110. Paul R. Ignatius, *On Board: My Life in the Navy, Government, and Business* (Annapolis, MD: U.S. Naval Institute Press, 2006), 163–64.

51. "Rickover," ZTT.

52. Ibid.

53. In *On Watch*, Bud noted that Rickover was the consummate infighter and strategist, which Bud had learned from time with Nitze as aide to the secretary of the navy. "Holland wanted some access to US Navy nuclear reactor technology, which Nitze thought should be rejected. Nitze asked for Rickover's opinion since he was the admiral in charge of nuclear propulsion. The reply came back not on Navy stationary, but on AEC stationary, allowing him to circulate copies to members of congress and avoid the navy chain of command."

54. Rickover (Box 37) interview, Jan. 27, 1977, by Rudy Abramsen of the *Los Angeles Times*, never published, Paul Nitze Papers, Library of Congress.

55. Ibid.

56. Bud insisted that he was not disappointed with never having a major command: "No, I really don't. I am pleased with the fact that I got the jobs that I got as a result of having been selected for rear admiral early. They were all fascinating jobs. I don't think in reflecting back that I, having ridden cruisers as my flagship when I was flotilla commander, that the skipper of a cruiser gets anything like the fun and zest that I had had in three previous commands. Nor, because he's so layered in, does he have the broadening that you get in the destroyer commands. I really believe that it's not harmful to skip that command." Stillwell, *Reminiscences of Admiral Zumwalt, Jr.*, 434.

57. A remarkable set of future leaders was in the E ring at this time. Dave Bagley was executive assistant to the undersecretary; Ike Kidd, executive assistant to the CNO; Jerry Miller, executive assistant to the VCNO; George Brown, military assistant to the secretary of defense; and Alexander Haig, military assistant to Joseph Califano.

58. I am indebted to Admiral Bill Thompson for providing me with a copy of chapters 18, 21, and 22 of the manuscript of his personal memoirs, later published as *Gumption: My Life—My Words*. It was copyrighted 2010 (self-published), and for allowing me to make extensive use of his recollections. "Once he determined that an associate or a subordinate had the experience, background and integrity to handle a task, he brought him/her into his confidence and shared the task and would work with them until fruition and then give him/her all the credit."

59. Ibid.

60. Ibid.; author interview with Bill Thompson. Thompson recalled that the first time he ever heard the term was "when listening to Bud addressing a convocation of the Naval District Commandants in Washington, I learned the term and after struggling with a dictionary, learned how to spell it."

61. Bill Thompson memoirs.

62. Bud was aware that he was on the selection list but was wary of counting it a done deal. Nitze had sworn him to silence.

63. "Paul Nitze," ZTT.

64. Ibid.

65. Ibid.

66. "It was almost five years to the day that Bud Zumwalt called me from Washington—I was at the Harvard Business School—to relate that he was nominated by the President to be the Navy's Chief of Naval Operations." Bill Thompson memoirs and interview with author.

67. Ibid.

CHAPTER 7: PATH TO VIETNAM

1. Plaque presented to Bud by the F-111B study team.
2. Stillwell, *Reminiscences by Staff Officers,* 432.
3. Ibid.
4. "Zumwalt Assumes Flotilla Command," *Tulare Advance-Register,* July 26, 1965.
5. Stillwell, *Reminiscences of Admiral Zumwalt, Jr.,* 440–45.
6. "We had a very busy social schedule, as a result. Mouza and I, I think, were out every night—at least to a cocktail party and sometimes to a dinner party. Our oldest son was then away at college, but the other three we were able to see quite a bit of, more so than in Washington duty." Ibid.
7. Ibid., 440.
8. Ibid., 445.
9. Ibid., 449.
10. Ibid., 450. See also David Vine, *Island of Shame: The Secret History of the U.S. Military Base on Diego Garcia* (Princeton, NJ: Princeton University Press, 2009).
11. Ignatius, *On Board: My Life in the Navy, Government, and Business,* 161; Bill Thompson memoir.
12. Charles J. DiBona stood second of the 681 graduates in the Naval Academy class of 1956 and subsequently studied as a Rhodes Scholar. In 1967, as a lieutenant commander, he resigned from active duty and became chief executive officer of the Center for Naval Analyses.
13. Author interview with Charles DiBona.
14. Stillwell, *Reminiscences of Admiral Zumwalt, Jr.,* 451.
15. Ibid., 520.
16. Ibid., 453.
17. A few months into his tour in OP-96, Bud made the decision to replace the Franklin Institute and hire the University of Rochester to manage the Office of Naval Analyses. As part of the process of getting Rochester to agree to manage the office, a powerful and vigorous president of the center was needed. Charlie DiBona was the only logical candidate.
18. The navy finally won, and the F-111 became an air force fighter-bomber that had little success, while the navy's substitute version, the F-14, a successful interceptor with its Phoenix missile system, remained in the inventory into the next century.
19. Stillwell, *Reminiscences of Admiral Zumwalt, Jr.,* 451.
20. Ibid.
21. *The Reminiscences of Vice Admiral Gerald E. Miller, U.S. Navy (Retired),* vol. 2, USNI Oral History Program (Annapolis, MD: U.S. Naval Institute, 1984), 594.
22. Stillwell, *Reminiscences of Admiral Zumwalt, Jr.,* 454.
23. Paul Stillwell asked Bud, "Did you study gas turbine propulsion?" Zumwalt answered, "Yes, and I became a firm devotee of it and insisted upon it when I was CNO. It was one of the significant sources of friction between me and Admiral Rickover. And I think that that's one area in which we've been proven to have been absolutely right, that they had proven to be magnificent." Ibid., 460.
24. About Rickover, Zumwalt wrote: "He's the only military man I know who can destroy a professional career willy nilly. He has more power over money than probably anybody in the executive branch—maybe even including the President—and he's got administrative control of what to do about the money really nailed down. Power to control the whole management method that goes probably beyond anything other than the Manhattan project itself." Zumwalt, *On Watch,* 456.
25. Ibid., 406. Bud delighted in telling the story of the running feud and vendetta between Rickover and the chief of naval personnel, Vice Admiral William R. Smedberg, who was "a great fighter for the interests of people and who found himself constantly in

battle with Admiral Rickover who put people a distinct second." CNO Fred Korth brought them both to his office, read them the riot act in four-letter words and told them he never wanted to see them in his office again on matters like this. As they left, Smedberg said, 'I've never been talked to by anybody in my life like that.' Rickover said, 'You deserved it.'"

26. Briefing 2, tape 30B, ZTT.

27. Polmar and Allen, *Rickover: Controversy and Genius*, 394–95.

28. Ibid., 406.

29. Stillwell, *Reminiscences of Admiral Zumwalt, Jr.*, 462–63.

30. All nineteen guerrillas were killed, as were four MPs, a marine guard, and a South Vietnamese embassy employee. See Larry Berman, *Lyndon Johnson's War: The Road to Stalemate in Vietnam* (New York: Norton, 1989).

31. Zumwalt's data indicated that "we could have been even more decisive in killing the F-111." ZTT 28, side A.

32. This account comes from Elmo Zumwalt, Jr., and Elmo Zumwalt III, *My Father, My Son* (New York: Dell, 1987).

33. Personal letter, Jan. 16, 1997, on the occasion of Nitze's ninetieth birthday. "Your graciousness in letting me listen to and participate in policy discussions over the years." ZFC.

34. Nick Thompson provided me with the following description of the farm: "A half mile paved road runs from the house past some corn fields and tobacco barns down to a tennis court and a swimming pool. There was also a trampoline down there when I was a child. There are horses stabled about two miles away that Z and N probably rode through the woods. Other things I remember: metal chairs out back, a beautiful full-wall mural in the dining room drawn by Nitze's old friend Charlie Child with foxes, snakes, and birds drawn into a beautiful nature scene on it, a piano that N used to play, a gravel driveway with magnolia trees in the middle, sheep just across the driveway from the front barn, pigs in another field, a large silo and a cow barn near to the house. It smelled and looked like a farm, and there were real farmers that worked it. But I don't think my grandfather was out there all that often milking the cows or sowing the corn crop."

35. "I developed a very high regard for Paul Ignatius. I did not have the close personal relationship with him that I had with Paul Nitze, who had, by that time, become a dear friend. But Paul Ignatius had obviously been left with a very good taste in his mouth about me by Paul Nitze, and by Worth Bagley, who stayed on with him. He was always extremely supportive. He had had a very good analytical background himself. He was Under Secretary of the Army and then an Assistant Secretary of Defense (I&L) and then he became Secretary of the Navy. So he'd had an opportunity to review a lot of studies, and he had a good analytical mind. He was very thoughtful; he didn't let himself get pushed around by the brass, but neither was he at all contemptuous of the military mind. Most of the time good work would persuade him, and sometimes good work would not, and bad work never got by him. I guess, the best epitaph I can put on that is that when it came time for me to go to Vietnam, he called me in and gave me his personal appreciation for the work I'd done and showed me a letter that he'd written to General Abrams, commending me to Abrams's attention. He'd worked with Abe closely when he was Under Secretary of the Army and had a deep reverence for him. He just kind of opened the door for me to General Abrams in a very helpful way." Stillwell, *Reminiscences of Admiral Zumwalt, Jr.*

36. Zumwalt and Zumwalt, *My Father, My Son*, 41.

37. Market Time, conducted by the navy's Task Force 115, consisted of offshore surveillance forces aimed at cutting off the waterborne flow of supplies to enemy forces in South Vietnam.

38. Bud had been tipped off by his detailer that he was being promoted to vice admiral and commander of naval forces, Vietnam. Howard Kerr noted that "those stars would also produce some strains in terms of bitterness on the part of others. He had jumped over an awful lot of people." Stillwell, *Reminiscences of Staff Officers*, and author personal interview with Kerr.

39. I am indebted to Leslie Julian Cullen, whose dissertation was completed under the supervision of Jim Reckner, "Brown Water Admiral: Elmo R. Zumwalt, Jr., and United States Naval Forces, Vietnam, 1968–1970," Texas Tech University, May 1998. I make extensive use of Cullen's interview notes, which are available in the Cullen Collection at the Vietnam Archive.

40. Never missing a beat, Jim closed with, "You know my personal feelings about the war in Viet Nam. You also know how I feel about the great social forces which engulf the suffering human race on this planet. We have often debated these issues, sometimes with great feeling. In spite of all the political differences, I have never stopped loving you as a brother, nor have I questioned your integrity or sincerity. If ever a military figure had compassion, you have it. I cannot think of another officer whom I would be more confident in than you. God protect you my dear brother and may He bestow on Mouza and your children the strength and forbearance and the patience which your absence will entail."

Chapter 8: Brown Water Navy

1. January 10, 2000, letter to Mouza Zumwalt and family following Bud's death. ZFC.
2. The first "Dear Clan" letter is dated October 23, 1968. Over Labor Day weekend, Bud went to the Naval Ordnance Test Station (NOTS), China Lake, California, for a briefing, but this merely provided cover to be given a party hosted by his proud father, Elmo, and Doris. Three days of parties in Tulare were intermixed with a trip to Carmel to visit with Saralee and her husband, Richard Crowe, and their son, Lieutenant Junior Grade Richard Crowe.
3. With longtime friend Harry Rowen, then president of the Rand Corporation, and with Rear Admiral William Heald Groverman, a commander in World War II and the Korean War, one of the leading Cold War experts on antisubmarine warfare and Soviet naval forces.
4. "I spent two years at the Fletcher School, and it was upon the completion of this tour that I first served with Admiral Zumwalt as his flag secretary and aide when he was assigned as Commander U.S. Naval Forces in Vietnam in late 1968." Stillwell, *Reminiscences by Staff Officers*.
5. Ibid.; and author interview with Howard Kerr.
6. Ibid.
7. *Frocking* refers to the practice of donning the uniform insignia and assuming the title of a higher rank before actually being promoted to it. Promotion would come when he took command in Vietnam.
8. Interview with Ann Zumwalt Coppola.
9. "He shared with us quite frequently at dinner time his thoughts, feelings, and worries." Ann Zumwalt Coppola e-mail to author.
10. October 23, 1968, "Dear Clan" letter. ZPP.
11. Jim was at the University of North Carolina and Elmo on his way to the USS *Ricketts*.
12. October 23, 1968, "Dear Clan" letter.
13. Stillwell, *Reminiscences of Admiral Zumwalt, Jr.*
14. Stillwell, *Reminiscences of Staff Officers*.
15. "My own personal belief at the time, based on what I knew from having sat in on, and done the studies about, the attrition of logistics coming down from North Vietnam,

and from my long bull sessions with Paul Nitze on his farm over the weekend, was that he was getting very disenchanted about our prospects in Vietnam. And of even more importance, he thought that the support around the country was running out. It was clear to me that support was running out." Stillwell, *Reminiscences of Admiral Zumwalt, Jr.*, 490.

16. Stillwell, *Reminiscences of Staff Officers*, and personal interview.
17. Ibid. This chapter also makes extensive use of William C. McQuilkin, "Operation SEALORDS: A Front in a Frontless War, an Analysis of the Brown-Water Navy in Vietnam," a thesis presented to the faculty of the U.S. Army Command and General Staff College, Fort Leavenworth, Kansas, 1997, as well as McQuilkin's interview notes and correspondence with Zumwalt. These are in the Cullen Collection at the Vietnam Archive. I also draw on Thomas J. Cutler's seminal contribution, *Brown Water, Black Berets* (Annapolis, MD: Naval Institute Press, 1988).
18. Stillwell, *Reminiscences of Staff Officers*.
19. Author interview with Kerr.
20. Ignatius, *On Board: My Life in the Navy, Government, and Business*, 160.
21. Paul Stillwell, unpublished interview with Vice Admiral Earl F. Rectanus, Nov. 19, 1982, ZPP.
22. Ibid.
23. Stillwell, *Reminiscences of Staff Officers*.
24. Leslie Cullen, Zumwalt interview notes, Leslie Cullen Collection at the Vietnam Archive.
25. Cutler, *Brown Water, Black Berets*, 249.
26. Edward J. Marolda, "Stories about Vance," www.ussvance.com/Vance/marketa.htm.
27. Howard Kerr told me that he was present when Bud made this promise in the Admiral's villa.
28. See Cutler, *Brown Water, Black Berets*, ch. 6, "SEALORDS."
29. Lewis Sorley, *Thunderbolt: From the Battle of the Bulge to Vietnam and Beyond—General Creighton Abrams and the Army of His Times* (New York: Simon & Schuster, 1992).
30. Stillwell, interview with Rectanus.
31. Lew Glenn to Mrs. Zumwalt, Oct. 1, 1968. ZPP.
32. "No eggs or milk; cereal with Apple Juice"—author interview with Dr. William Narva.
33. From: COMNAVFOR Vietnam, To: NAVFOR Vietnam, Subject: CHANGE OF COMMAND, Oct. 1, 1968, ZPP.
34. Cullen, Zumwalt interview notes.
35. Ibid.
36. Stillwell, *Reminiscences of Staff Officers*. Glenn named his son Russell after Bud; Bud was godfather to Howard Kerr's daughter Heather.
37. Stillwell, interview with Rectanus, 63; Dan Lawrence, "Passing of VADM Earl F. 'Rex' Rectanus," Naval Intelligence Professionals, http://navintpro.net/?p=134.
38. "VADM Emmett H. Tidd," USS *Richard B. Anderson* DD-786, www.vietnamproject.ttu.edu/dd786/tidd.html.
39. McQuilkin interview with Zumwalt transcript, 17.
40. On June 23, 1969, Joe Rizzo left and Tidd arrived. When Tidd pointed out to Bud that there were four staff officers senior to him, Zumwalt replied, "They may be senior to you, but they are not senior to me."
41. Chick Rauch spent thirty years in the U.S. Navy before retiring in the rank of rear admiral. He was commanding officer of two nuclear submarines, instructor in reactor engineering at the Navy's Nuclear Power School, and senior naval advisor to the Vietnamese Navy. Chick also served as assistant chief of naval personnel for human resource development, during which time his office developed the navy's early programs in race relations and equal opportunity, women's rights, alcohol and drug abuse rehabilitation and control, and cross-cultural relations.

42. Author interview with Bob Powers. See Robert C. Powers, *Save the Belknap: A True Story—a Night to Forget That Will Never Be Forgotten* (New York: Simon & Schuster, 2011).

43. Zumwalt tape, "History of Jogging," ZTT.

44. Bud got his name and address and sent him a thank-you note on his official letterhead when he got back home, so the man would know that Bud wasn't pulling his leg. See "I Run the United States Navy," *Noble Perspective*, vol. 32, July 1999, 1. I am grateful to Bud's son Jim for clarifying some facts in the published version.

45. Stillwell, *Reminiscences of Staff Officers*.

46. Vice Admiral Earl Rectanus, Mar. 3, 2008, the Naval Intelligence Organization Vietnam, ZPP.

47. Cutler, *Brown Water, Black Berets*, 253–55; see Cullen interviews with Zumwalt, 4.

48. Ibid.

49. Ibid.; Elmo Zumwalt, Jr., and Elmo Zumwalt III, *My Father, My Son* (New York: Dell, 1987).

50. Stillwell, interview with Rectanus.

51. The PBRs were being used as backups for the PCFs offshore. "We'll bring the PBRs inland, and use the PCFs along the coast . . . in the past that had been discouraged. Zumwalt put these boats in place." Cutler, *Brown Water, Black Berets*.

52. McQuilkin interview notes, Cullen Collection.

53. "And when he endorsed it as bold and risky, but doable, I felt very comfortable about going forward with it. And I guess I would also say that, in essence, by giving me those assets, Bob Salzer was giving me the battleships of the brown water Navy; the Swift boats, representing the cruisers; and Art Price's forces, the PBRs, representing the destroyers of the brown water Navy."

54. Tom Glickman, "Memories," unpublished manuscript provided to the author, and interview with Tom Glickman. Glickman spent the evening at the villa, having dinner and breakfast with the admiral and staff.

55. Ibid.

56. Cutler, *Brown Water, Black Berets*, 252.

57. Ibid., 252–53. See Glickman account as well.

58. Glickman, "Memories."

59. Ibid.

60. Personal letter, Sept. 22, 1995, Rex Rectanus to Pricilla Barry, on the occasion of the Zumwalts' fiftieth wedding anniversary.

61. Author interview with Bob Powers.

62. Author interview with Chick Rauch.

63. Quoted in J. A. Davidson, USN, "The 'Z' Meets His Recruiters," Navy News Release #21-74, NHHC.

64. Stillwell, interview with Rectanus, 29–30.

65. Bud told Paul Stillwell the same story. "I got the idea that we could fly boats by these huge helicopters from one river to another. If we could get enough of the helicopters, you could suddenly have a ten-boat convoy dropped in on a river in an area where the Vietcong weren't expecting it and give them some surprise. Well, it turned out that when people did the numbers and calculations, there weren't enough helicopters available to get more than one or two in at a time, and it would be a pretty hairy operation to get it there. So after just a little bit of staffing, we decided that was a downer." Stillwell, *Reminiscences of Admiral Zumwalt, Jr.*, 489.

66. Stillwell, interview with Rectanus.

67. Salzer Oral History, 653.

68. General Abrams requested that the navy brief Major General George Eckhardt, the IV Corps senior advisor, about the scope of the plan.

69. Letter from Bud to the "Clan," July 30, 1969, ZPP.

70. Ibid.; See Ralph Christopher, *Duty, Honor, Sacrifice* (Bloomington, IN: AuthorHouse, 2007), and Ralph Christopher, *River Rats* (Bloomington, IN: AuthorHouse, 2005). I am indebted to Ralph for taking the time to discuss these issues with me at Texas Tech.
71. Stillwell, interview with Rectanus.
72. Reckner was the senior advisor of this group and can attest to the fact that they were there. Author interview with Jim Reckner.
73. "Naval Intelligence Field Operations Vietnam (NAVINTFOV), "The NILO Program: Tailoring Naval Intelligence to Fit the War," Presentation, Texas Tech University Vietnam Symposium, Mar. 14, 2008; see also H. Lawrence Serra, *NILO Ha Tien: A Novel of Naval Intelligence in Cambodia* (Bloomington, IN: AuthorHouse 2009). I am indebted to Larry Serra for providing the insights and details of the NILO narrative. There are no available records of the NILOs' actual post assignments or intelligence operations—only monthly summary reports of their combat deaths, capture, or participation in significant operations. There is an alumni group consisting of most of the two hundred living NILOs, IOs (intelligence officers), and Collection Branch officers.
74. Only a handful were experienced intelligence officers trained in agent handling, air intelligence, photo interpretation, communications security, and such. More than 75 percent were unrestricted surface line officers who had come from blue-water commands where they had served as qualified officers of the deck, navigators, gunfire director officers skilled in placing naval gunfire on enemy positions, and small-boat officers skilled in the operation of small craft in shallow coastal and riverine areas.
75. There were combat mishaps: NILO Ken Tapscott was killed in a firefight on the Song Ong Doc River (in Operation Sea Float) in August 1970; Rectanus's friend and protégé IO Jack Graf was shot down several times, eventually captured, and killed in attempting to escape captivity in the delta. All in all, the NILOs sustained casualty rates similar to the boat forces with whom they operated, a result of Zumwalt's aggressive combat strategy of taking the fight to the enemy.
76. Serra, *NILO Ha Tien*.
77. The NILO enthusiasm and audacity was infectious and spread to those around them. After the departure of one NILO by medevac, a local ATSB (advanced tactical support base) naval officer executed a covert coast-watching mission for enemy infiltration from a jungle hilltop on one of the Cambodian Pirate Islands, a mission originally conceived by the NILO. (See H. Lawrence Serra, "Pirate Islands, Cambodia," *The Monk, and Other Stories* (Bloomington, IN: AuthorHouse, 2012).
78. See James G. Zumwalt, *Bare Feet, Iron Will: Stories from the Other Side of Vietnam's Battlefields* (Jacksonville, FL: Fortis, 2010), 197.
79. Stillwell, *Reminiscences of Staff Officers*.
80. Ibid.; Lew Glenn.
81. Ibid.; and author interview with Howard Kerr.
82. This account appears in Sorley, *Thunderbolt*, as well as in Cullen, Zumwalt interview notes.
83. At the Oct. 29, 1968, meeting in the Cabinet room, Abrams told Johnson, "I am blessed with four good men: Goodpaster, Brown, Zumwalt, and Cushman. They don't belong to any service. They belong to the U.S. Government."
84. Stillwell, *Reminiscences of Admiral Zumwalt, Jr.*, and also Stillwell, *Reminiscences of Staff Officers*.
85. McQuilkin interview with Zumwalt, 10.
86. Naval historian Richard L. Schreadley, a critic of Admiral Zumwalt, noted that the admiral used the opportunity presented by Abrams's rejection of the air force plan to give a "virtuoso performance" in which "no one seemed to notice the lack of slick

charts and graphs. . . . The new vice admiral had scored big." Richard L. Schreadley, *From the Rivers to the Sea: The U.S. Navy in Vietnam* (Annapolis, MD: Naval Institute Press, 1992), 9.

87. McQuilkin interview with Zumwalt, 10.
88. Bud especially enjoyed the Saturday-morning meetings where the top commanders and senior staff officers met to dissect the war strand by strand "and then put it back together again analytically." He was "intellectually vitalized by these meetings."
89. See Dale Van Atta, *With Honor: Melvin Laird in War, Peace, and Politics* (Madison: University of Wisconsin Press, 2008).
90. Nov. 29, 1968, ZPC.
91. Ibid.
92. Larry Oswald, letter to Mouza recalling event after Bud's passing, ZFC.
93. Stillwell, interview with Rectanus, 66.
94. Author interview with Mike Spiro, who recalled the name as Ensign Billups.
95. Stillwell, *Reminiscences of Staff Officers*.
96. In a letter to the family, Glenn observed, "It was a real experience for me to be part of this Christmas holiday, and to share in the conviviality and the jocular repartee that characterizes the Zumwalts. They make up for their separations with a closeness, when together, that is a marvelous thing to observe."
97. Oswald, personal letter to Mouza, ZFC.
98. Jim Morgan, patrol officer, senior patrol officer, and operation officer, River Division 593, e-mail to author.
99. Powers, *Save the Belknap*, 57–58.
100. Stillwell, *Reminiscences of Admiral Zumwalt, Jr.*, 506.
101. Starry was recovered and eventually became a four-star general. Cullen interview transcript, 31.
102. Jimmy R. Bryant, *Man of the River* (Fredericksburg, VA: Sergeant Kirkland's, 1998), 96–97; and author personal correspondence with Bryant.
103. "About John C. 'Bubba' Brewton," USS *Brewton* DE-1086, www.ussbrewton.com /brewton.htm.
104. Aug. 2, 1971, letter, ZPP.
105. Bud was in attendance in 1968 when Elmo was commissioned an ensign in Naval ROTC at the University of North Carolina.
106. Elmo also wrote, "Your shadows of success and influence are a part of why I believe I must get out. . . . Dad, I realize the last letter I wrote you hit hard. I am not acting with a 'sense of exaltation.' I am doing what I feel I must. Captain Mullane agreed that if you were in my place that you would be doing the same." ZFC.
107. "To Jock," enclosing Elmo's letter, saying he was proud of his son, ZFC and ZPP.
108. Letter to Nitze, Feb. 10, 1970, ZFC.
109. Undated letter from Elmo to his mother, ZFC.
110. Zumwalt and Zumwalt, *My Father, My Son*, 87.
111. Letter, Nov. 17, 1969, ZFC.
112. Letter, Jan. 20, 1970, ZFC.
113. Zumwalt and Zumwalt, *My Father, My Son*, 81; Harvey Miller, radarman.
114. Letter, Dec. 11, 1986. Bud asked Do to advise Elmo not to take so many risks. "Captain Kiem, talk to him, he might listen to you." ZPP.
115. Stillwell, *Reminiscences of Admiral Zumwalt, Jr.*, 523.
116. Confidential message, COMNAVFORV to CTF 115, 116, 117, Jan. 17, 1969 [Task Force].
117. Stillwell, *Reminiscences of Admiral Zumwalt, Jr.*, 524.
118. "Vietnam," ZTT 24, side A.
119. Stillwell, *Reminiscences of Admiral Zumwalt, Jr.*, 500.

120. These are synthetic chemical compounds developed to kill weeds in order to increase crop yield. In Vietnam, the military cooked a fifty-fifty mixture of 2,4-D and 2,4,5-T at higher temperatures than the normal process and named it Agent Orange. Dioxin is a highly toxic by-product of military-grade 2,4,5-T. As Zierler documents in his pathbreaking book, cooking it at higher temperatures made the dioxin levels even more deadly. See David Zierler, *The Invention of Ecocide: Agent Orange, Vietnam, and the Scientists Who Changed the Way We Think about the Environment* (Athens: University of Georgia Press, 2011).

121. Zumwalt and Zumwalt, *My Father, My Son*, 94.

122. Kirkwood wrote, "I later found out this chemical was a carcinogen. . . . I believe I died up there on that deck and I didn't even know it." Ron Kirkwood, "Fear of Living (Vietnam)," manuscript excerpt sent to Zumwalt, ZFC, online at http://parker5nc .tripod.com/story/fear.html.

123. They became close friends, and Wages later served as personal aide to Zumwalt in the Office of Chief of Naval Operations.

124. Wages's river patrol group was awarded the Presidential Unit Citation by President Nixon for extraordinary heroism.

125. Letter to clan, Nov. 30, 1969, ZPP.

126. Letter, Feb. 10, 1970, ZPP.

127. July 30, 1969, letter to clan, ZPP.

128. Letter to Nitze, Feb. 10, 1970, ZPP.

129. He retired eight years later, on Nov. 1, 1974, at the age of fifty-four.

130. "He was a traditional Buddhist, very devout Buddhist—a man of very high ethical and moral principles. He lived very poorly, and not very many of the flag and general officers on the Vietnamese side did," recalled Bud to Paul Stillwell. Stillwell, *Reminiscences of Admiral Zumwalt, Jr.*, 512.

131. Letter, "Dear Clan," July 30, 1969, ZPP.

132. Ibid.

133. Letter, Nov. 30, 1969, ZPP.

134. All quotes are from Bud's letters home to Saralee and the "Clan," ZPP.

135. Schreadley, *From the Rivers to the Sea*, 367.

136. Stillwell, interview with Rectanus, 81.

137. A letter, "Dear Zumwalts," Jan. 12, 1970, provides Bud's review of their family Christmas, the first time in eighteen months that the entire family was together. Elmo left the boat in the southern tip of Ca Mau Peninsula, Jim flew in from the University of North Carolina and went fifteen pounds over his wrestling weight, Kathy, their future daughter-in-law, flew in from Manila, and they all went to Baguio, the Philippines. ZPP.

138. Sept. 30, 1970, ZPC and NHHC.

139. Chafee interview, ZTT 5, side A, transcript, 6.

140. Bud told Paul Stillwell, "I went down and in 15 minutes briefed Laird on our operations and on the turnover proposal. . . . Mel Laird . . . seemed both intrigued and pleased by the briefing, asked a couple of questions. . . . And that was all that was heard from Mel Laird. I had the sensation that he was pleased." Stillwell, *Reminiscences of Admiral Zumwalt, Jr.*

141. Bud briefed Laird on the evolution of naval power against the North Vietnamese. Their logistical system had been kept to trawlers, and the communists had been forced to develop the trail, and this, combined with what they were able to bring into Sihanoukville and across the sanctuary of Cambodia to the borders of South Vietnam represented their principal logistical lines. The application of sea power had been accomplished by the use of air and surface elements of the Seventh Fleet supplemented from 1966 to 1970 with a hundred Swift Boats (PCFs) and about two hundred river patrol boats (PBRs), plus about a hundred armored amphibious boats. The

Swifts were used for inshore coastal patrol to prevent infiltration of small craft and to provide Coast Guard–type functions. The PBRs patrolled the major branches of the Mekong in the delta, and the armored boats were married to the U.S. Ninth Division to form the mobile riverine flotilla bringing support to the cities of the delta. ZTT 5, side A.

142. Telephone conversation, Laird and Kissinger, Mar. 10, 1970.
143. Telephone conversation, Laird and Kissinger, Mar. 13, 1970.

CHAPTER 9: THE WATCH BEGINS

1. Letter on the occasion of Bud's retirement, June 25, 1974, oo. (These double-zero files refer to Zumwalt's CNO papers, "oo" being his designation. When cited, they are from NHHC.)
2. Zumwalt, *On Watch*, 44.
3. Ibid.; "The Summons," ZTT.
4. Ibid.
5. Ibid.
6. "Jerome H. King Jr.," Arlington National Cemetery, www.arlingtoncemetery.net /jhking-03.htm.
7. Admiral Harry DePue Train II, *Oral History* (Annapolis, MD: U.S. Naval Institute, 1997), 237. Excerpts and printed volumes of USNI oral histories are available at www .usni.org/heritage/oral-history-catalog.
8. The other recommended candidates included Admiral Ralph Wynne Cousins and Vice Admiral Bush Bringle.
9. Moorer told Bud that he felt he was coming up too early and should wait four years, but he made it clear that he would work with him and be supportive. Bud said that Warner spilled his guts that night and acknowledged that there had been much infighting over the nomination. Warner had favored Chick Clarey. "The Summons," ZTT.
10. Clarey served in that position from from Dec. 5, 1970, through Sept. 30, 1973.
11. Captain Samuel Gravely would be the first one promoted to admiral.
12. Van Atta, *With Honor: Melvin Laird in War, Peace, and Politics*, 329.
13. Ibid., 143–44, and Van Atta's interview with Warner, Nov. 10, 1999.
14. June 2, 1972. Bud inscribed a book on riverine warfare "To John Chafee with deep admiration for your leadership, wise counsel and friendship as Secretary of the Navy." On Oct. 24, 1973, Bud wrote a handwritten P.S.: "What a great opportunity to serve you gave me. I shall always be grateful." This was in response to Chafee's Oct. 2, 1973, "Congratulations on the Trident victory. I know you put tremendous energy and effort into that, and the results must have been most satisfying. Please don't make the votes any closer." On June 23, 1974, Bud wrote Chafee that "it will be tough to come up with another job which approaches the challenge and opportunity of this one." Executive Correspondence, NHHC
15. Telephone conversation, Apr. 13, 1970, National Archives.
16. Admiral Stansfield Turner Oral History, p. 335.
17. He first called his father but neglected to tell him not to say anything. The president had yet to announce his appointment.
18. "The Appointment," ZTT. See also "The Admiral's Wife."
19. *Washington Post*, Apr. 16, 1970.
20. *Wall Street Journal*, Apr. 16, 1970.
21. Letter, Apr. 15, 1970. In reply Bud wrote, "As I look back at all California has done for me and for the navy, I am indeed proud and grateful to be her native son." Letter, Apr. 16, 1970, ZFC.

22. Executive Correspondence, Apr. 16, 1970, NHHC.

23. Robert Powers, Executive Correspondence, Apr. 27, 1970, NHHC.

24. On the tapes, Bud says that he called Abrams from Moorer's home to tell him when he was returning to Saigon, and Abrams did not congratulate him. "I was quite surprised and a little hurt but did not comment on it." ZTT.

25. Letter, Apr. 29, 1970, responding to Jim's letter of Apr. 23, ZPP.

26. Letter, Apr. 9, 1970, Robert Powers, Executive Correspondence, NHHC.

27. William J. Crowe, Jr., *The Line of Fire: From Washington to the Gulf, the Politics and Battles of the New Military* (New York: Simon & Schuster, 1993), 87. Moreover, retirements and losses had not occurred at anticipated levels, and budget reductions led to the loss of flag positions. To increase the flow of younger officers into flag rank, a policy of non-reversion for vice admirals was adopted by asking three-star officers to retire at the end of their tours when no additional vice admiral billet was available.

28. Letter, Apr. 27, 1970, NHHC.

29. May 2, 1970. Nicholson was indeed later recruited to head one of the Mod Squad units.

30. Letter, Apr. 14, 1970, NHHC.

31. Letter, Apr. 16, 1970, NHHC.

32. Letter, June 25, 1970, NHHC.

33. U.S. Code, title 10, chapter 509, section 5081, subsection D.

34. Stillwell, *Reminiscences by Staff Officers.*

35. The oral history of Salzer, who retired as a vice admiral, is in the Naval Institute collection, www.usni.org/heritage/oral-history-catalog.

36. Stillwell, *Reminiscences of Admiral Zumwalt, Jr.*

37. "Jerome H. King Jr.," Arlington National Cemetery, www.arlingtoncemetery.net/jhking-03.htm; Stillwell, *Reminiscences of Admiral Zumwalt, Jr.*

38. Nomination of Admiral Elmo R. Zumwalt, Jr., USN, to Be Chief of Naval Operations, Hearings before the Committee on Armed Services, United States Senate, Apr. 16, 1970.

39. Bud's farewell staff party in Saigon was held on May 4 at Top of the Duc. It was a raucous affair during which his team offered an adaptation of Hungarian-born American composer Sigmund Romberg's operetta *The Student Prince.*

40. In his speech, Bud identified the sources of inspiration in his life—his family; the seamanship of his officers; the Vietnamese navy, whose personnel, along with Chon, "have helped us to understand and come to love the Vietnamese people"; General Abrams, "tough, demanding, compassionate, and understanding . . . a great military captain in war"; and his Brown Water Navy for "their sacrifices and heroism."

41. Letter, Apr. 21, 1970, ZPP.

42. On Aug. 8, 1970, Bud wrote Chon that Truc was doing quite well at the academy, excelling in certain areas. "He is the best of the foreign students, no doubt." ZFC.

43. Zumwalt and Zumwalt, *My Father, My Son.* Elmo was not technically a state resident, and had low LSAT scores with a few Ds on his transcript. Elmo did not qualify under the out-of-state quota. Caldwell was able to get him admitted as an in-state student because Bud had no fixed residence and had taught there and because both boys had been undergraduates there.

44. May 7, 1970, ZFC.

45. Aug. 4, 1970, ZPP.

46. Author interview with Dr. Narva.

47. ZTT 30, side A, 8, addition to summons chapter.

48. Author interview with Dr. Narva. Bud remained in the hospital for seven days under the care of Dr. Narva, who would accompany him on almost 90 percent of his overseas trips as CNO.

49. Zumwalt tape 26, side A, ZTT.

50. Much of the discussion here benefits from the work of Malcolm Muir, Jr., *Black Shoes and Blue Water: Surface Warfare in the United States Navy, 1945–1975* (Washington, DC: Department of the Navy, 1996; Honolulu: University Press of the Pacific, 2005); Edgar F. Puryear, Jr., *American Admiralship: The Art of Naval Command* (Minneapolis: Zenith Press, 2005); George W. Baer, *One Hundred Years of Sea Power: The U.S. Navy, 1890–1990* (Stanford, CA: Stanford University Press, 1993); James C. Bradford, *Quarterdeck and Bridge: Two Centuries of American Naval Leaders* (Annapolis, MD: U.S. Naval Institute Press, 1997); Robert William Love, Jr., *The Chiefs of Naval Operations* (Annapolis, MD: U.S. Naval Institute Press, 1980); and Thomas C. Hone, *Power and Change: The Administrative History of the Office of the Chief of Naval Operations, 1946–1986* (Washington, DC: Naval Historical Center, 1989).

51. Change-of-command remarks, ZPP.

52. Chafee wanted the vice chief of naval operations to "truly be your alter ego" and able to enunciate the CNO's position on all matters of policy. "You will be tied up so much in the JCS that it's most helpful to have someone who can speak for you with authority." Bud told Chafee his VCNO would be an "alter ego" and "I have the fullest confidence in Chick Clarey to speak for me on any Navy matter."

53. Stansfield Turner Oral History, 336.

54. Letter, June 18, 1970, ZPP.

55. B. A. Clarey, president of Line Flag Selection Board, Apr. 8, 1971. Executive Correspondence, NHHC.

56. Memo from CNO, July 1, 1970, NHHC.

57. Jeffrey Sands, *On His Watch: Admiral Zumwalt's Efforts to Institutionalize Strategic Change* (Alexandria, VA: Center for Naval Analyses, 1993), www.cna.org/sites/default/files/research/Zumwalt's%20Efforts%20to%20Institutionalize%20Strategic%20Change%202793002200.pdf; and David Alan Rosenberg, "Project 'Sixty': Landmark or Landmine—an Evaluation," unpublished draft, July 23, 1982. OP-965. I have relied on both Sands and Rosenberg throughout the discussion of Project Sixty. I have also used Zumwalt tape transcipts titled Project Sixty.

58. "To some it was the highly idiosyncratic creation of an entirely idiosyncratic individual." David Alan Rosenberg, "Project 'Sixty.' "

59. Elting E. Morison, "A Case Study of Innovation," *Engineering and Science* 13, no. 7 (Apr. 1950): 5–11.

60. Stansfield Turner Oral History, 365.

61. Stansfield Turner Naval Institute Oral History is cited in David Alan Rosenberg, "Project 60: Twelve Years Later," unpublished manuscript; and Elmo Russell Zumwalt, Jr., "After *On Watch*," unpublished manuscript.

62. Sands, *On His Watch*, 20.

63. Rosenberg, "Project 60: Twelve Years Later."

64. Zumwalt, *On Watch*, 64. "For twenty years Rickover has been working successfully toward a super ship navy, and so it is partly his doing that for twenty years the Navy has been getting smaller except of course in the item of nuclear powered submarines."

65. ZTT 2, side A, part 6–7.

66. Polmar and Allen, *Rickover: Controversy and Genius*.

67. Zumwalt, *On Watch*, 65.

68. Letter, Nov. 18, 1970, to Stansfield Turner, at the time a rear admiral and commander of Cruiser-Destroyer Flotilla Eight. "Ps . . . sorry it took me so long to record my great pride in the work you did for me. We have much left to do." ZFC.

69. Rosenberg, "Project 60: Twelve Years Later"; Sands, *On His Watch*.

70. In "Project 60: Twelve Years Later," David Alan Rosenberg identified the accomplishments of Project 60:

Explicit missions and rationale for justifying the navy were developed, which served as an overarching philosophy for the navy.

Nuclear attack submarines were employed for the first time as carrier battle-group escorts.

Minesweeping helicopters were developed; the first RH-53Ds entered service in 1973.

PGs (patrol gunboats) were assigned to the Mediterranean to serve as task-force escorts and protect against Soviet tattletale activity.

The first interim sea-control ship, the USS *Guam*, operated with V/STOL (vertical or short takeoff and landing) aircraft and helicopters from 1972 to 1974. The sea-control ship was canceled, and this was a major setback for Project Sixty.

Major innovations on conduct of fleet exercises were implemented.

Major improvements were made in Sixth Fleet readiness and logistics, including electronic warfare, surveillance, and antisubmarine warfare.

The first interim surface-to-surface missile, the standard ARM (antiradiation missile, which homes in on active radars and radios), was converted and deployed.

Twenty UH-2 Seasprite utility helicopters were converted to SH-2D LAMPS-I (light airborne multipurpose system) antisubmarine helicopters.

Development of the basic point-defense missile system was accelerated, and widespread installation soon began.

The CV (carrier variant) concept was adopted, and the S-3 Viking was procured to help implement it.

Marine air squadrons began overseas deployments on attack carriers to fulfill both force-multiplier and cross-training requirements.

The CVN-70, the USS *Carl Vinson*, was funded in fiscal year 1974.

The first PFG (patrol frigate, guided missile), the USS *Oliver Hazard Perry*, was funded after a long struggle in fiscal year 1973.

The first PHM (patrol hydrofoil guided missile ship), the USS *Pegasus*, was authorized in fiscal year 1973.

The undersea long-range missile system became reality with the Trident program.

The first Trident submarine, the USS *Ohio*, was authorized in fiscal year 1974.

Initiatives that became effective after Zumwalt's term included production of the Harpoon antiship missile.

71. Richard J. Levine, "A Final Z-Gram from Zumwalt," *Wall Street Journal*, May 13, 1974.

72. Sands, *On His Watch*; Rosenberg, "Project 'Sixty': Twelve Years Later."

73. President Carter's decision not to fund the SCS was significant because "it was the most revolutionary naval technology in all of Project 60 and appeared to a number of naval officers to represent the wave of the future that would enable the surface Navy to survive and prevail against Soviet high-speed torpedo armed nuclear attack submarines." Ibid.

74. The president's foreign policy declaration of February 1970 promised that "our interests, our foreign policy objectives, our strategies and our defense budgets are being brought into balance—with each other and with our overall national priorities."

75. Letter, Clarey to Zumwalt, Apr. 22, 1971, NHHC.

76. Ibid.

77. Bud decided to bring the Joint Strategic Target Planning Staff to Washington so that the JCS and senior military authority could work closely on weapons-system development. He was unsuccessful. Unnumbered ZTT side A, parts 1–2.

78. After the briefing, Bud shared the view of another JCS member, who thought the Soviets would "soon be able to blackmail and we will have a Cuba in reverse." Ibid.

79. ZTT unnumbered side B, part 9.

80. Presidential briefing, Aug. 18, 1970, NHHC.

81. In previous crises—Lebanon, the Dominican Republic, Quemoy and Matsu, the Chinese incursion into India, and the Cuban Missile Crisis—the United States had demonstrated the capability to control the seas and project its forces.

82. Toward the end of his term, Bud became more outspoken about classified studies bearing on war outcomes. On Jan. 23, 1974, Senator William Proxmire wrote James Schlesinger that he had been informed that the CNO "has privately told several members of Congress, while lobbying for Navy programs, that had the US and USSR navies come to battle in the Mediterranean during the recent crisis there, the US fleet would either have been forced to retire or accept huge losses. The point was made emphatically and without qualification." Proxmire charged "deliberate deception," since "quite the opposite conclusion was reached by the Chairman of the Joint Chiefs on national television." Executive Correspondence, NHHC.

83. Bud's first six weeks on the job coincided with the final six weeks of SALT II meetings, which ended Aug. 14, 1970.

84. SALT, ZTT.

85. Zumwalt, *On Watch*, 283.

86. Sept. 10, 1970, letter to Abrams, ZPP.

87. On Sept. 18, 1970, Abrams responded with a positive note that they need to accept the budget realities and "devote our best judgment and imagination to making it all turn out right." ZPP.

CHAPTER 10: ZINGERS

1. Letter, Jan. 1, 1970.

2. The goal for reenlistments after a first hitch had been 35 percent. "In 1970 the actual figure was 9½%." Zumwalt, *On Watch*, 167.

3. *Washington Post*, Dec. 10, 1972.

4. Letter from CNO, July 22, 1970, NHHC.

5. "Morale," undated speech by Bud Zumwalt, oo Files, NHHC.

6. The first flurry of sixty-nine was followed by twenty-three in the next six months, then by ten and eleven in Bud's third and fourth half years.

7. "Zingers is about the best name I've heard yet for the Z-Grams." Letter to Lew Glenn, Sept. 21, 1970, Executive Correspondence, NHHC.

8. Letter, Sept. 9, 1970, Executive Correspondence, NHHC.

9. Letter, Sept. 1, 1970, Executive Correspondence, NHHC.

10. Rear Admiral Dave Bagley (older brother of Worth) headed the new office. He would be replaced by Charles "Chick" Rauch.

11. NAVOP-091457Z.

12. Cited in Van Atta, *With Honor: Melvin Laird in War, Peace, and Politics*, 254.

13. Stockdale note, Mar. 25, 1973, NHHC and ZFC.

14. ZFC.

15. Letter, Jan. 15, 1971, NHHC.

16. Letters, Aug. 3, 1971, and Oct. 19, 1971, from Blanche Seaver. "Have you completely given up on the great heritage left us by our Founding Fathers who fought and bled and died, giving us the great uncompromising principles, which have made our country so enviable." NHHC.

17. Letter, Nov. 13, 1972, to Mrs. Seaver, NHHC.

18. Letter, Feb. 3, 1971, NHHC.

19. Letter, Dec. 23, 1970, NHHC. He must have heard incorrectly. The Battle of Trenton occurred on Christmas Day, and the "British" defeated there were Hessians—German mercenaries.

20. Letter, Jan. 6, 1971, NHHC.

21. Letter, Nov. 16, 1970, NHHC.

22. Chafee told *Pacific Stars and Stripes* that he fully supported Zumwalt's liberalization policies and Z-grams.

23. Letter, Dec. 3, 1970.
24. For example, Z-04 gave thirty days' leave between assignments.
25. Issued July 29, 1971.
26. Issued Dec. 23, 1970.
27. Dec. 26, 1970, NHHC.
28. Holloway letter, Executive Correspondence, Dec. 23, 1971, NHHC.
29. Letter, July 19, 1971, NHHC.
30. Letter, Jan. 12, 1976, to Jack Connery, NHHC. First-term reenlistments were less than 10 percent, and on aircraft carriers, less than 4 percent. After just thirteen days, Petty Officer Allen Buckalew thanked Z, for "in every Z gram you instilled and reinforced a new courage and pride in our Navy."
31. Wolfgang Saxon, "Adm. John Hyland, 86, Dies; Championed Naval Air Power," *New York Times*, Nov. 1, 1998.
32. Paul Stillwell interview, p. 488.
33. Hyland reply, Sept. 27, 1970. Hyland was replaced by Admiral Bernard A. "Chick" Clarey.
34. Letter, Oct. 15, 1970, personal, Sensitive, Eyes Only, NHHC.
35. Miller Oral History, 591, USNI oral histories, www.usni.org/heritage/oral-history-catalog.
36. Ibid.
37. Letter, Aug. 8, 1970. Bud told Chon that he was trying to help on a few fronts, like getting a U.S. headquarters for Operation Helping Hand going and making every effort to increase the caliber of advisors being sent. ZPP.
38. Author interview with Admiral Charles "Chick" Rauch.
39. Letter to Harvard Law School Admissions Department, Dec. 30, 1970, ZPP.
40. Letter, Dec. 10, 1970. Antle possessed "an invaluable capacity to assimilate and correlate facts in a logical decision matrix." ZPP.
41. Zumwalt, *On Watch*, 171; and Halperin interview with Rice, ZPP.
42. Letter, July 21, 1970, ZPP.
43. Zumwalt, *On Watch*, 172.
44. Letter, Jan. 19, 1972, to "Howie," ZPP.
45. Letter, Jerry Carr, AMS2 [Aircraft Intermediate Maintenance Department], ZFC.
46. Dick Nicholson, letter, Sept. 4, 1970. He gives the example of Lieutenant Commander Rolf Clark, the group's analyst, "who has changed his mind about a career."
47. Jan. 11 (year is obscured, but circa 1970–1973); ZFC; confirmed in author interview with Halperin.
48. Zumwalt, *On Watch*, 168.
49. John Darrell Sherwood, *Black Sailor, White Navy: Racial Unrest in the Fleet during the Vietnam War Era* (New York: New York University Press, 2007), 11. I have drawn extensively from Sherwood's pathbreaking work.
50. ZTT 21, side A, part 3.
51. Stillwell, *Reminiscences by Staff Officers*, 291.
52. Sherwood, *Black Sailor, White Navy*.
53. Ibid., 47.
54. See Wallace Terry, *Bloods: An Oral History of the Vietnam War by Black Veterans* (New York: Random House, 1984). Also see Norman and Zumwalt accounts of this session on Zumwalt tapes, "Retention," ZTT.
55. Norman account, ZTT, "Retention."
56. Sherwood, *Black Sailor, White Navy*.
57. ZTT 35, side A. The action list was presented at their Nov. 9, 1970, meeting.
58. Letter, Taylor Branch, Nov. 18, 1970, ZPP.
59. Zumwalt, *On Watch*, 43.

60. Ibid
61. Bill Thompson memoirs.
62. Sherwood, *Black Sailor, White Navy*, 45.
63. Bill Norman tapes, tape 36, side A, ZTT.
64. Gerald Astor, *The Right to Fight: A History of African Americans in the Military* (Novato, CA: Presidio Press, 1998; New York: Da Capo, 2001), 466.
65. David Halperin was initially the only white person involved.
66. Letter, Dec. 19, 1970, NHHC.
67. Astor, *The Right to Fight*, 453–54.
68. Sherwood, *Black Sailor, White Navy*, 48.
69. Bill Norman tapes, tape 36, side A, ZTT.
70. Ibid.
71. Sherwood, *Black Sailor, White Navy*, 49.
72. Bill Norman tapes, tape 36, side A, ZTT.
73. Letter for the record upon Norman's resignation, NHHC.
74. Another program paying exceptional dividends was the Personal Services Division, which focused on sports, travel, and other opportunities. On May 9, 1973, Lew wrote in his second situation report (SITREP), "Let me express the appreciation of every member of DESRON [Destroyer Squadron] TWELVE for your quick action in resolving the problem of landing our liberty parties." NHHC.
75. Letter, June 2, 1973, NHHC.
76. Letter, Mar. 10, 1972, NHHC.
77. Letter, Apr. 24, 1972, NHHC.
78. *Time*, Mar. 1, 1982.
79. Zumwalt, *On Watch*, 265.
80. Letter to Anna Cox, Nov. 8, 1994, ZFC.
81. Zumwalt, *On Watch*, 262.
82. A 1978 Supreme Court ruling declared unconstitutional the law blocking assignment of women to ships. *Owens v. Brown*, 455 F.Supp. 291 (1978).
83. Rosemary Mariner, "Adm. Zumwalt Changed My Life," *Washington Post*, Jan. 9, 2000, B07. She closed by saying, "Cheers and thanks, Admiral."
84. June 1972, NHHC.
85. *Ms.*, Aug. 1982.
86. Letter, June 29, 1974, ZPP.
87. Letter, Oct. 10, 1995, ZFC.
88. Hazard retirement speech, Sept. 1972, provided to author by Admiral Hazard.

CHAPTER 11: ROUGH SEAS

1. ZTT 21, side B, part 6. Thermidor refers to the coup of 9 Thermidor 2 (July 27, 1794), the end of the Reign of Terror, after which Maximilien Robespierre was guillotined. For historians of revolutionary movements, Thermidor has come to mean the phase in some revolutions when the political pendulum swings back toward something resembling a prerevolutionary state, and power slips from the hands of the original revolutionary leadership. See, e.g., www.wordiq.com/definition/Thermidor.
2. Letter, Feb. 17, 1972.
3. It was difficult to comprehend or understand other people's perceptions of discriminatory practices. "We frequently are taken aback by the intensity of feeling revealed over some issue or practice we had long assumed to be perfectly innocuous or unrelated to Equal Opportunity affairs," wrote Bud. ZPP.

4. There was a concerted effort by antiwar organizations to get literature aboard ships encouraging mutiny as part of the effort to stop the war machine. See Gregory Freeman, *Troubled Water: Race, Mutiny, and Bravery on the USS* Kitty Hawk (New York: Palgrave Macmillan, 2009), 258; and Leonard F. Guttridge, *Mutiny: A History of Naval Insurrection* (Annapolis, MD: U.S. Naval Institute Press, 1992), 255.

5. Four days after the *Kitty Hawk*, the crew of the USS *Hassayampa*, a fleet oiler docked at Subic Bay Naval Base, was informed by the executive officer that the ship would not sail unless money stolen from the wallet of a black sailor was returned. Five white sailors were assaulted, and eleven black sailors were put ashore. The news coverage suggested that Z-66 and Z-113 had contributed to an attitude that protest and mutiny were permissible. Sherwood, *Black Sailor, White Navy*.

6. Mark Clodfelter, *The Limits of Air Power: The American Bombing of North Vietnam* (New York: Free Press; London: Collier Macmillan, 1989).

7. Stillwell, *Reminiscences of Admiral Elmo Zumwalt, Jr.*, 554.

8. What followed was the most successful use of air power during the war and one of the largest aerial bombardments in world history. Targeting roads, bridges, rail lines, troop bases, and supply depots, the attack utilized precision-guided laser bombs for the first time in modern aerial warfare.

9. Sherwood, *Black Sailor, White Navy*, 52.

10. Bill Thompson, memoir provided to author.

11. Sherwood, *Black Sailor, White Navy*, 53.

12. Thompson memoir, 393.

13. Sherwood, *Black Sailor, White Navy*, 53.

14. John F. Lehman, Jr., *On Seas of Glory: Heroic Men, Great Ships, and Epic Battles of the American Navy* (New York: Touchstone, 2001); and John F. Lehman, Jr., *Command of the Seas* (Annapolis, MD: U.S. Naval Institute Press, 1988).

15. Sherwood, *Black Sailor, White Navy*, 56.

16. Ibid.

17. Ibid., 58.

18. Captain Cloud had this to say about the reluctance of whites to enter black areas: "People in those areas were sort of off limits because that's where the minorities chose to live and chose to live in happiness. So if they're happy, leave them alone." Ibid., 58.

19. Sherwood, *Black Sailor, White Navy*, 85.

20. Guttridge, *Mutiny*, 265.

21. Sherwood, *Black Sailor, White Navy*, 91.

22. Ibid.

23. Ibid., 101.

24. Ibid., 91.

25. Forty-seven men, all but six of them white, were treated for injuries; three required medical evacuation.

26. Sherwood, *Black Sailor, White Navy*, 93.

27. Ibid., 94.

28. Ibid., 95.

29. Ibid.

30. Ibid.

31. Zumwalt, *On Watch*, 222.

32. Ibid.; and Sherwood, *Black Sailor, White Navy*, 140.

33. Sherwood, *Black Sailor, White Navy*, 142.

34. Ibid., 143.

35. Guttridge, *Mutiny*, on the beach detachment, 275.

36. Sherwood, *Black Sailor, White Navy*, 155.

37. Ibid.

38. Ibid., 146.

39. Ward had been able to identify 15 men who were "making a definite effort to distort actions of the command just to create hate and dissension," but 132 members (120 blacks and 12 whites) of the dissident group (all but 2 nonrated) were established as a shore detachment and moved to barracks at NAS North Island. Eventually, 24 would be returned to ship. In all, 120 were removed from the roster and investigated, resulting in 46 discharges, 36 honorable; 74 were given new assignments. Earl Caldwell, "Kitty Hawk Back at Home Port; Sailors Describe Racial Conflict," *New York Times*, Nov. 29, 1972.

40. Sherwood, *Black Sailor, White Navy*, 158.

41. Ibid., 157.

42. Ibid., 155.

43. All three wanted the protesters treated evenhandedly—neither threatening them nor kowtowing to them.

44. Letter, Nov. 4, 1972, from Rear Admiral Draper L. Kauffman, NHHC and ZPP.

45. Cited in Sherwood, *Black Sailor, White Navy*, 159.

46. Memorandum of conversation, Nov. 9, 1972, ZPP.

47. Memorandum of conversation, Nov. 8, 1972. The shore party had been staying in barracks, which Bud ordered closed so there would be no place to stay except the ship. Doing so "might help some of them come back rather than going and buying a hotel room." ZPP.

48. Memorandum of conversation, Nov. 9, 1972. "I gave this to John Warner after he refused to act. He would not keep it, insisted on returning it, but did finally approve my recommendation." ZPP.

49. Ibid.

50. "I had first met John Warner when he came to South Vietnam in 1969 to visit our forces there. He appeared to be a man of charm, young looking for his age, and handsome. One's first impression of him was quite favorable." ZTT.

51. Memorandum of conversation, Nov. 7, 1972, ZPP.

52. Memorandum of conversation, Nov. 10, 1972, "Sensitive," ZPP.

53. Sherwood, *Black Sailor, White Navy*, 168.

54. Letter from Humphrey, Nov. 15, 1972.

55. Letter, Nov. 14, 1972.

56. Nov. 15, 1972 [column 46], "Zumwalt Sets Navy on Course." The new seven-point program to end racial discrimination and enforce compliance through the Office of the Inspector General offered a promising new beginning.

57. Response dated Dec. 2, 1972, ZFC and NHHC.

58. Thompson memoirs, 394.

59. Robert Salzer Oral History Interview, 653.

60. Ibid., 653.

61. H. R. Haldeman, *The Haldeman Diaries: Inside the Nixon White House* (New York: Putnam, 1994), 533.

62. ZTT 31, sides A and B.

63. Sherwood, *Black Sailor, White Navy*, 170.

64. Bud told Murphy that Kissinger went from "hostile to moderately hostile" once he showed a little muscle.

65. Telephone conversation, Nov. 11, 1972, ZPP.

66. Zumwalt, *On Watch*, 308.

67. ZTT, undated and unnumbered.

68. Executive Correspondence, Aug. 19, 1970, and Dec. 1, 1970, NHHC.

69. June 7, 1971, Executive Correspondence, Zumwalt Papers, NHHC.

70. Henry Kissinger, *White House Years* (Boston: Little, Brown, 1979), 810.

71. "I have heard him express privately his great disenchantment with the President but while in the presence of the President playing the role of total sycophant." ZTT, "Henry Kissinger."

72. ZTT, "Kissinger"; also see the "Thermidor" tape.

73. Memorandum of conversation with Clarey, Nov. 11, 1972. Bud said that he hadn't spoken or heard from Warner and his staff "is worried about his uncertainty." Bud wanted to speak with John Ehrlichman, whom he had gotten to know. "That would be dangerous," said Murphy. "He would not trust him as far as he could throw him, and he thinks the CNO would be courting trouble." ZPP.

74. Telephone conversation, Nov. 13, 1972, ZPP.

75. Birthday greeting note, Nov. 29, 1972, NHHC.

76. L-gram (Laird memo), Nov. 29, 1971, NHHC.

77. Handwritten note, Executive Correspondence, Nov. 24, 1972, NHHC.

78. Telephone conversation, Nov. 14, 1972, ZPP.

79. Arleigh Burke Oral History, 312.

80. Telephone conversation, Nov. 14, 1972, ZPP.

81. Telephone conversation, Nov. 17, 1972, ZPP.

82. ZTT 17, side B.

83. Ibid.

84. Anderson Papers, box 62, Apr. 3, 1974. In a letter to journalist Hanson Baldwin, Anderson noted, "I am still shocked every time I see the appearance of our enlisted men and many junior officers in the Mediterranean area." Anderson Papers, box 62, Sept. 12, 1973, NHHC.

85. Anderson Papers, Nov. 7, 1972, NHHC.

86. Anderson Papers, May 1, 1973, NHHC.

87. ZPP.

88. Zumwalt responded on Dec. 22, 1972, "God bless you for your courage, your Americanism and your support." Handwritten note, Nov. 19, 1972, NHHC.

89. Letter, Nov. 30, 1972; reply, Dec. 12, 1972, NHHC.

90. A further sampling of the hate mail: "Spending $600,000 for race relations to force the 'niggers' further on to the whites is paralleling that of our school systems that are brainwashing the young and innocent to eventually bed down with the black. How do you personally feel about eating with these nauseating Bastards and having your daughter go to bed with them. It is the nature of these arrogant slobs to take a mile for every inch given them. ooo, box 86, folder 4. Undated letter from another racist: "Zumwalt: How the hell any white man can sell his own race of people 'down the river' to the niggers is hard to understand." ooo, box 86, folder 4. From F. B. Ward, Nov. 15, 1972: You are "a traitor to your race and the position of leadership you hold." From Columbus, Georgia, a group of concerned citizens wrote, "This is a white man's country and no negro is going to take it away from us." Mrs. Albertina Pimental, San Pedro, California, Nov. 11, 1972, referring to Zumwalt's dressing down of flag officers: ". . . slaps all whites in the face. . . ." A letter, Nov. 15, 1972, signed "class of 34": "Zumwalt you have reduced the greatest Navy the world has ever known to the laughing stock of the world. You and your black will never reduce the great traditions of our great Navy to your level. Name withheld for my protection, not from you, but from your blacks. We are taking action to get rid of you. You can expect to be moved in the near future." All from NHHC.

91. Nov. 11, 1972, ooo, box 86, folder 4, NHHC.

92. Nov. 15, 1972, ooo, box 86, folder 4, NHHC.

93. Nov. 26, 1972, ooo, box 86, folder 5, NHHC.

94. Dec. 3, 1972, ooo, box 86, folder 5, NHHC.

95. Nov. 23, 1972, ooo, box 86, folder 5, NHHC.

96. Nov. 24, 1972, 000, box 86, folder 4, NHHC.

97. Undated, 000, box 86, folder 4, NHHC.

98. White House tapes, Kissinger, Nov. 13, 1972.

99. White House tapes, Kissinger, tape 827, conversation 10, Dec. 20, 1972.

100. Zumwalt, *On Watch.*

101. Nov. 13, 1972.

102. Sherwood, *Black Sailor, White Navy,* 172.

103. Ibid.

104. "Carl Vinson: A Legend in His Own Time," http://georgiainfo.galileo.usg.edu/c-vinson.htm.

105. Thompson memoir, manuscript, 394.

106. On Nov. 21, 1972, Lew wrote that in Vietnam Bud had demonstrated "brilliant leadership and perseverance." NHHC. Operations Giant Slingshot, SEALORDS, and Barrier Reef "required absolute discipline at every level from Task Group Commander to PBR Boat Captain." The close combat on the inland rivers and canals of Vietnam required resourcefulness, innovation, courage, and total discipline. Progressive thinking was being misconstrued as permissiveness.

107. Telephone conversation, Nov. 15, 1972, ZPP.

108. Sherwood, *Black Sailor, White Navy,* 67.

109. ZTT 12, no. 3 of 3.

110. Ibid.

111. Author interview with Admiral Roberta Hazard.

112. Telephone conversation with Murphy, Nov. 15, 1972, ZPP.

113. Ibid.

114. Telephone conversation, Nov. 13, 1972, ZPP.

115. Zumwalt requested that Hare read the text of the speech that he delivered to the flag officers, as well as the related Z-grams, and "find it in his conscience to see his way clear to talk with Carl Vinson."

116. Sherwood, *Black Sailor, White Navy,* 177.

117. ZTT 31, side A.

118. Nov. 27, 1972, ZPP.

119. Ibid. Washington: Government Printing Office, 1969.

120. Conference call, conversation among lawyers, Nov. 21, 1972, summary by Hicks. 09/Jag/ (judge advocate general). Robbie Robertson led the discussion. ZPP.

121. Sherwood, *Black Sailor, White Navy,* 186.

122. Telephone conversation, Nov. 24, 1972. John Stennis shared Pirnie's view. "He is concerned with the trouble we are having with the blacks and the whole thing of discipline—you have a discipline problem that the blacks are a part of . . . you might want to be more selective on the ones we take in."

123. Sherwood, *Black Sailor, White Navy,* 183.

124. Nov. 29, 1972, Executive Correspondence, NHHC.

125. He had not heard from Haig or the White House since the "stupid call about getting everyone thrown out." Murphy asked "if he had patched it up with John Warner." He also told Bud that Vinson was a good friend of Laird and that they have a "father-son relationship."

126. Letter, Nov. 22, 1972, Executive Correspondence, NHHC.

127. Telephone conversation, Nov. 24, 1972, ZPP.

128. McNamara letter, ZFC.

129. Orr Kelly, "Riding Out a Navy Storm," Dec. 11, 1972, NHHC.

130. Joining the president were: Dr. Henry A. Kissinger, assistant to the president for national security affairs; General Alexander M. Haig, Jr., deputy assistant to the president for national security affairs; Melvin R. Laird, secretary of defense; Kenneth Rush, deputy secretary of defense; Admiral Thomas H. Moorer, chairman of

the Joint Chiefs of Staff; Admiral Elmo R. Zumwalt, Jr., chief of naval operations; General Creighton W. Abrams, army chief of staff; General Robert E. Cushman, Jr., commandant of the Marine Corps; General Horace M. Wade, vice chief of staff for the air force.

131. See Larry Berman, *No Peace No Honor: Nixon, Kissinger and Betrayal in Vietnam* (New York: Free Press, 2001), and Zumwalt, *On Watch*.

132. Dec. 16, 1972, ZPP.

133. Bud told this to Dan Murphy, ZPP.

134. 000, box 86, folder 7, NHHC.

135. 000, box 86, folder 5, NHHC.

136. 000, box 86, folder 7, NHHC.

137. Howard Kerr, letter, Dec. 15, 1972, Executive Correspondence, NHHC.

138. Dec. 17, 1972, from Subic Bay Naval Base, ZPP.

139. Dec. 7, 1972, ZPP.

140. *Disciplinary Problems in the Navy*, House Armed Services Committee. Headlines in the nation's newspapers captured the uncertainty of the times: "Navy's 'Old Guard' Out to sink Zumwalt," *Chicago Daily News*, Jan. 2, 1973; "Navy's Zumwalt Irks Racists," *Chicago Tribune*, Jan. 5, 1973; "Rivalries Imperil Zumwalt's Career," *Chicago Tribune*, Jan. 3, 1974; and "Navy Boss Is a National Disaster," which repeated Max Rafferty's charge that the navy was "rotten with rebellion, palsied with permissiveness and disintegrating with disobedience."

141. *Disciplinary Problems in the Navy*, House Armed Services Committee, vol. 5, 489, 1050. See Zumwalt Papers, "Hicks Committee and Critical Analysis of Hicks," NHHC.

142. Feb. 12, 1973, NHHC.

143. Zumwalt, *On Watch*.

144. Aug. 7, 1973, ZPP.

Chapter 12: The Zumwalt Intelligence Service

1. White House Tapes, Tape 308–13, National Archives.

2. Zumwalt, *On Watch*, xii.

3. Zumwalt to JCS historians, Historical Division, Joint Secretariat, *The Joint Chiefs of Staff and National Security Policy*, vol. 10: 1969–1972, 9. See Editorial Note, Document 159, *Foreign Relations of the Unites States, 1969–1976*, vol. 2, 328, cited in Peter W. Rodman, *Presidential Command: Power, Leadership, and the Making of Foreign Policy from Richard Nixon to George W. Bush*, 5th ed. (New York: Knopf, 2009), 67.

4. Kissinger, *White House Years*, 722.

5. Rodman, *Presidential Command*, 56; and White House Tapes, June 13, 1971, 3:09 p.m., cassette 825, conversations 5–590.

6. *Foreign Relations of the United States, 1969–1976*, vol. 2, 336.

7. "Nixon Travels—China," United States History, www.u-s-history.com/pages/h1877 .html; and Richard Nixon, *RN: The Memoirs of Richard Nixon* (New York: Grossett and Dunlap, 1978), 533.

8. Rodman, *Presidential Command*, 53.

9. Ivo H. Daalder and I. M. Destler, *In the Shadow of the Oval Office: Portraits of the National Security Advisers and the Presidents They Served—from JFK to George W. Bush* (New York: Simon & Schuster, 2009), 80.

10. Rodman, *Presidential Command*, 68.

11. Daalder and Destler, *In the Shadow of the Oval Office*, 80.

12. Zumwalt, *On Watch*, 397.

13. Ibid.

14. Bruce Lambert, "Seymour Weiss, Long an Adviser on Military Policy, Is Dead at 67," *New York Times*, Sept. 25, 1992; and Zumwalt, *On Watch*, 398ff.

15. Zumwalt, *On Watch*, 348.

16. Ibid., 348.

17. Ibid., 348–49.

18. Zumwalt CNO Papers, July 24, 1972. "He was very mad about the DPRC [Defense Program Review Committee] agenda because he felt that they were really trying to usurp his authority. He was mad about the supply line on project ENHANCE." NHHC.

19. ZTT 3, side A, 29–30.

20. Told to Secretary of Defense Schlesinger in conversation, June 24, 1974, ZPP.

21. Zumwalt, *On Watch*, 320.

22. Ibid., 321.

23. ZTT, "Henry Kissinger."

24. Quoted in unpublished "Rice" interview for Zumwalt, ZPP .

25. Letter to Jack Connery, July, 13, 1996, ZPP.

26. Rodman, *Presidential Command*, 65.

27. Ibid., 66; and Van Atta, *With Honor*, 224–25. I draw extensively on Van Atta's detailed study, especially because he had access to Laird's papers that are still closed to scholars.

28. Rodman, *Presidential Command*, 66; and Van Atta, *With Honor*, 224–25.

29. Van Atta, *With Honor*, 217.

30. Rodman, *Presidential Command*, 66.

31. Ibid., 67.

32. Seymour Hersh, *The Price of Power: Kissinger in the Nixon White House* (New York: Simon & Schuster, 1984), 68. I am indebted to Hersh for telling me that while he could not disclose sources because some were still alive, that I was on the right trail and knew what he knew.

33. Walter Issacson, *Kissinger: A Biography* (New York: Simon & Schuster, 2005), 202.

34. Sept. 13, 1995. Bud was Halperin's proposer for membership. ZFC. Kissinger added, "Although I understood and supported his desire to pursue a career in law after a couple of grueling years in the White House, he was hard to replace." ZFC.

35. Jan. 6, 1988, ZFC.

36. Memorandum for the record, phone conversation between Admiral Zumwalt and H.K., ZPP.

37. Wharton alumni magazine, *The Ties That Bind*, http://beacon.wharton.upenn.edu /whartonmagazine/files/2012/03/am01fal.pdf

38. DC debriefing, June 19, 1972. All of the DC briefings are available in the Vietnam Archive. ZPP.

39. ZTT. Henry Kissinger and Al Haig.

40. Zumwalt on Haig, ZTT.

41. DC debriefing, Mar. 20, 1973, ZPP.

42. DC debriefing, June 4, 1973, ZPP.

43. Zumwalt, *On Watch*, 375–76.

44. "Meeting with Kissinger," ZTT, undated side B, part 9.

45. On June 7, 1971, Bud wrote Kissinger, "You were most thoughtful to share your high regard for Rear Admiral Robinson, as expressed in your 22 May letter, with me. I have insured your letter is placed in his official record." Kissinger name file, Executive Correspondence, NHHC.

46. Executive Correspondence, NHHC.

47. "The Radford Affair," ZTT; and Mark Feldstein, *Poisoning the Press: Richard Nixon, Jack Anderson, and the Rise of Washington's Scandal Culture* (New York: Farrar, Straus & Giroux, 2010). I draw extensively from Feldstein's pathbreaking research.

48. Hearings before the Committee on Armed Services, United States Senate, Feb. 20–21, 1974; also see *Foreign Relations of the United States, 1969–1976*, vol. 2, *Organization and Management of Foreign Policy, 1969–1972*, Documents, 164–66.

49. Nov. 19, 1982. Stillwell, *Reminiscences by Staff Officers*, 136.

50. Hersh, *Price of Power*, 460. Readers will note that I draw extensively on Hersh's research, who confirmed in an e-mail that, while he still could not identify his sources, my documentation was consistent with what he knew.

51. Henry Kissinger, *Years of Upheaval* (Boston: Little, Brown, 1982; New Haven, CT: Phoenix Press, 2000), 806–7; *Foreign Relations of the United States, 1969–1976*, vol. 2, 334.

52. John Prados, *Keeper of the Keys: A History of the National Security Council from Truman to Bush* (New York: William Morrow/HarperCollins, 1991), 316.

53. "Sometime after returning from the trip with Dr. Henry Kissinger, Captain [Arthur] Knoizen walked through the office and said, 'Radford, keep up the good work.' I knew what he meant. Nothing else was said."

54. "Again I returned with copies of several documents of interest to Admiral Welander. In each case copies were made, or mental notes made, and in each case these copies, or mental notes, were given to my admiral, Robinson or Welander. No documents were given to anyone else."

55. Hersh, *Price of Power*, 467.

56. Ibid., 467, 470.

57. Asaf Siniver, *Nixon, Kissinger, and U.S. Foreign Policy Making: The Machinery of Crisis* (New York: Cambridge University Press, 2008), 149.

58. Robert Dallek, *Nixon and Kissinger: Partners in Power* (New York: Harper, 2007), 340; Richard Reeves, *President Nixon: Alone in the White House* (New York: Simon & Schuster, 2002), 391; and Feldstein, *Poisoning the Press*, 156.

59. Hersh, *Price of Power*, 457.

60. Minutes of senior staff meeting, India-Pakistan, Dec. 6, 1971, cited in Van Atta, *With Honor*, 302.

61. Zumwalt, *On Watch*, 367–68; see also Siniver, *Nixon, Kissinger*, 150.

62. Van Atta, *With Honor*, 302.

63. Years later Zumwalt gave a series of lectures at the Indian War College in New Delhi. At dinner, Admiral Sardari Mathradas Nanda, who had commanded the Indian Navy in 1971, said he needn't have worried, because "my instructions to the Indian Navy were that if they encountered US Navy ships, invite their skippers aboard your ships for a drink."

64. Siniver, *Nixon, Kissinger*, 178.

65. John Ehrlichman, *Witness to Power: The Nixon Years* (New York: Simon & Schuster, 1982), 302; cited in Feldstein, *Poisoning the Press*, 154.

66. Radford testimony, Hearings before the Committee on Armed Services, United States Senate, Feb. 20–21, 1974.

67. Haldeman, *The Haldeman Diaries: Inside the Nixon White House*, 386.

68. Feldstein, *Poisoning the Press*, 178; and memo, W. Donald Stewart to Martin Hoffman, Jan. 21, 1974.

69. The Defense Department investigation of the unauthorized disclosures to columnist Jack Anderson found that the wives of Radford and Anderson were close friends who had shopped together and shared an interest in genealogy. See Feldstein, *Poisoning the Press*.

70. Cited in Feldstein, *Poisoning the Press*, 179.

71. Ibid.

72. Anderson maintained to his death that Radford was not his source.

73. Len Colodny and Robert Gettlin, *Silent Coup: The Removal of a President* (New York: St. Martin's, 1992). This is the seminal work on the spy ring. See http://www.silentcoup

.com/. I started my research on this topic with *Silent Coup* and have drawn extensively from it.

74. Hersh, *Price of Power*, 472.
75. National Archives, Nixon Presidential Materials, White House Central Files, President's Daily Diary.
76. Feldstein, *Poisoning the Press*, 181. Ehrlichman, *Witness to Power*.
77. Feldstein, *Poisoning the Press*, 181.
78. Nixon wanted Welander fired. "Can him. Can him. Can him. Get him the hell out of here." White House Tapes, tape 189, Dec. 21, 1971, conversation no. 639–30.
79. Oval Office meeting, Dec. 21, 1971.
80. Zumwalt's name comes up immediately on the Dec. 21, 1971, tape.
81. Feldstein, *Poisoning the Press*, 189.
82. Ibid., 188; and Ehrlichman, *Witness to Power*, 129.
83. Feldstein, *Poisoning the Press*, 182; White House Tapes, tapes 640–45, Dec. 22, 1971.
84. Feldstein, *Poisoning the Press*, 183.
85. Nixon agreed: "Let the poor bastards stew over Christmas, and then crack 'em." "New Evidence Confirms Pentagon Stole and Leaked Top Secret Documents from Nixon White House," NixonTapes.org, http://nixontapes.org/welander.html, Dec. 21, 1971, tape.
86. National Archives, Nixon Presidential Materials, White House Tapes, Recording of conversation among Nixon, Mitchell, Haldeman, and Ehrlichman, Oval Office, conversation no. 639–30.
87. Haldeman, *Diaries*, 385–86.
88. Nixon, *RN*, 532.
89. Feldstein, *Poisoning the Press*, 197.
90. National Archives, Nixon Presidential Materials, White House Tapes, Conversation between Nixon and Mitchell, Dec. 24, 1971, 5:33 p.m., White House Telephone, conversation no. 17–37.
91. This account is from ZTT, "The Radford Affair."
92. Ibid.
93. Van Atta, *With Honor*, 303; J. Fred Buzhardt, "Interim Report of Investigation of Recent Unauthorized Disclosure of Classified material to Columnist Jack Anderson and the Use of Unauthorized Communications Channels between the National Security Council Staff and the Office of the Joint Chiefs of Staff," memorandum to the secretary of defense, Jan. 10, 1972.
94. Van Atta, *With Honor*, 304.
95. Ibid.
96. Zumwalt, *On Watch*, 370.
97. White House Tapes, conversation no. 641-10, Dec. 23, 1971.
98. The CNO was never apprised of the allegations that Welander had improperly provided information from Kissinger/NSC to JCS until CJCS briefed the Joint Chiefs after the first news story on Radford/Welander.
99. ZTT 21, side A, part 6.
100. Moorer to JCS, Jan. 4, 1972, ZTT, "The Radford Story," side B, part 3.
101. Hearings before the Committee on Armed Services, United States Senate, Feb. 20–21, 1974; also see *Foreign Relations of the United States, 1969–1976*, vol. 2, *Organization and Management of Foreign Policy, 1969–1972*, Documents, 164–66.
102. "Radford/Welander Matter," Jan. 26, 1974, Sensitive—Eyes Only Memo for the record, telephone conversation between CNO and RADM Welander.
103. James F. McHugh, memo to CNO, "Moorer/Welander/Radford Matter," Feb. 1, 1974.
104. ZTT, "The Radford Story."
105. Ibid.

106. Author interview with Burt Shepherd, March 2012.

107. Paul Stillwell, interview with Rectanus, Oral History, Nov. 19, 1982, 136.

108. White House Tapes, conversation no. 308–13, Dec. 22, 1971; and Feldstein, *Poisoning the Press*, 197.

109. See ZTT 10, side B, part 9. "K urged me not to go to Haig because he did not trust Haig. Not to cut him out and deal directly with the President."

CHAPTER 13: RUFFLES AND FLOURISHES

1. ZTT A, part 4.

2. Zumwalt, *On Watch*, 492.

3. Schlesinger had worked previously as the Rand Corporation's director of strategic studies from 1967 to 1969. One of his major papers at Rand had dealt with the role of systems analysis in political decision making. At Rand he had been deeply involved in efforts to rethink the strategic nuclear doctrine. Schlesinger left Rand in 1969, joining the Nixon administration as assistant director of the bureau of the budget (BOB), specializing in military and international programs. He was also responsible for forming the administration's energy policy relating to air and water pollution. He was soon given the title of acting deputy director of BOB, which he held until 1970. When BOB was reorganized as the Office of Management and Budget (OMB), Schlesinger became assistant director. In OMB he took the lead on a detailed study for restructuring the nation's intelligence community. He next served as chairman of the Atomic Energy Commission, where his dual interests in atomic energy and concern for the environment manifested themselves. Faced with trying to reconcile the opposing interests of conservationists and nuclear power advocates, he began by announcing that the AEC would no longer take the traditional position of championing the rights of nuclear energy above all others. He remained at the AEC until, in rapid succession, he was appointed director of the CIA (confirmed January 23, 1973) and then secretary of defense, confirmed June 28, 1973.

4. Cockell to Admiral Zumwalt, Sensitive—Eyes Only, July 3, 1973, ZPP.

5. Schlesinger, unlike Bud, was "very much a private man, not given to socializing . . . unpretentious, likes plain living and disdains creature comforts; wears off-the bargain-rack suits and drives a retirement car. A tweedy, pipe smoking perfectionist and 'Harvard rustic,' he is a careful planner who seldom leaves home without consulting a road map," wrote Cockell, NHHC.

6. Rowen report forwarded by Cockell to Zumwalt, NHHC.

7. Sherry Sontag, Christopher Drew, and Annette Lawrence Drew, *Blind Man's Bluff: The Untold Story of American Submarine Espionage* (New York : Public Affairs, 1998), 190.

8. Nitze, *From Hiroshima to Glasnost*, 334.

9. Sontag, *Blind Man's Bluff*, 191.

10. Robert Kaufman, *Henry M. Jackson: A Life in Politics* (Seattle: University of Washington Press, 2000), 242.

11. Zumwalt, *On Watch*, 422.

12. Nitze, *From Hiroshima to Glasnost*, 335.

13. Michael Krepon, "The Jackson Amendment," Arms Control Wonk, http://krepon .armscontrolwonk.com/archive/2414/the-jackson-amendment.

14. See Paul Nitze, "Nuclear Strategy in an Era of Detente," Sept. 5, 1975, Working file, Nitze Papers, Library of Congress.

15. Schlesinger, "Director of DOD Salt Task Force," Sept. 15, 1973, NHHC.

16. Nitze, *From Hiroshima to Glasnost*, 335.

17. Kaufman, *Henry M. Jackson*, 273, based on Kaufman interview with Schlesinger in 1994.

18. Abba Solomon Eban, *Abba Eban: An Autobiography* (New York: Random House, 1977), 515.

19. Tad Szulc, *The Illusion of Peace: Foreign Policy in the Nixon Years* (New York: Viking, 1978), 734. I draw extensively from Szulc's account.

20. Matloff interview, Office of the Secretary of Defense Historical Office, ZPP.

21. Kaufman, *Henry M. Jackson*, 275.

22. Cited in ibid., 276.

23. Kissinger tape, Nixon Library, Oct. 11, 1973.

24. Kaufman, *Henry M. Jackson*, 276; and Thompson, *The Hawk and the Dove*, 234.

25. Nitze, *From Hiroshima to Glasnost*, 338.

26. Ibid., 338.

27. Ibid., 340.

28. Conversation between Fred Wikner and Haig, June 2, 1974, ZPP.

29. Jan. 6, 1988, ZPP.

30. Tape 21, side A, "The Problem of Succession," ZTT.

31. Zumwalt, *On Watch*, 476.

32. Ibid., 474.

33. Tape 21, side A, "The Problem of Succession," ZTT.

34. Ibid., "Ike Kidd." ZTT.

35. W. A. Cockell, memorandum for the record, Dec. 17, 1973, conversation with PHN. "In view of the subject, I have typed this memo myself, and made no copies." ZPP.

36. DC debriefing, May 20, 1974, ZPP.

37. Cited in Zumwalt, *On Watch*, 399. "Kissingerology had a 'Haigological' branch," wrote Zumwalt (p. 397).

38. Tape 21, side A, "The Problem of Succession," ZTT.

39. Michael Getler, "Holloway Nominated to Succeed Zumwalt," *Washington Post*, Mar. 29, 1974, A3.

40. *Chicago Tribune*, June 29, 1974, D12.

41. Mar. 28, 1974, ZPP.

42. Memorandum of conversation, June 6, 1974, ZPP.

43. May 28, 1974, Executive Correspondence, NHHC.

44. Zumwalt, *On Watch*.

45. May 30, 1974, ZPP.

46. Clements also told Zumwalt that he viewed Peet as "a question mark with me for some time and I think you know that," ZPP.

47. Richard Reston, "Nixon Warns Against Attempts to Sway Soviet Domestic Affairs," *Los Angeles Times*, June 6, 1974, A1.

48. ZTT, "Change of Command."

49. Telephone conversation, CNO and CJCS, June 6, 1974, speaking about that day's *New York Times* article; Zumwalt wonders who the source might be. ZPP.

50. Ibid.

51. ZTT unnumbered, side B, 87.

52. ZTT 15, side B, part 9.

53. June 20, 1974, ZPP.

54. Telephone conversation, CNO and Captain Cockell, June 21, 1974, ZPP.

55. In a subsequent memorandum for the record, Zumwalt recorded Moorer's debriefing from the June 21 NSC meeting.

56. Zumwalt was told by Cockell that the president said he and not Zumwalt was in charge. Bud "was proud that he got honorable mention 3 times at meeting." ZPP. Cockell also reported that the president was upset that "Moorer is spying on him."

57. Welander, assistant deputy chief of naval operations (plans and policy), memorandum for the record, debriefing June 21, 1974, of NSC meeting of June 20, 1974.

58. Memorandum for the record, June 24, 1974, based on a "sensitive—eyes only conversation with Holloway and Bagley." ZPP.

59. Zumwalt, *On Watch*, 486. I also benefited from Howard Kerr's account in his letter of remembrance following Bud's passing. Provided to author by Kerr.

60. June 25, 1974, ZFC.

61. June 25, 1974, NHHC.

62. SSBN-640 Gold, June 4, 1974, NHHC.

63. July 5, 1974, NHHC.

64. June 24, 1974, NHHC.

65. June 25, 1974, NHHC.

66. June 26, 1974, NHHC.

67. Harold E. Shear oral history interview, USNI, 313–14.

68. David Woodbury, "Musings of an Ancient Mariner, on Being a Naval Aide," unpublished manuscript.

69. Jerry Wilson, "Petticoat Politics," Sunday, July 1974, *Palestine* (Texas) *Herald Press*, 3, Executive Correspondence, Wilson Files, NHHC.

70. June 27, 1974, ZFC.

71. Author interview with Howard Kerr.

72. "Zumwalt Navy Era Ends with Warning," *Washington Star News*, June 30, 1974.

73. July 3, 1974, NHHC.

74. Aug. 15, 1974, Executive Correspondence, NHHC.

75. "'Well Done,' Bud Zumwalt," *Salt Lake Tribune* headline on June 29, 1974, 16.

76. "Zumwalt Navy Era Ends with Warning," *Washington Star News*, June 30, 1974.

77. "Zumwalt Says He Turned Down VA Post, Cites 'Domestic Politics,'" *Washington Post*, July 1, 1974, A8.

78. Unnumbered tape transcript, side B, 85.

79. Tape 14, side A, ZTT.

80. Senate Foreign Relations Committee, China, July 15, 1974.

81. Elmo Russell Zumwalt, Jr, "After *On Watch*," unpublished manuscript, Sept. 9, 1999.

82. Tape 21, side A, ZTT.

Chapter 14: The Watch Never Ends

1. Letter, June 2, 1995, ZFC.

2. Personal letter, July 8, 1974.

3. Elmo Russell Zumwalt, Jr, "After *On Watch*," unpublished manuscript, Sept. 9, 1999. Bud was planning another update for 2010, when he would have been ninety.

4. July, 12, 1974, ZFC.

5. Zumwalt, "After *On Watch*."

6. Ibid.

7. This was the period Bud set aside for writing his memoirs, allowing him to reenter the arena.

8. Correspondence shows that Bud was also under consideration for the presidency of the University of California system.

9. Ibid.

10. Zumwalt, "After *On Watch*," 3.

11. Ibid.

12. Rowland Evans and Robert Novak, "Adm. Zumwalt Lays His Political Groundwork," *Washington Post*, Dec. 22, 1974, B7.

13. Jay Mathews, "Zumwalt Seeks Support for Race Against Byrd," *Washington Post*, May 20, 1975, C1; also see Bill McAllister, "Admiral Zumwalt Seeks Byrd's Seat," *Washington Post*, Feb. 24, 1976, B1.

14. Zumwalt, "After *On Watch*."

15. Berman, *No Peace, No Honor*.

16. Sheehan interview with Negroponte, Sheehan papers, Negroponte folder, Library of Congress.

17. Berman, *No Peace, No Honor*; and Gregory Tien Hung Nguyen and Jerrold L. Schecter, *The Palace File: The Remarkable Story of the Secret Letters from Nixon and Ford to the President of South Vietnam and the American Promises That Were Never Kept* (New York: Harper & Row, 1986).

18. President Gerald Ford issued an official statement: "The Government of the Republic of Vietnam has surrendered. Prior to its surrender, we have withdrawn our Mission from Vietnam. Vietnam has been a wrenching experience for this nation. . . . History must be the final judge of that which we have done or left undone, in Vietnam and elsewhere. Let us calmly await its verdict."

19. Interview with author; see Tran Van Chon, "Back Home from Hell," unpublished paper, ZPP.

20. Khue letter to Admiral Zumwalt, May 6, 1975, ZPP.

21. The two brothers were separated for seven years.

22. See Murrey Marder, "Jackson: Hill, Saigon Mislead," *Washington Post*, May 2, 1975; see also: "The Watergate Connection," *Time*, May 5, 1975, and "War-Watergate Tie Is Seen by Zumwalt," *New York Times*, May 6, 1975.

23. Paul G. Edwards, "Zumwalt Goes on TV, Attacks Sens. Byrd, Scott," *Washington Post*, Sept. 8, 1976.

24. Paul G. Edwards, "Zumwalt's Campaign Languishes," *Washington Post*, Oct. 10, 1976.

25. Zumwalt, "After *On Watch*."

26. May 10, 1976, Nitze Papers, Manuscript Division, Library of Congress.

27. Zumwalt, "After *On Watch*."

28. After practicing law in Virginia Beach for three years, Finchem served in the White House as deputy advisor to the president in the Office of Economic Affairs in 1978 and 1979. In the early 1980s, Finchem cofounded the National Strategies and Marketing Group in Washington, D.C. He then served as commissioner of the PGA Tour.

29. Megan Rosenfeld, "Zumwalt in Uphill Race," *Washington Post*, Oct. 12, 1976.

30. Paul G. Edwards, "Zumwalt Readies Attack on Byrd," *Washington Post*, May 10, 1976.

31. Dec. 12, 1975, ZPP.

32. Memorandum for the record, Apr. 2, 1976, Nitze Papers, Library of Congress.

33. Ibid. He reported it to Captain Davy and Ann Zumwalt.

34. Nicholas Thompson, "Did Henry Kissinger Really Plan 'An Accident' for Bud Zumwalt?" Sept. 14, 2009, Danger Room, *Wired*, www.wired.com/dangerroom/2009/09/did-henry-kissinger-really-plan-an-accident-for-bud-zumwalt.

35. *Virginian-Pilot*, Jan. 12, 1976.

36. Bill McAllister, "Zumwalt Urges Defense Cuts," *Washington Post*, July 14, 1976, A15.

37. The speech was covered by all three networks, but it was so late that the venue was half empty by the time he spoke.

38. "We cannot lose sight of the fact that our nation's essential task is the promotion of better lives, not bigger guns."

39. Paul G. Edwards, "Zumwalt's Campaign Languishes," *Washington Post*, Oct. 10, 1976.

40. Megan Rosenfeld, "Zumwalt: A 'Passed Over' Politician," *Washington Post*, Nov. 4, 1976.

41. Douglas Feith, *War and Decision: Inside the Pentagon at the Dawn of the War on Terrorism* (New York: Harper, 2008). "Zumwalt went out of his way to promote my career, inviting me to dinners with his friends and recommending me for speeches and membership in groups such as the Council on Foreign Relations," 27.

42. Feb. 1, 1977, ZFC.

43. June 29, 1977, Nitze Papers, Library of Congress.

44. Aug. 10, 1977, handwritten note, "Dear Boss," Nitze Papers, Library of Congress.

45. May 7, 1970. Bud thanked Dr. James Caldwell at the University of North Carolina for all the help he has given Elmo and the family.

46. With his father dead, Bud took over the job of writing an annual letter to the entire clan, March 18, 1974, ZFC.

47. James Zumwalt, Award of the Order of the Silver Rose, Nov. 11, 1997, ZPP.

48. My thanks to the Zumwalt family for providing me with a set of Elmo's medical records. Some of these records are also in ZPP.

49. *US News and World Report*, Jan. 17, 2000.

50. *Parade*, Oct. 7, 1984, 16, interview with Walter Anderson, editor.

51. May 13, 1983, ZPP.

52. June 9, 1983, ZPP.

53. Dec. 12, 1983, ZPP.

54. Apr. 17, 1984, was the day he put feet on American soil. ZPP.

55. May 5, 1984, ZPP.

56. June 18, 1984, ZPP.

57. Oct. 2, 1984, ZPP.

58. July 30, 1984, handwritten note, ZFC.

59. From Aunt Saralee, Jan. 13, 1986, ZFC.

60. Until Mouzetta was safely out of danger because of the difficulty during the procedure. "Her bones were stronger than most and the needle punctures needed to get 1000 ccs of marrow had to be doubled. She had a rough time." Mar. 6, 1986, progress report, ZFC.

61. Cited in Walter Anderson, *Courage Is a Three-Letter Word* (New York: Random House, 1986), 168, 171.

62. Mar. 26, 1986, update, ZFC and ZPP.

63. The May 27, 1986, update is the last one.

64. Sept. 14, 1987, ZFC.

65. Sept. 13, 1987, ZFC.

66. Dec. 30, 1987, ZFC.

67. After seven years as chairman, Bud gave it up in 1994, but continued to serve in emeritus capacity.

68. Aug. 19, 1987, ZPP.

69. Ibid.

70. Oct. 2, 1987, ZPP.

71. Nov. 30, 1987, Truc wrote with photos, ZPP.

72. Dictated to Dad on Aug. 12, 1987, and signed book July 26, 1988.

73. James G. Zumwalt, vignette, "The Valiant Warrior," ZFC.

74. See Zumwalt and Zumwalt, *My Father, My Son*, as well as *Dateline* interview with Elmo.

75. Tim Larimer, interview with Admiral Zumwalt, *Washington Post*, ZPP, date obscured.

76. Letter, Aug. 16, 1990, to Earl Collins, Austin, Texas, ZPP.

77. Richard Ehrlich, "Zumwalt Haunted After Spraying Vietnam with Agent Orange," Oct. 1994, ZPP.

78. ZFC.

79. ZFC.

80. Author interview with Dr. William Narva.

81. Elmo died on August 13, 1988. "The tears and the mourning will never cease. But over time the remembering of our years together brings joy and laughter amidst those tears," said Bud.

82. James G. Zumwalt, *Bare Feet, Iron Will: Stories from the Other Side of Vietnam's Battlefields* (Jacksonville, FL: Fortis, 2010), 8.

83. George Esper, "Man with a Mission," *Cape Cod Times*, Jan. 22, 1995.

84. On July 17, 1974, Zumwalt wrote, "Dear Ed, I sincerely appreciate your kind remarks and insertion into the Congressional Record of Bill Anderson's farewell column marking the end of my tour. In my view any enduring success we attained is due in large part to the efforts of our great Navy team and of Congressional leaders such as you who courageously supported the changes that would improve Navy life and navy strength in the years ahead. I am most grateful for that support." ZPP.

85. Oct. 11, 1989, ZPP.

86. On Jan. 19, 1990, Bud wrote the FOI officer that he'd been designated by Derwinski, and "I need information from your files on the following": He was looking for all available data on where Agent Orange had been used in Vietnam and the troops deployed in those areas.

87. Bud testified before the House Human Resources and Intergovernmental Relations Subcommittee of the Committee on Government Operations, "Links Between Agent Orange, Herbicides, and Rare Diseases," June 26, 1990. See also his "Report to the Secretary of the Department of Veterans Affairs on the Association Between Adverse Health Effects and Exposure to Agent Orange," May 5, 1990. See also Statement by Admiral E. R. Zumwalt, Jr., USN (Ret), Chairman of the Agent Orange Coordinating Council before the House Subcommittee on Energy and Environment, Dec. 13, 1995.

88. Ibid., 4. Statement by Admiral E. R. Zumwalt, Jr., USN (Ret.), Chairman of the Agent Orange Coordinating Council before the House Subcommittee on Energy and Environment, Dec. 13, 1995.

89. Zumwalt, "After *On Watch*."

90. Sept. 13, 1989, ZPP.

91. Oct. 8, 1989, ZPP.

92. Interview with author.

93. See Draft Report on Manipulations and 1988 letter to Sen. Tom Daschle.

94. The best book on this subject is David Zierler, *The Invention of Ecocide: Agent Orange, Vietnam, and the Scientists Who Changed the Way We Think about the Environment* (Athens: University of Georgia Press, 2011).

95. See Bill McAllister, "Ex-Admiral Zumwalt Claims Manipulation on Agent Orange," *Washington Post*, June 27, 1990.

96. Jan. 17, 1984, ZPP.

97. Had Carter been reelected "and had his defense instincts continued, the entire Middle East would have ended up in Soviet hands. By the year 1990 they would have had control of the oil flow and minerals of Africa." *Time*, Mar. 1, 1982.

98. Ibid.

99. Ronald Reagan to Elmo, on his fortieth birthday, ZFC.

100. Aug. 15, 1988, ZFC.

101. At about same time the Agent Orange Scientific Task Force commissioned by the American Legion identified a number of diseases that passed the "as likely as not" test.

102. Aug. 9, 1990. Weiss Committee Report and Statement by Admiral E. R. Zumwalt, Jr., USN (Ret.), Chairman of the Agent Orange Coordinating Council before the House Subcommittee on Energy and Environment, Dec. 13, 1995.

103. Associated Press, "Agent Orange Study Obstruction Charged," *Houston Chronicle*, Aug. 10, 1990.

104. Letter to Kirkwood, ZFC.

105. Letter to Jerry Wages, Oct. 12, 1993, ZPP.

106. Wednesday, Dec. [date obscured] 1992, 8:30 p.m., handwritten notes. "If I live, we'll get some more." ZPP.

107. ZFC.

108. Zumwalt, *Bare Feet, Iron Will.*
109. Ibid.
110. On his way home, in Bangkok, *Newsweek* magazine caught up with Bud. He was asked by Ron Moreau (*Newsweek*, Sept. 26, 1994) if it was a mistake to use Agent Orange. "In my judgment we'd have to do it again in identical circumstances."
111. Sept. 29, 1994, ZPP.
112. Oct. 27, 1994 flyer. Following Robb's victory, Bud handwrote a note of congratulations, attaching a letter he had sent to President Reagan, saying, "I believe that it led to Nancy Reagan's decision to blast Ollies follies!" ZPP.
113. Nov. 14, 1994, ZPP.
114. Warner handwritten note, Dec. 12, 1994, ZPP.
115. July 31, 1996, ZFC.
116. Celeste Bohlen, "Zumwalt's Arms Nomination Is Now in Trouble," *Washington Post*, Dec. 10, 1982, B4.
117. Bud Zumwalt, "If These Were My Last Words," Jan. 2, 1999, author interview with Philip Lader, ZFC.
118. Nov. 15, 1994, ZPP.
119. Nov. 1, 1996, ZPP.
120. Ibid.
121. Author interview with Mike Spiro.
122. Oct. 3, 1988, ZPP.
123. Oct. 12, 1990, ZPP.
124. Jan. 1, 1991, ZPP.
125. Paper written by Chon at Evergreen Valley College, 1993, Political Science 1 class. ZPP.
126. *Capitol Conversations*, a tape in ZFC.
127. Jan. 26, 1998, ZFP.
128. Oct. 5, 1995, ZFP.
129. Final report, Special Oversight Board, Dec. 2000, ZFC.
130. May 7, 1998, ZPP.
131. See George V. Galdorisi and Scott C. Truver, "The *Zumwalt*-Class Destroyer: A Technology 'Bridge' Shaping the Navy after Next," *Naval War College Review* 63, no. 3 (Summer 2010): 63–72.
132. "As you celebrate this milestone in the ship's building, my thoughts are with Bud's children, Jim, Ann, and Mouzetta, and their families. As dedicated as Bud was to his beloved Navy, his life revolved around all of you, who reminded him every day what protecting our country was really about. I know he would be moved that the keel will bear your initials and those of your late brother, Elmo."
133. Steve Vogel, "Remembering a Hero—and Friend: At Zumwalt Grave, Vietnamese Admiral Honors Man Who Saved Him," *Washington Post*, July 30, 2003.
134. Ibid.; see also Sept. 13, 2003, letter to Mouzetta Zumwalt, ZFC.

Author's Research Note

1. Conversation with Turkish foreign minister Melih Esenbel, Mar. 10, 1975, nsarchive .files.wordpress.com/2010/11/kiss-foia.pdf.
2. "Book in Progress: History of Vietnam," C-SPAN Video Library, www.c-spanvideo .org/program/123213-1.
3. Matthew M. Aid, ed., "Declassification in Reverse: The U.S. Intelligence Community's Secret Historical Document Reclassification Program," National Security Archive, Feb. 21, 2006, www.gwu.edu/~nsarchiv/NSAEBB/NSAEBB179.

4. Section 3161(b) of the Strom Thurmond National Defense Authorization Act for Fiscal Year 1999 (Public Law 105-261; 112 Stat. 2260; 50 U.S.C. 435 note), National Defense Authorization Act for Fiscal Year 2000, Section 1068, Declassification of Restricted Data and Formerly Restricted Data, Section 3161(b), www.fas.org/sgp/news/1999/02 /lottamend.html.

5. Michael Dobbs, "Our History, Off-Limits," *Washington Post*, June 10, 2008, www .washingtonpost.com/wp-dyn/content/article/2008/06/09/AR2008060902241.html.

6. Larry Berman, Plaintiff-Appellant, v. Central Intelligence Agency, Defendant-Appellee, No. 05-16820, argued and submitted July 10, 2007, and Sept. 04, 2007, www .gwu.edu/~nsarchiv/news/20070905/bermanopinion.pdfhttp://www.silha.umn .edu/news/fall2007.php?entry=198955.

7. Harry Train Oral History, USNI.

Index

Page numbers beginning with 443 refer to notes.

Larry Berman has written four previous books on the war in Vietnam: *Planning a Tragedy: The Americanization of the War in Vietnam*; *Lyndon Johnson's War: The Road to Stalemate in Vietnam*; *No Peace, No Honor: Nixon, Kissinger, and Betrayal in Vietnam*; and *Perfect Spy: The Incredible Double Life of Pham Xuan An,* Time *Magazine Reporter and Vietnamese Communist Agent.* He has been featured on C-SPAN Book TV, *The Public Mind with Bill Moyers,* and David McCullough's *American Experience.* He has been a Guggenheim fellow and a fellow in residence at the Woodrow Wilson International Center for Scholars. He received the Bernath Lecture Prize for contributions to our understanding of foreign relations and the Vice Admiral Edwin B. Hooper Research Grant. Berman is professor emeritus at the University of California, Davis, and founding dean of the Honors College at Georgia State University. He lives in Atlanta, Georgia.